Springer Proceedings in Earth and Environmental Sciences

Series Editor

Natalia S. Bezaeva, The Moscow Area, Russia

The series Springer Proceedings in Earth and Environmental Sciences publishes proceedings from scholarly meetings and workshops on all topics related to Environmental and Earth Sciences and related sciences. This series constitutes a comprehensive up-to-date source of reference on a field or subfield of relevance in Earth and Environmental Sciences. In addition to an overall evaluation of the interest, scientific quality, and timeliness of each proposal at the hands of the publisher, individual contributions are all refereed to the high quality standards of leading journals in the field. Thus, this series provides the research community with well-edited, authoritative reports on developments in the most exciting areas of environmental sciences, earth sciences and related fields.

More information about this series at http://www.springer.com/series/16067

Sergey Glagolev
Editor

14th International Congress for Applied Mineralogy (ICAM2019)

Belgorod State Technological University
named after V. G. Shukhov,
23–27 September 2019,
Belgorod, Russia

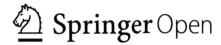

Editor
Sergey Glagolev
Belgorod State Technological University
Belgorod, Russia

ISSN 2524-342X   ISSN 2524-3438   (electronic)
Springer Proceedings in Earth and Environmental Sciences
ISBN 978-3-030-22976-4   ISBN 978-3-030-22974-0   (eBook)
https://doi.org/10.1007/978-3-030-22974-0

This Springer imprint is published by the registered company Springer Nature Switzerland AG
The registered company address is: Gewerbestrasse 11, 6330 Cham, Switzerland

Foreword

Dear Participants of the 14th International Congress for Applied Mineralogy!

On behalf of the Presidium of the Russian Mineralogical Society, let us congratulate you with this important scientific event and express our hope for the successful and productive work at its sessions and symposia.

Challenges that will be considered at the Congress, which is held in Russia for the first time, are exceptionally broad. They coincide with many directions that were recently developed by the Russian Mineralogical Society at its Commissions on technological mineralogy, environmental mineralogy and geochemistry, crystal chemistry and spectroscopy, and organic mineralogy.

In 2017, the Russian Mineralogical Society (RMS) celebrated its 200th anniversary, and today, it unites more than twelve hundred members from different fields and disciplines. RMS sections are present in many scientifically important cities and mining regions of our country. RMS members are actively solving important problems of development and production of raw materials for nuclear, diamond, rare metal, and oil industries.

The continuing progress in microscopy (electron, tunnel, atomic force) brought mineralogical studies to the micro- and nanolevels. Intervention into micro- and nanoworlds is an important new stage of development of modern mineralogy and mineralogical crystallography. Due to the success in these fields, a range of new possibilities was opened for:

prospecting and elaboration of novel types of mineral deposits,

solution of the remediation problems for mining regions and the rational use of mine tailings,

deeper understanding of the nature of oil fields, organic minerals and biominerals,

creation of principally new geo-inspired materials with improved functional properties.

The methods of mineralogical sciences are still in the focus of problems that are faced by Earth sciences, rational use of natural resources, conservation of natural diversity, and cultural heritage of the humankind.

We are sure that the Congress will demonstrate a broad spectrum of important scientific discoveries, will develop new directions for research and exploration in applied mineralogy, and, last but not least, will provide a lavish basis for new professional contacts and collaborations.

Wishing you great inspiration, new discoveries, and achievements.

Yury B. Marin
President of the Russian Mineralogical Society,
Corresponding Member of the Russian Academy of Sciences

Sergey V. Krivovichev
Vice-President of the Russian Mineralogical Society,
Corresponding Member of the Russian Academy of Sciences

Preface

Dear Participants of the Congress!

I welcome you to Belgorod at the 14th International Congress for Applied Mineralogy ICAM 2019!

Considering a special significance and prestige of the events of the International Council on Applied Mineralogy and the key role of our country as one of the world leaders in the industrial sectors associated with mineralogy, it is gratifying to note that the Congress is held on the base of Belgorod Technological University, which is a flagship university of Russia, an educational and scientific-innovative center of attraction for talents and formation of a professional regional elite, the personnel base and potential of enterprises of mining and processing industries.

Main scientific trends of the forthcoming Congress are devoted to modern methods of studying rocks and minerals, geometallurgy and processing, biomineralogy, functional materials, hydrocarbon raw materials, and environmental problems of the industry. Undoubtedly, this large-scale and productive meeting of budding and leading specialists of the industry will be an important milestone in the history of applied mineralogy, will determine promising areas of industrial development of this science in Russia and abroad, and will contribute to overcoming significant technological barriers, creating new and improving existing industry collaborations.

Colleagues, I wish you successful reports and presentations, constructive dialogue, recognition of colleagues and like-minded people, strengthening the deserved authority of domestic science in the global educational and professional community!

Sergey N. Glagolev
Editor
Rector of Belgorod State Technological University named after V. G. Shukhov,
Doctor of Economics, Professor,
Co-chairman of the National Organizing Committee

14th ICAM

National Organizing Committee

Evgeny Savchenko (Chairman, Corresponding Member RAS)
Sergey Glagolev (Co-chairman)
Alexandr Gliko (Academician RAS)
Valery Maslennikov (Corresponding Member RAS)
Dmitry Pushcharovsky (Academician RAS)
Nina Zaitseva
Dmitry Rundkvist (Academician RAS)
Igor Burtsev
Valeria Strokova
Natalia Timonina

National Scientific Committee

Leonid Vaisberg (Chairman, Academician RAS)
Nikolay Bortnikov (Co-chairman, Academician RAS)
Olga Kotova (Executive Secretary)
Askhab Askhabov (Academician RAS)
Yury Marin (Corresponding Member RAS)
Nikolay Sobolev (Academician RAS)
Tamara Matveeva
Valentin Chanturiya (Academician RAS)
Elena Ozhogina
Sergey Krivovichev (Corresponding Member RAS)
Vladimir Shchiptsov
Olga Frank-Kamenetskaya (Associate Member)

ICAM Council

Saverio Fiore (President), Italy
Olga Kotova (Vice-President), Russia
Dieter Rammlmair (Secretary), Germany

Supporting from IMA-Commission for Applied Mineralogy

Maarten A. T. M. Broekmans (Chairman IMA-CAM), Norway
Jan Elsen (Secretary IMA-CAM), Belgium

International Board of Reviewers

Askhab Askhabov
Nikolay Bortnikov
Maarten A. T. M. Broekmans
Igor Burtsev
Valentin Chanturiya
Emin Ciftci
Faqin Dong
Saverio Fiore
Olga Frank-Kamenetskaya
Olga Kotova
Sergey Krivovichev
Yury Marin
Valery Maslennikov
Tamara Matveeva
Elena Ozhogina
Dmitry Pushcharovsky
Dieter Rammlmair
Vladimir Shchiptsov
Nikolay Sobolev
Valeria Strokova
Shiyong Sun
Natalia Timonina
Leonid Vaisberg

14th ICAM Organizers

 MINISTRY OF SCIENCE AND HIGHER EDUCATION

 *Russian Academy of Sciences*

    БГТУ им. В. Г. Шухова

 РААСН

General Sponsor

 Металлоинвест

Information Support

 *Вестник*

 РАЗВЕДКА И ОХРАНА НЕДР

 ОБОГАЩЕНИЕ РУД

 И РУДЫ МЕТАЛЛЫ

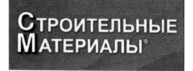 СТРОИТЕЛЬНЫЕ МАТЕРИАЛЫ®

 Научно-теоретический журнал *Вестник* Белгородского государственного технологического университета им. В.Г. Шухова ISSN 2071-7318

Contents

Geometallurgy, Technological Mineralogy and Processing of Mineral Raw

Scherrer Width and Topography of Illite as Potential Indicators for Contrasting Cu-Recovery by Flotation of a Chilean Porphyry Cu (Mo) Ore 3
G. Abarzúa, L. Gutiérrez, U. Kelm, and J. Morales

Ore Mineralogical Study of Cerattepe Au-Cu (±Zn) VMS Deposit (Artvin-Turkey) 7
İ. Akpınar and E. Çiftçi

Correlation Value of the Mineralogical Composition of Tills in the North of European Russia 11
L. Andreicheva

Comparative Gold Deportment Study on Direct Leaching and Hybrid Process Tails of Oxide Ores from Mayskoye Au Deposit (Chukotka, Russia) 15
I. Anisimov, A. Dolotova, A. Sagitova, M. Kharitonova, B. Milman, and I. Agapov

Mineral-Geochemical Criteria to Gold and Silver Recovery for Geometallurgical Sampling Campaign on Primorskoe Gold-Silver Deposit 19
I. Anisimov, A. Sagitova, O. Troshina, and I. Agapov

Mineralogical-Geochemical Criteria for Geometallurgical Mapping of Levoberezhnoye Au Deposit (Khabarovsk Region, Russia) 24
I. Anisimov, A. Sagitova, M. Kharitonova, A. Dolotova, and I. Agapov

**Mineralogical Reasons of Au Recovery Variability from
North-Western Pit of Varvara Au-Cu Mine (Kazakhstan)
and Criteria for Geometallurgical Mapping** 29
I. Anisimov, A. Dolotova, A. Sagitova, M. Kharitonova, and I. Agapov

Mineralogical Breakthrough into Nanoworld: Results and Challenges ... 33
A. Askhabov

**Case Study: Geochemistry and Mineralogy of Copper Mine Tailings
in Northern Central-Chile** 37
K. Berkh, D. Rammlmair, M. Drobe, and J. Meima

**Application of Fluoride Technology for Processing of Off-Grade
Aluminum Raw Materials** 41
I. Burtsev, I. Perovskiy, and D. Kuzmin

Applied Mineralogy for Complex and Profound Mineral Processing ... 45
V. Chanturiya and T. Matveeva

**Mineralogical and Technological Features of Tin Minerals
at Pravourmiysky Deposit (Khabarovsk Region)** 49
T. Chikisheva, S. Prokopyev, E. Kolesov, V. Kilin, A. Karpova,
E. Prokopyev, and V. Tukuser

**Ore Mineralogy of High Sulfidation Çorak-Taç Epimesothermal
Gold Deposit (Yusufeli-Artvin-Turkey)** 53
K. Diarra, E. Sangu, and E. Çiftçi

**Mineralogical and Technological Aspects of Phosphate
Ore Processing** 59
A. Elbendari, V. Potemkin, T. Aleksandrova, and N. Nikolaeva

Properties and Processing of Ores Containing Layered Silicates 66
A. Gerasimov, V. Arsentyev, and V. Lazareva

Applied Mineralogy of Anthropogenic Accessory Minerals 70
A. Gerasimov, E. Kotova, and I. Ustinov

**Crystal-Chemical and Technological Features of the KMA
Natural Magnetites** 75
T. Gzogyan and S. Gzogyan

**Practical Application of Technological Mineralogy on the Example
of Studying of Sulphidization in the KMA Ferruginous Quartzites** 80
S. Gzogyan and T. Gzogyan

**Ore Mineralogy of Kirazliyayla (Yenişehir-Bursa-Turkey)
Mesothermal Zn-Pb-(±Cu) Deposit: Preliminary Results** 84
F. Javid and E. Çiftçi

Use of Borogypsum as Secondary Raw   90
A. Khatkova, L. Nikitina, and S. Pateyuk

**New Approaches in X-ray Phase Analysis of Gypsum Raw Material
of Diverse Genesis**   94
V. Klimenko, V. Pavlenko, and T. Klimenko

**Nanotechnologies in Mineral-Geochemical Methods for Assessing
the Forms of Finding of Gold, Related Elements, Technological
Properties of Industrial Ores and Their Tails**   99
R. Koneev, R. Khalmatov, O. Tursunkulov, A. Krivosheeva,
N. Iskandarov, and A. Sigida

Applied Mineralogy of Mining Industrial Wastes   103
O. Kotova and E. Ozhogina

**High-Tech Elements in Minerals of Massive Sulfide Deposits:
LA-ICP-MS Data**   107
V. Maslennikov, S. Maslennikova, N. Aupova, A. Tseluyko, R. Large,
L. Danyushevsky, and U. Yatimov

Absolutely Pure Gold with High Fineness 1000‰   111
Z. Nikiforova

New Data on Microhardness of Placer Gold   115
Z. Nikiforova

**Modern Methods of Technological Mineralogy in Assessing
the Quality of Rare Metal Raw Materials**   119
E. Ozhogina, A. Rogozhin, O. Yakushina, Yu. Astakhova,
E. Likhnikevich, N. Sycheva, A. Iospa, and V. Zhukova

**Topochemical Transformations in Sodium-Bismuth-Silicate System
at 100–900 °C**   123
A. Pavlenko and R. Yastrebinskiy

**Ag-Bearing Mineralization of Nevenrekan Deposit
(Magadan Region, Russia)**   127
E. Podolian, I. Shelukhina, and I. Kotova

**Th/U Relations as an Indicator of the Genesis of Metamorphic
Zircons (On the Example of the North of the Urals)**   129
Y. Pystina and A. Pystin

**Phenomenon of Microphase Heterogenization by Means
of Endocrypt-Scattered Impurity of Rare and Noble Metals
as a Result of Radiation by Accelerated Electrons of Bauxites**   133
I. Razmyslov, O. Kotova, V. Silaev, and L. A. Gomze

Gold Extraction to Ferrosilicium, Production of Foam Silicate from Processing Tails of the Olimpiada Mining and Processing Complex Gold Processing Plant (Russia, Krasnoyarsk Territory) 136
A. Sazonov, V. Pavlov, S. Silyanov, and E. Zvyagina

Predictive Assessment of Quality of Mineral Aggregates Disintegration 140
S. Shevchenko, R. Brodskaya, I. Bilskaya, Yu. Kobzeva, and V. Lyahnitskaya

Development of Methods for Anti-filtration Formations Destruction Inside a Heap Leach Pile 143
H. Tcharo, M. Koulibaly, and F. K. N. Tchibozo

Microtomographic Study of Gabbro-Diabase Structural Transformations Under Compressive Loads 146
L. Vaisberg and E. Kameneva

Process Mineralogy as a Basis of Molybdoscheelite Ore Preparation ... 152
L. Vaisberg, O. Kononov, and I. Ustinov

Crystallomorphology of Cassiterite and Its Practical Importance 157
I. Vdovina

Modal Analysis of Rocks and Ores in Thin Sections 162
Yu. Voytekhovsky

Quality Assurance Support (QA/QC System) of Mineralogical Analysis 167
O. Yakushina, E. Gorbatova, E. Ozhogina, and A. Rogozhin

Mineral Preparation in Geological Research 172
T. Yusupov, A. Travin, S. Novikova, and D. Yudin

Industrial Minerals, Precious Stones, Ores and Mining

Impact Diamonds: Types, Properties and Uses 179
V. Afanasiev, N. Pokhilenko, A. Eliseev, S. Gromilov, S. Ugapieva, and V. Senyut

Authentic Semi-precious and Precious Gemstones of Turkey: Special Emphasis on the Ones Preferred for Prayer Beads 183
E. Çiftçi, H. Selim, and H. Sendir

Biooxidation of Copper Sulfide Minerals 189
Yu. Elkina, E. Melnikova, V. Melamud, and A. Bulaev

Genetic Problem of Quartz in Titanium Minerals in Paleoplacers of Middle Timan 192
I. Golubeva, I. Burtsev, A. Ponaryadov, and A. Shmakova

**Gold and Platinum Group Minerals (PGM) from the Placers
of Northwest Kuznetsk Alatau (NWKA) (South Siberia, Russia)** 195
V. Gusev, S. Zhmodik, G. Nesterenko, and D. Belyanin

**Noble Metal Mineralization of the PGM Zone "C" of the East-Pana
Layered Intrusion (Kola Peninsula)** 198
O. Kazanov, G. Logovskaya, and S. Korneev

Shungites and Their Industrial Potential 201
V. Kovalevski and V. Shchiptsov

**Gold-Silver Natural Alloy of Chromitites from the Kamenushinsky
Massif (The Middle Urals)** 205
A. Minibaev

**Microbial Processes in Ore-Bearing Laterite at the Tomtor Nb-REE
Deposit: Evidence from Carbon Isotope Composition in Carbonates** ... 208
V. Ponomarchuk, E. Lazareva, S. Zhmodik, N. Karmanov, and A. Piryaev

Peridot: Types of Deposits and Formation Conditions 212
S. Sokolov

**Mineralogical Analysis of Glacial Deposits and Titanium
Paleoplacers of the East European Part of Russia** 214
N. Vorobyov and A. Shmakova

Oil and Gas Reservoirs, Including Gas Hydrates

**A Bench Scale Investigation of Pump-Ejector System
at Simultaneous Water and Gas Injection** 219
S. Karabaev, N. Olmaskhanov, N. Mirsamiev, and J. Mugisho

Integrated Use of Oil and Salt Layers at Oil Field Development 221
V. Malyukov and K. Vorobyev

**Oil and Gas Reservoirs in the Lower Triassic Deposits
in the Arctic Regions of the Timan-Pechora Province** 223
N. Timonina

**Associated Petroleum Gas Flaring: The Problem
and Possible Solution** 227
A. Vorobev and E. Shchesnyak

**Innovative Technology of Using Anti-sand Filters at Wells
of the Vankor Oil and Gas Field** 231
K. Vorobyev and A. Gomes

Analytical Methods, Instrumentation and Automation

Thermometry of Apatite Saturation (The Kozhym Massif,
The Subpolar Urals).................................... 235
Y. Denisova, A. Vikhot, O. Grakova, and N. Uljasheva

Studies of Structural Changes in Surface and Deep Layers
in Magnetite Crystals After High Pressure Pressing............... 239
P. Matyukhin

The Potential of Lacquer Peel Profiles and Hyperspectral Analysis
for Exploration of Tailings Deposits.......................... 244
W. Nikonow and D. Rammlmair

Methods of Extraction of Micro- and Nanoparticles of Metal
Compounds from Fine Fractions of Rocks, Ores
and Processing Products................................... 248
A. Smetannikov and D. Onosov

Advanced Materials with Improved Characteristics, Including Technical Ceramics and Glass

Efficiency Evaluation for Titanium Dioxide-Based Advanced
Materials in Water Treatment 255
M. Harja, O. Kotova, S. Sun, A. Ponaryadov, and T. Shchemelinina

The Use of Karelia's High-Mg Rocks for the Production of Building
Materials, Ceramics and Other Materials with Improved Properties ... 259
V. Ilyina

Kinetic Features of Formation of Supramolecular Matrices
on the Basis of Silica Monodisperse Spherical Particles 263
D. Kamashev

Three-Cation Scandium Borates $R_xLa_{1-x}Sc_3(BO_3)_4(R = Sm, Tb)$:
Synthesis, Structure, Crystal Growth and Luminescent Properties 267
A. Kokh, A. Kuznetsov, K. Kokh, N. Kononova, V. Shevchenko,
B. Uralbekov, A. Bolatov, and V. Svetlichnyi

Rational Usage of Amorphous Varieties of Silicon Dioxide in Dry
Mixtures of Glass with Specific Light Transmittance 272
N. Min'ko and O. Dobrinskaya

Peculiarities of Phase Formation in Artificial Ceramic Binders
for White-Ware Compositions 277
I. Moreva, E. Evtushenko, O. Sysa, and V. Bedina

Experimental Modeling of Biogeosorbents 281
T. Shchemelinina, O. Kotova, E. Anchugova, D. Shushkov,
G. Ignatyev, and M. Markarova

**Heating Rate and Liquid Glass Content Influence on Cement
Brick Dehydration** 286
V. Strokova and D. Bondarenko

**Structure and Surface Reactivity Mediated Enzymatic
Performances of Clay-Based Nanobiocatalyst**.................... 290
S. Sun, K. Wang, F. Dong, B. Ma, T. Huo, Y. Zhao, H. Yu, Y. Huang,
and J. Huang

**Structural-Phase Stabilization of Clay Materials
in Hydrothermal Conditions**.................................. 292
O. Sysa, E. Evtushenko, I. Moreva, and V. Loktionov

**Phase Changes in Radiation Protection Composite Materials Based
on Bismuth Oxide**.. 296
S. Yashkina, V. Doroganov, E. Evtushenko, O. Gavshina, and E. Sysa

**Development of Technology for Anti-corrosion Glass Enamel
Coatings for Oil Pipelines**.................................... 300
E. Yatsenko, A. Ryabova, and L. Klimova

Building Materials

**Optimization of Formulations of Cement Composites Modified
by Calcined Clay Raw Material for Energy Efficient Building
Constructions** 307
A. Balykov, T. Nizina, V. Volodin, and D. Korovkin

Santa Maria Clays as Ceramic Raw Materials.................. 311
Â. Cerqueira, C. Sequeira, D. Terroso, S. Moutinho, C. Costa,
and F. Rocha

**Alkaline Activation of Rammed Earth Material – "New Generation
of Adobes"** 313
C. Costa, D. Arduin, C. Sequeira, D. Terroso, S. Moutinho, Â. Cerqueira,
A. Velosa, and F. Rocha

**Structurization of Composites When Using 3D-Additive Technologies
in Construction**.. 315
M. Elistratkin, V. Lesovik, N. Chernysheva, E. Glagolev, and P. Hardaev

**Influence of Flow Blowing Agent on the Properties of Aerated
Concrete Variable Density and Strength** 319
V. Galdina, E. Gurova, P. Deryabin, M. Rashchupkina, and I. Chulkova

**Structuring Features of Mixed Cements on the Basis
of Technogenic Products** 323
M. Garkavi, A. Artamonov, E. Kolodezhnaya, A. Pursheva,
and M. Akhmetzyanova

Use of Slags in the Production of Portland Cement Clinker 327
V. Konovalov, A. Fedorov, and A. Goncharov

**Geopolymerization and Structure Formation in Alkali Activated
Aluminosilicates with Different Crystallinity Degree** 331
N. Kozhukhova, V. Strokova, I. Zhernovsky, and K. Sobolev

**Matrix Instruments for Calculating Costs of Concrete
with Multicomponent Binders** 335
T. Kuladzhi, S.-A. Murtazaev, S. Aliev, and M. Hubaev

Characterisation of Perovskites in a Calcium Sulfo Aluminate Cement ... 339
G. Le Saout, R. Idir, and J.-C. Roux

**Geonics (Geomimetics) as a Theoretical Basis for New
Generation Compositing** 344
V. Lesovik, A. Volodchenko, E. Glagolev, I. Lashina, and H.-B. Fischer

**Regularities in the Formation of the Structure and Properties
of Coatings Based on Silicate Paint Sol** 348
V. Loganina, E. Mazhitov, and V. Demyanova

**Influence of Sodium Oxide on Brightness Coefficient
of Portland Cement Clinker** 352
D. Mishin and S. Kovalyov

Production of Bleached Cement 356
D. Mishin and S. Kovalev

Multicomponent Binders with Off-Grade Fillers 360
S.-A. Murtazaev, M. Salamanova, M. Saydumov, A. Alaskhanov,
and M. Khubaev

**High-Quality Concretes for Foundations of the Multifunctional
High-Rise Complex (MHC) «Akhmat Tower»** 365
S.-A. Murtazaev, M. Saydumov, A. Alaskhanov, and M. Nakhaev

**Designing High-Strength Concrete Using Products of Dismantling
of Buildings and Structures** 369
T. Murtazaeva, A. Alaskhanov, M. Saidumov, and V. Hadisov

**Estimation of Rheo-Technological Effectiveness of Polycarboxylate
Superplasticizer in Filled Cement Systems in the Development
of Self-compacting Concrete for High-Density Reinforced
Building Constructions** 372
T. Nizina, A. Balykov, V. Volodin, and D. Korovkin

Parameters of Siliciferous Substrate of Photocatalytic Composition Material as a Factor of Its Efficiency 376
Y. Ogurtsova, E. Gubareva, M. Labuzova, and V. Strokova

Properties Improvement of Metakaolin-Zeolite-Diatomite-Red Mud Based Geopolymers 381
F. Rocha, C. Costa, W. Hajjaji, S. Andrejkovičová, S. Moutinho, and A. Cerqueira

Features of Production of Fine Concretes Based on Clinkerless Binders of Alkaline Mixing 385
M. Salamanova, S.-A. Murtazaev, A. Alashanov, and Z. Ismailova

Impact of Thermal Modification on Properties of Basalt Fiber for Concrete Reinforcement 389
V. Strokova, V. Nelyubova, I. Zhernovsky, O. Masanin, S. Usikov, and V. Babaev

Activation of Cement in a Jet Mill 393
S. Titov and A. Kazakov

The Law of Similarity and Designing High-Performance Composites ... 395
A. Tolstoy, V. Lesovik, E. Glagolev, and L. Zagorodniuk

Genesis of Clay Rock of the Incomplete Stage of Mineral Formation as a Raw Material Base for Autoclaved Materials 399
A. Volodchenko and V. Strokova

Abnormal Mineral Formation in Aluminate Cement Stone 403
I. Zhernovsky, V. Strokova, V. Nelyubova, Yu. Ogurtsova, and M. Rykunova

Structural Transformations of Low-Temperature Quartz During Mechanoactivation 407
I. Zhernovsky and V. Strokova

Biomimetic Materials on a Mineral Basis, Biomineralogy

Effect of Earthquake on the Landscape of Jiuzhaigou-Huanglong Travertine and Its Restoration 413
F. Dong, Q. Dai, Q. Li, F. Wang, and Y. Luo

Microbial Colonies in Renal Stones 415
A. Izatulina, M. Zelenskaya, and O. Frank-Kamenetskaya

Fabrication of ZnO/Palygorskite Nanocomposites for Antibacterial Application 419
Y. Kang, A. Hui, and A. Wang

Bacterial Oxidation of Pyrite Surface 423
S. Lipko, I. Lipko, K. Arsent'ev, and V. Tauson

**Biomimetic Superhydrophobic Cobalt Blue/Clay Mineral Hybrid
Pigments with Self-cleaning Property and Different Colors** 427
B. Mu, A. Zhang, and A. Wang

Silicon Dioxide in Mineralized Heart Valves 432
A. Titov, V. Zaikovskii, and P. M. Larionov

**Preparation of Macroporous Adsorbent Based on Montmorillonite
Stabilized Pickering Medium Internal Phase Emulsions** 436
F. Wang, Y. Zhu, W. Wang, and A. Wang

Environment and Energy Resources

Depletion of the Land Resources and Its Effect on the Environment . . . 443
M. Abou Zahr Diaz, M. A. Alawiyeh, and M. Ghaboura

**Geochemical Behavior of Heavy Metals During Treatment by
Phosphoric Fertilizer at a Dumping Site in Kabwe, Zambia** 445
H. Kamegamori, K. Lawrence, T. Sato, and T. Otake

**Murataite-Pyrochlore Ceramics as Complex Matrices for Radioactive
Waste Immobilization: Structural and Microstructural Mechanisms
of Crystallization** 447
S. Krivovichev, S. Yudintsev, A. Pakhomova, and S. Stefanovsky

**Cs Leaching Behavior During Alteration Process of Calcium Silicate
Hydrate and Potassium Alumino Silicate Hydrate** 451
K. Kuroda, K. Toda, Y. Kobayashi, T. Sato, and T. Otake

Environmental Pollution Problems in the Mining Regions of Russia . . . 453
E. Levchenko, I. Spiridonov, and D. Klyucharev

**Environmental Solutions for the Disposal of Fine White
Marble Waste** 457
I. Shadrunova, T. Chekushina, and A. Proshlyakov

**Security Test of New Technology in View of Increased Performance
of Oil Platforms Without Increasing Environmental Risks** 461
E. M. Tanoh Boguy and T. Chekushina

**Calcite Mineral Generation in Cold-Water Travertine
Huanglong, China** 463
F. Wang, F. Dong, X. Zhao, Q. Dai, Q. Li, Y. Luo, and S. Deng

**Optimization of the Natural-Technical System "Iron Ore Quarry"
Management Based on the Algorithm of the Rock Mass
Stability Ensuring**   466
L. Yarg, I. Fomenko, and D. Gorobtsov

**Utilization of Associated Oil Gas: Geo-ecological Problems
and Modernization of the State**   471
L. Z. Zhang and H. Y. Sun

Cultural Heritage, Artifacts and Their Preservation

**Identifying the Decorative Stone Samples from the Mining
Museum's Collection: First Results**   475
N. Borovkova and M. Machevariani

**Monitoring of the State of St. Petersburg Stone Monuments
and the Strategy of Their Preservation**   479
O. Frank-Kamenetskaya, D. Vlasov, V. Rytikova, V. Parfenov,
V. Manurtdinova, and M. Zelenskaya

Ceramics Sugar Jars Pieces from Aveiro Production   483
S. Moutinho, C. Costa, Â. Cerqueira, C. Sequeira, D. Terroso, J. Nobre,
P. Morgado, A. Velosa, and F. Rocha

Author Index   485

Geometallurgy, Technological Mineralogy and Processing of Mineral Raw

Scherrer Width and Topography of Illite as Potential Indicators for Contrasting Cu-Recovery by Flotation of a Chilean Porphyry Cu (Mo) Ore

G. Abarzúa[1], L. Gutiérrez[2], U. Kelm[1(✉)], and J. Morales[3]

[1] Instituto de Geología Económica Aplicada (GEA),
Universidad de Concepción, Concepción, Chile
ukelm@udec.cl, ued.kelm@gmail.com
[2] Departamento de Ingeniería Metalúrgica, Facultad de Ingeniería,
Universidad de Concepción, Concepción, Chile
[3] Departamento de Geología, Universidad de Salamanca, Salamanca, Spain

Abstract. Two contrasting feeds in terms of copper recovery from a Cu (Mo) porphyry deposit but with similar overall mineralogy have been characterized by X-ray diffraction for their <1 µm fraction illite crystallinity, Scherrer width and by atomic force microscopy for surface roughness. The unfavorable feed displayed slightly higher crystallinity, larger Scherrer width and surface roughness factors, than the feed with good Cu recovery. As Scherrer width is an easy and cheaply to determine parameter it is suggested as complementary information to particle size distribution analyses when dealing with feeds where illite may affect pulp viscosity or gangue adhesion to bubbles during flotation.

Keywords: Illite · Scherrer width · Surface roughness · Flotation · Sericite · Atomic force microscopy

1 Introduction

For over half a century, illite crystallinity has been used as an indicator of mineral maturity in metasediments between the transition of diagenesis to very low temperature metamorphism and the incipient low-grade metamorphism or epizone (Frey 1999). Illite crystallinity (later Kübler Index) measured at full width half medium height (FWHM) of the basal XRD-reflection is also an indirect indicator for the size of the jointly diffracting illite sheets, also known as Scherrer width, which has been directly visualized with the widespread availability of Transmission Electron Microscopy (Frey 1999). Superimposition of metamorphic and hydrothermal alteration processes, paired with time consuming analytical routines, has limited the application to alteration halos of ore deposits; an exception is the study by Beaufort et al. (2005) on the East Alligator River Uranium deposit in the Northern Territory, Australia, due to its abundance of chlorite and illite gangue. Sericitic alteration (muscovite/illite) also represents widespread gangue for Andean Cu (Mo) porphyry ore deposits, but systematic studies of phyllosilicate crystallinity or Scherrer width within ore deposit areas are not available.

© The Author(s) 2019
S. Glagolev (Ed.): ICAM 2019, SPEES, pp. 3–6, 2019.
https://doi.org/10.1007/978-3-030-22974-0_1

Cheng and Peng (2018) suggest negative effects for low crystallinity kaolinite rich ore, much of the work being based on artificial ore-gangue mixtures. However Jorjani et al. (2011) single out illite and vermiculate as key gangue affecting flotation for the Iranian Sarcheshmeh porphyry copper deposit. The present exploratory study has been sparked by the effort to develop a formula for blending ore based on mineralogical-chemical parameters of a giant porphyry copper deposit in Chile. Ore with similar sericitic alteration and chalcopyrite dominated ore phases, were identified by the concentrator operation as favorable (F) and unfavorable (UF) floating feed, the latter entering the concentrator only as blend. Given this overall mineralogical similarity, it was decided to characterize the contrasting feeds based on their illite crystallinity – Scherrer width and concomitant surface roughness.

2   Methods and Approaches

Triplicate samples of favorable (F) and unfavorable (UF) floating ore were prepared for X-ray diffraction (XRD) analysis of the clay size fraction (>0.45 <1 µm) based on the recommendations of Moore and Reynolds (1997). XRD measurements were carried out on a Bruker D4 diffractometer operated with Ni-filtered Cu-radiation. Illite crystallinity and Scherrer width were determined on the 001 basal reflection following Lorentzian adjustment using the Origin 8.5 program. Atomic Force Microscope (AFM) measurements of topography were carried out with an AIST-NT equipment in contact mode on a 5×5 µm surface. WSxM5.0 software was used for calculating Ra (arithmetic average) and Rrms (root mean square) roughness factors (Horcas et al. 2007; Erinosho et al. 2018).

3   Results and Discussion

Illite crystallinity values of both samples (F: 0.15 Δ° 2Θ, UF: 0.12 Δ° 2Θ) correspond to epizone values, and for the unfavorable feed are at the sensitivity limit of this method (Frey 1999). Though values for the favorable feed are marginally lower, nevertheless this difference is expressed in an increased Scherrer width or crystallite size (F: 48.8 nm, UF: 62.6 nm) for the unfavorable feed.

Topographic images of sample surfaces show different roughness, being the favorable feed (F) the smoother. Roughness factors were calculated for different surfaces scanned by AFM. As a mean, 10 sample surfaces were measured and statistically compared, giving values of Ra = 57 ± 20 and Rrms = 66 ± 23 for the favorable feed and Ra = 68 ± 18 and Rrms = 86 ± 19 for the unfavourable feed sample. Undoubtedly, different scales of images analysed imply changes in the surface parameters. To avoid this, images with the same size have been compared. As it can be observed in the Fig. 1, roughness parameters confirm the XRD results.

Despite the difficulties of differentiating between illite generations in rocks with sericitic overprint in porphyry (and other) ore deposits, this simple XRD measurement permits concomitant calculation of coherently diffracting particle sizes or Scherrer width for a given geometallurgical unit. The correspondence observed for this

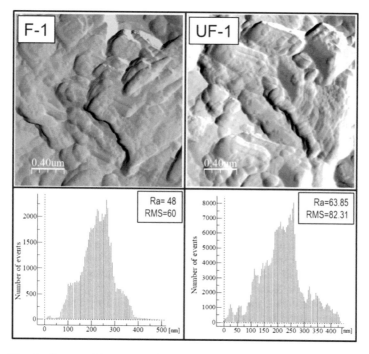

Fig. 1. Topographic aspects of F-1 sample (left) and UF-1 sample (right)

exploratory study between increased Scherrer width and higher AFM measured surface roughness, recommends the XRD based value as an easily available tool to assess difference in flotation for samples where traditional methods like optical microscopy, automated mineralogy, and semi-quantitative whole rock mineralogy do not reveal obvious mineralogical differences between feeds with contrasting flotation behavior. Farrokhpay and Ndlovu (2013) discussed the effect of clay particle size on pulp rheology; here the X-ray coherent Scherrer width is suggested as a complementary indicator to the particle size distribution measurements by laser diffraction in the clay size range. The scale of surface roughness as a factor impacting on particle adhesion to bubbles has been studied by Nikolaev (2016). However, direct AFM measurements are still no routine procedures to define geometallurgical units within an ore deposit, whereas XRD information can be generated faster and in a more standardized fashion.

4 Conclusions

Illite crystallinity, Scherrer width and AFM-determined surface roughness have been determined for two flotation-feed of a Cu (Mo) porphyry copper deposit with contrasting Cu-recovery. Samples did not display any mineralogical difference allowing a straightforward explanation of this difference. For this exploratory study case, Scherrer width is considered an easy to obtain parameter that points to differences in surface roughness of illite particles in the <1 μm fraction and thus may influence pulp viscosity and/or particle adhesion to bubbles during flotation.

Acknowledgements. Dr Manuel Melendrez, Departamento de Ingeniería de Materiales, Universidad de Concepción is thanked for access to the AFM equipment.

References

Beaufort D, Patrier P, Laverret E (2005) Clay alteration associated with proterozoic unconformity-type uranium deposit in the East Alligator Rivers uranium fields, Northern Territory, Australia. Econ Geol 100:515–536

Chen X, Peng Y (2018) Managing clay minerals in froth flotation. A critical review. Miner Process Extr Metall Rev 39(5):289–307

Erinosho MF, Akinlabi ET, Johnson OT (2018) Characterization of surface roughness of laser deposited titanium alloy and copper using AFM. Appl Surf Sci 435:393–397

Farrokhpay S, Ndlovu B (2013) Effect of phyllosilicate minerals on the rheology, colloidal and flotation behaviour of chalcopyrite mineral. In: Australasian conference on chemical engineering, Chemeca 2013, Challenging Tomorrow, p 733

Frey M (1999) Very low-grade metamorphism of clastic sedimentary rocks: in Low Temperature Metamorphism. Blackie and Sons, Glasgow, pp 9–58

Horcas I, Fernández R, Gomez-Rodriguez JM, Colchero JWSX, Gómez-Herrero JWSXM, Baro AM (2007) WSXM: a software for scanning probe microscopy and a tool for nanotechnology. Rev Sci Instrum 78:013705

Jorjani E, Barkhordari HR, Khorami MT, Fazeli A (2011) Effects of aluminosilicate minerals on copper–molybdenum flotation from Sarcheshmeh porphyry ores. Miner Eng 24(8):754–759

Moore DM, Reynolds RC Jr (1997) X-ray diffraction and the identification and analysis of clay minerals. Oxford University Press, Oxford

Nikolaev A (2016) Flotation kinetic model with respect to particle heterogeneity and roughness. Int J Miner Process 155:74–82

Ore Mineralogical Study of Cerattepe Au-Cu (±Zn) VMS Deposit (Artvin-Turkey)

İ. Akpınar[1(✉)] and E. Çiftçi[2]

[1] Department of Geological Engineering, Faculty of Engineering and Natural Science, Gumushane University, Gumushane, Turkey
hiakpinar@gumushane.edu.tr
[2] Department of Geology, Faculty of Mines, Istanbul Technical University, Istanbul, Turkey

Abstract. The Cerattepe mine, one of the volcanogenic massive sulfide deposits in northeastern Turkey, is hosted within the late Cretaceous volcanic, intrusive and sedimentary rocks. Deposit's main ore body contains high-grade massive copper sulfides and a gold-silver and barium rich oxide zone, characterized by dense alteration stages, is situated on top of it. Replacement, cataclastic, breccia, dissemination, dendritic, concentric growth, colloform, and framboidal textures were identified. Pyrite, sphalerite, marcasite, chalcopyrite, bornite, galena, tennantite-tetrahedrite, gold, covellite, digenite, chalcocite, cuprite and cubanite constitute the mineral paragenesis where quartz, calcite and barite account for the gangue minerals. Limonite, hematite, lepidocrocite, malachite, azurite and jarosite developed in the oxidation zone.

Keywords: Eastern Pontide · Cerattepe · VMS deposit · Framboidal pyrite · Bird's eye-texture

1 Introduction

The eastern Pontide orogenic belt of Turkey is an important segment of the Tethyan-Eurasian Metallogenic Belt. This belt carries a special importance in metallogeny of Turkey and hosts numerous VMS deposit. (Akıncı 1984; Çiftçi and Hagni 2005; Güven 1993; Revan et al. 2013; Zaykov et al. 2006; Yiğit 2005). The most of the eastern Pontide VMS deposits show some similarities in many aspects to the Kuroko deposits of Japan (Çiftçi and Hagni 2005; Ciftci 2000; Pejatoviç 1979). The Cerattepe Deposit is a Kuroko-type VMS deposit located in late Cretaceous age volcanic, intrusive and sedimentary rocks. It is distinguished by an unusual basal zone of high-grade copper sulfides and an overlying Au-rich oxide zone from the other VMS deposits of northeastern Pontides.

2 Methods and Approaches

A total of 46 samples representing oxide (14), sulfide (28) and stringer zones (4) of Cerattepe ore deposit were collected from drill cores, underground audits and surface outcrops. Polished sections were prepared and Nikon Eclipse LV100 reflected light

© The Author(s) 2019
S. Glagolev (Ed.): ICAM 2019, SPEES, pp. 7–10, 2019.
https://doi.org/10.1007/978-3-030-22974-0_2

microscopy was employed for examination. The ore minerals and the paragenesis were identified on the basis of their petrographical features and their textural relationships, respectively. Electron Probe Micro Analysis (EPMA) and Secondary Electron Microscopy-Energy Dispersive Spectroscopy (SEM-EDS) were used for chemistry of sulfide minerals.

3 Results and Discussion

The mineral paragenesis (Fig. 1) of Cerattepe VMS deposit comprises of pyrite, sphalerite, marcasite, chalcopyrite, bornite-idaite, galena, covellite, chalcocite, cubanite, cuprite including sulfosalts (mainly tennantite and lesser tetrahedrite) gold, silver, arsenopyrite and bournonite.

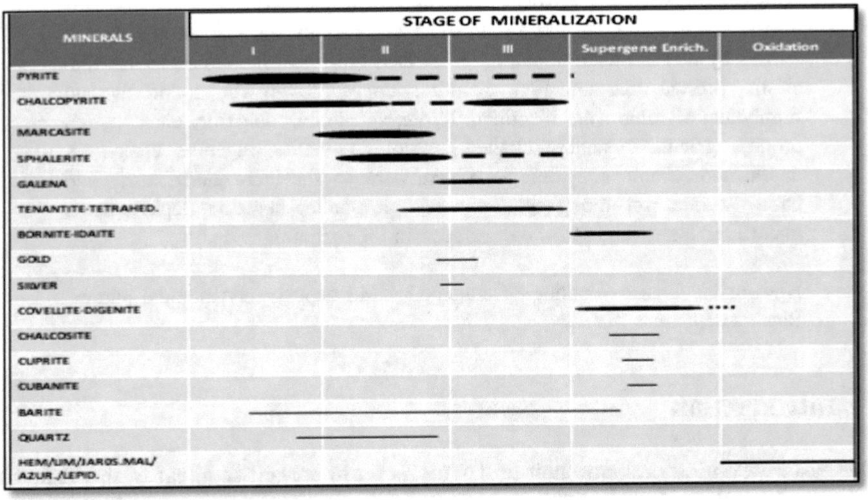

Fig. 1. Ore mineral paragenetic sequence of Cerattepe Au-Cu (±Zn) VMS deposit

Gangue minerals include barite, quartz, gypsum, anhydrite and calcite. Hematite, limonite, lepidocrocite, malachite and azurite, and jarosite are the oxidation minerals. Four generation of pyrite is specified. Pyrite I is represented by colloform-concentric textured (Fig. 2) grains formed from initial solution reached the seafloor from the chimney. The colloform textured pyrites are progressively overgrown later coarse crystalline grains, which are the second generation pyrites.

These are later extensively altered to marcasite. The framboidal-pelletal textured collomorphic grains, seen sometimes coeval with sphalerite and galena, are the third generation pyrites. The fourth generation pyrites are small sized euhedral, subhedral grains observed on transition zone of sulfide ore- footwall rock and veinlets of stringer zone.

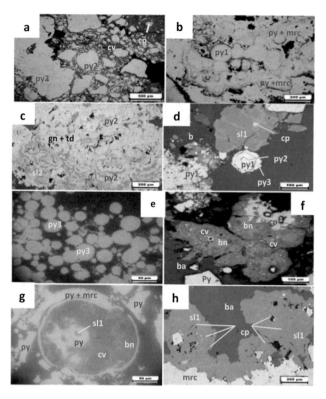

Fig. 2. Photomicrographs of selected polished sections showing observed textures in Cerattepe deposit. (a) Cataclastic texture of pyrite grains. (b) Colloform textured pyrite and marcasite intergrowth surrounded by sphalerite "Bird's eye texture"; (c) Colloform textured pyrite aggregate with a spongy textured intergrowth zones comprised of sphalerite, galena and tetrahedrite; (d) Three different generation of pyrites, (e) Pelletal pyrite framboids with minor sphalerite and galena in the cracks; (f) Bornite and chalcopyrite replaced by covellite; (g) collomorphically banded sulfides in melnikovite pyrite; (h) Dissociation texture of chalcopyrite in sphalerite replaced by marcasite

Most of the minerals are small - fine grained, and the larger grains of the major minerals are in the order of 100–800 μm up to 1.2–2 mm in size and some rare pyrite grains have a size of 5 mm. Most of the minor and trace minerals are much smaller, typically in the order of 1–20 μm in size. Majority of ore minerals are anhedral with the exception of pyrite, quartz, barite and some sphalerite and galena occur as euhedral to subhedral crystals. Observed ore textures are dissemination and veinlet textures in the stockwork and siliceous ore zones, whereas replacement, overgrowth, concentric and colloform textures become prevalent in the massive ore, particularly in the center of main deposit.

Banded textures of black and yellow ore are seen with polymetallic sulfides. In the outer part of the main ore body, at lateral zones, clastic or fragmental ore textures are present.

4 Conclusions

It is concluded that the ore mineral assemblage and textures observed in Cerattepe VMS deposit are comparable to those of other VMS deposits occur in north eastern Pontide region, Kuroko deposits of Japan and also comparable modern seafloor–seamount VMS deposits in the world.

Acknowledgements. This study was financially supported by BAP Project unit (No: 39615) of Istanbul Technical University in Turkey. We are thankful to Mr. Ünsal Arkadaş and staff of Etibakır A.Ş. in Artvin, Turkey for their courteous support for the field work.

References

Akıncı ÖT (1984) The geology and the metallogeny of the Eastern Pontides (Turkey). In: International world geological congress abstracts, pp 197–198

Ciftci E (2000) Mineralogy, paragenetic sequence, geochemistry, and genesis of the gold and silver bearing upper cretaceous mineral deposits, Northeastern Turkey. Ph.D. Thesis, University of Missouri-Rolla, Rolla, MO, USA (unpublished)

Çiftçi E, Hagni RD (2005) Mineralogy of the Lahanos deposit a Kuroko-type VMS deposit from the eastern Pontides (Giresun, NE Turkey). Geol Bull Turk 48(1):55–64

Güven İH (1993) Artvin Kafkasör Sahası Maden Jeolojisi Raporu No 2600, MTA, Ankara. metallogeny. Ore Geol Rev 147–179

Pejatovic S (1979) Metallogeny of the Pontid type massive sulphide deposits, Ankara, 98 p

Revan MK, Genç Y, Maslennikov V, Ünlü T, Delibaş O, Hamzaçebi S (2013) Original findings on the ore-bearing facies of volcanogenic massive sulphide deposits in the eastern black sea region (NE Turkey). Bull MTA 147:73–89

Yigit O (2005) Gold in Turkey - a missing link and metallogenic features of the in Tethyan

Zaykov V, Novoselov K, Kotlyarov V (2006) Native gold and tellurides in the Murgul and Çayeli volcanogenic Cu deposits (Turkey)

Correlation Value of the Mineralogical Composition of Tills in the North of European Russia

L. Andreicheva[✉]

Institute of Geology KomiSC UB RAS, Syktyvkar, Russia
andreicheva@geo.komisc.ru

Abstract. The material composition of the tills, including its mineral component, is formed during of the exaration-accumulative activity of the cover glacier and depends on the composition of the rocks of the glaciation centers, transit areas, local underlying rocks and on the relief of the preglacial bed. Thus the composition of the till is conditioned by the total influence of source glacier provinces and represents an average sample of the rocks on the way of the glacier.

Keywords: Mineral composition · Correlation · Till · Petrographic composition

1 Introduction

The studies were conducted on a vast and geologically heterogeneous territory of the Timan-Pechora-Vychegda region, which determined the variability of lithological and mineralogical parameters of the tills in term of the regional plan. This restricts their use for correlations. But taking into account the factors of the glacial sedimentation genesis and the complex study of the tills, the stratigraphic confinement of moraine horizons is established very confidently. The features of the mineral-petrographic composition of the tills and the rocks of the source provinces (remote, transit and local) are discussed in detail in the papers by Andreicheva (1994, 2017).

2 Results and Discussion

Pomusovski (Okskii) till in the lower Pechora and Laya is characterized by a low content of a heavy fraction (0.26–0.4%) and amphibole (10–12%) - garnet (12–19%) - epidote (35–41%) mineral association. The concentration of titanium minerals (rutile, titanite, leucoxene) is 7–9%, the total amount of pyrite and siderite does not exceed 10–11%. Eastward (in the valleys of the Kolvaand Pechora Rvers - in borehole 301-Kushshor) the yield of the heavy fraction is significantly higher - 0.8–1.16%, but the epidote content decreases (to 26–30%), amphiboles (to 5–7%) and the total concentration of pyrite and siderite increases sharply to 37%. In borehole 301-Kushshor the

© The Author(s) 2019
S. Glagolev (Ed.): ICAM 2019, SPEES, pp. 11–14, 2019.
https://doi.org/10.1007/978-3-030-22974-0_3

number of garnets (up to 23%) and the content of titanium minerals (up to 11%) increase.

The decrease in the content of amphiboles in the Pomusovskitill eastward with the underlying Aptian and Albianrocks. They are characterized by the epidote-amphibole mineral association with a distinct W-E decreasing content of amphiboles. The enrichment of the heavy fraction with pyrite and siderite in the of Kolvavalley is associated with assimilation of these minerals from the Mesozoic sediments. The petrographic composition of the fragments and their south-eastward orientation testifies to the formation of the pomusovskytill in the region due to source from Fennoscandinavia and the Northern Timan.

The composition of the heavy fraction of Pechora(Dneprovian) till is also characterized by areal variability, reflecting the specifics of the mineral composition of the rocks of remote, transit and local source provinces. The minimum yield of the heavy fraction is 0.33%, noted in Pechoratill in the middle Pechora River, the maximum - 0.86%, in the SeydaRver. The mineral association is represented by epidote, garnets, amphiboles, pyrite, siderite and ilmenite. In some areas, the role of titanium minerals increases, most often due to an increase in the amount of leucoxene, sometimes the content of metamorphogenic minerals is increased: kyanite, staurolite, sillimanite.

Epidote prevails in the Pechoratill, its quantity varies from 14 to 37%. Maximum concentrations are recorded in sections of the middle Pechora, the lowest - in the ChernayaRver. The number of garnets is increased to 21% in the ShapkinaRver, reduced to 5% - on the valley of the SeydaRver. The content of amphiboles is low: its maximum concentration (16%) is marked in the VychegdaRver, and the lowest - on the right bank of the UsaRver, where is only 0.6%, which is associated with the complete absence of amphiboles in the underlying Mesozoic rocks. The amount of pyrite also varies significantly: its minimum content (3%) is in the basin of the ShapkinaRver, the maximum in the SeydaRver- 27%. Siderite in the LayaRver is 8%, in the sections in the ChernayaRver reaches of 27%, pyrite and siderite are constantly present. Their total concentrations are different, but the ratio is stable: almost everywhere siderite dominates pyrite. In the petrographic thin sections of the pechorskytill, glauconite is contained in significant quantities (up to 60 grains in a standard petrographic section), which, like pyrite, siderite, is characteristic of local Mesozoic rocks. The composition of the minerals of the heavy fraction testifies to the participation of underlying Triassic, Jurassic and Cretaceous rocks in the formation of the mineral spectrum of the till (Andreicheva and Nikitenko 1989). Single reference boulders are constantly present - pink crinoid-bryozoan Novaya Zemlya limestones. Another characteristic feature of the Pechora glaciation is a steady south-western trend of ice movement, which is consistent with the petrographic composition of rock fragments in the till. The obtained data testify to the Paykhoy-Ural-Novaya Zemlya center of glaciation in the Pechora time throughout the region.

The peculiarity of the paleogeography of Vychegda (Moskovian) glaciation is the existence in the west and in the east of the region of the ice covers of various centers. Accordingly, the material composition of the till was formed by the material of different source provinces. In the western and central parts of the region, the heavy fraction of till, which ranges from 0.42% in the lower Pechora River to 0.95% in the Vychegda valley, is represented by an amphibole-garnet-epidote mineral association with increased total pyrite and siderite contents (up to 32% in the ChernayaRver valley in the north of the Bolshezemelskaya tundra). Sometimes the amounts of ilmenite and titanium minerals are increased. In the south of the region, the association of heavy minerals contains less epidote than in the Pechoratill, and there are more amphiboles and garnets, sometimes amphiboles reach 57% of the heavy fraction. Pyrite and siderite make up the first percentages of the weight of the heavy fraction, and ilmenite concentrations are increased (9-12%). The amount of glauconite in Vychegdatill is 3–4 times lower than in Pechoratill: up to 15–20 grains per standard petrographic section. Heteroagedtills particularly sharply differ from one another by the composition of heavy minerals in the southern regions of the region. The orientation of the fragments in sector 270–3600 and the presence of indicator rocks of the North-West Terrigenous Mineralogical Province also indicate the migration of material from the west and from the north-west. Another feature of this till is the predominance of light colored limestones in the group of carbonate rocks.

In the northeast of the region, the mineral composition of the heavy fraction of the Vychegdatill is variable and consists of ilmenite (12%), garnet (15%), pyrite (15%), siderite (25%), epidote (32%). The content of titanium minerals is slightly increased – 10%, and amphiboles – is decreased (4–12%). The petrographic spectrum of fragments represents Uralian rocks. The fragments of rocks are oriented in sub latitudinal and SW directions, i.e. material for the formation of Vychegdatill came from the east-northeast.

The main mineral association of the Polar (Ostashkovian) till contains epidote (19–27%), garnets (14–20%) and amphibole (11–16%). The total content of siderite and pyrite is 13–35% with dominating of siderite. The crinoid-bryozoanNovaya Zemlya limestones are present. The orientation of detrital material from NNE to SSW confirms the onset of the cover glacier in polar time from Pay-Khoy-Novaya Zemlya.

3   Conclusions

Despite the variability of lithologic and mineralogical parameters of tills in terms of the regional plan, the data of the mineral composition of tills can be used with confidence for the dismemberment of Quaternary sections and wide spatial correlations of glacial deposits, taking into account the factors of glacial sedimentation genesis, as well as the petrographic composition of rock fragments and their orientation.

Acknowledgements. This research was supported by UB RAS project № 18-5-5-50 and project AAAA-A17-117121140081-7.

References

Andreicheva LN (2017) Correlation of neopleistocene tills of the north of the Russian plain in petrographic composition. Lithol Miner (1):82–94. (in Russian)

Andreicheva LN (1994) The source provinces and their influence on the formation of the composition of the moraines of the Timan-Pechora-Vychegda region. Lithol Miner (1):127–131. (in Russian)

Andreicheva LN, Nikitenko IP (1989) The mineral composition of the fine earths of the main moraines of the Timan-Pechora-Vychegda region. In: Mineralogy of the Timan-Northern Ural region, Syktyvkar, proceedings of the institute of geology, Komi SC UB RAS, no 72, pp 52–62. (in Russian)

Comparative Gold Deportment Study on Direct Leaching and Hybrid Process Tails of Oxide Ores from Mayskoye Au Deposit (Chukotka, Russia)

I. Anisimov[✉], A. Dolotova, A. Sagitova, M. Kharitonova,
B. Milman, and I. Agapov

Science and Technology Research Division, Polymetal Engineering,
Saint-Petersburg, Russia
anisimovis@polymetal.ru

Abstract. Mayskoye gold deposit is located in Chukotka, Russia. Ore bodies are mineralized brecciation zones composed of vein-quartz, argillic altered rocks (siltstone and carbonaceous silts) with fine disseminated and veinlet gold-bearing arsenopyrite and pyrite. Two main technological types of ores were distinguished at the deposit: primary and oxide. The main reserves were represented by primary ores, which are classified as refractory. Oxide ore had a quartz-micaceous composition with minor feldspars, kaolinite and sulfides. Beside native visible and colloidal gold, other main carriers of gold in oxide ore are arsenopyrite, pyrite, minerals of scorodite group and stibnite. Tested oxide ore sample showed low recoveries, according to the existing flow sheet in the CIL plant. Cyanidation tests showed some preg-robbing effect on organic matter. Flotation of carbonatious matter with consecutive leaching of flotation tails proved to have better recoveries than direct leaching by reducing preg-robbing on carbon three times.

Keywords: Refractory gold · Preg-robbing · Invisible gold ·
Surface contamination

1 Introduction

Mayskoye Au deposit is located in Chaunsky region in Chukotka, Russia. The ore was formed by quartz-pyrite-arsenopyrite shear opening veins in terrigenous sequence of carbonaceous siltstone and sandy-siltstones. Ore bearing sequence altered to micaceous-carbonate-albite rocks (of beresite formation). Thus, the ore had quartz-micaceous composition with minor feldspars and sulphide material. Major part of gold was refractory and concentrated in arsenopyrite and partially pyrite. Thus primary ores are treated with flotation recovering around 90% of gold with sulfide concentrate, which was shipped to Amursky POX plant for oxidation and cyanide tank leaching.

Part of the carbonaceous matter is preg-robbing gold. Upper part of the deposit contains oxide ore, which mostly was not refractory and free leaching. The carbon in pulp plant treats oxide ore during the warm season.

© The Author(s) 2019
S. Glagolev (Ed.): ICAM 2019, SPEES, pp. 15–18, 2019.
https://doi.org/10.1007/978-3-030-22974-0_4

The studied oxidized ore sample had low gold recovery in CIL plant, thus directed to gold deportment analysis and lab test work. The sample mineral composition was quartz-micaceous with minor kaolinite and secondary minerals scorodite, jarosite, tripuhyite and others. Natural coal content was 1,4%. Sulfides were represented by pyrite, marcasite, arsenopyrite and stibnite. Oxidized ore appeared to be partially refractory. Most of the refractory gold was represented by "invisible" gold. It was contained both in the native form, and was dispersed in liberated ore minerals: pyrite, arsenopyrite, stibnite and scorodite, and their binaries with gangue. The form of occurrence of gold in the ore and its connection with mineral carrier was important to determine effective methods for metal extraction. Such study in plant products was necessary to chase the issue and find optimization in reagent scheme or the flow-sheet.

2 Samples and Methods of Study

Two samples of initial oxidized ores from 1st and 2nd ore zones of Mayskoe gold deposit and four plant products: direct cyanidation tails, flotation tails and combined leach tails of flotation tails leaching product. The examination methodology included the following stages: classification of samples into narrow size classes, separation of material in heavy liquids with density of 2,90 g/cm^3 and 2,5 g/cm^3. Gold concentrate was panned from the heavy concentrate. The following separation products were obtained: gold gravity concentrate, heavy concentrate, rock-forming lights and carbonaceous fraction. Bulk mineral composition of the initial size classes and lights was studied by powder X-ray phase analysis using the Rietveld refinement. The mineral composition of the gold concentrates and heavy concentrates were studied by optical microscopy and the mineral identification of the carbonaceous fraction was done with stereo microscope.

Gold content in the products was analyzed by fire assays; arsenic, antimony and other elements by XRF in all products, except for gold concentrates. In the gold tips, the quantitative finding of gold was carried out by optical and electron microscopy.

Chemical composition of minerals-carrier of gold and other ore minerals was studied with SEM-EDX.

3 Results and Discussion

Quartz and phyllosilicates (muscovite, illite, smectite, kaolinite, dickite) were the main minerals in the samples; scorodite, jarosite, sulfides and other minerals were accessories. The main minerals-carrier of gold are represented by arsenopyrite, pyrite, marcasite and stibnite, oxidized forms of arsenic - scorodite, and antimony - tripuhyite, stibiconite, cervantite and valentinite. Native carbon content in the ore was 0,1–1,4%.

Visible native gold in samples had high fineness (902–914‰). The distribution pattern of visible native gold particles in ore zone 1 and 2 were slightly different. 70% of the visible native gold from head sample from the first ore zone was larger than 90 μm. 65% of the visible native gold from the second ore zone head sample was coarser than 70 μm. All visible gold particles in the cyanidation tails of first ore zone were less than 45 μm. Gold particles coarser than 45 μm dominated in cyanidation tails of the sample from the second ore zone. Native gold observed in the samples of direct leach tails were found in free particles with the surface shielded or in binaries with oxidized antimony forms. Rarely binaries of gold with pyrite, stibnite and gangue minerals were observed. Fineness of visible gold in cyanidation tails was 10–15% lower than in initial products with higher silver content. Besides natural gold the metal was in dispersed form in the mineral-carriers: pyrite, arsenopyrite, stibnite, scorodite, antimony oxidized forms and bound with carbon (Fig. 1).

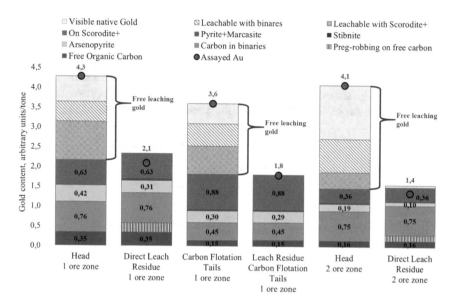

Fig. 1. Gold deportment in the plant feed, direct leach tailings, flotation tails and flotation tails leach residues of oxidized ores from 1st and 2nd ore zones of Mayskoe gold deposit

Free leaching gold accounted for 49% in head sample from the first ore zone and 64% in head sample from the second ore zone. The rest of the gold was on carbon (about 22–26%), in sulfides (about 5–10%) and scorodite (about 9–15%). Visible gold extraction by direct cyanidation was in range of 96–99% and gold from scorodite only 53–61%. The losses of visible gold in leaching occurred due to the blocking of the surface with compounds of oxidized antimony, as well as the slower cyanidation of gold with lower fineness. Coal flotation reduced gold losses by 22%.

4 Conclusions

Gold in the head samples was represented by both visible high-grade gold, most of which was a of gravity size and invisible/colloidal gold (<0,5 micron), forming more than half of the metal of the original ore. Lower grade gold particles leached slower.

The oxide ore of zones 1 and 2 of Mayskoye deposit was partially refractory, since most of the gold was invisible and dispersed in arsenopyrite, scorodite and with carbon.

The main gold losses during cyanidation were associated with invisible gold enclosed in sulfides, scorodite and organic carbon.

Based on gold deportment results obtained, the following suggestions were made:

1. Introduction of gravity in the grinding cycle would help reduce the loss of coarse gold and extract some of the refractory gold with large particles of arsenopyrite and stibnite. Increase in recovery can vary as coarse gold content and be as much as 2–5%.
2. Addition of lead nitrate in cyanidation process may increase the dissolution of low-grade gold particles.
3. Flotation of carbon with consecutive tails leaching proved to have better recoveries than direct leaching. Partial coal removing reduced preg-robbing on carbon for three times and reduced gold losses by 22% from zone 1 ore.
4. The most effective objective to recovery would be liberation of minerals-carriers of gold (mainly scorodite and arsenopyrite) from binaries with rock forming minerals.
5. Leaching of gold from scorodite can be enhanced by NaOH pretreatment with recovery increase of 10–15% for ore zone 1 and lesser effect for ore zone 2.

Mineral-Geochemical Criteria to Gold and Silver Recovery for Geometallurgical Sampling Campaign on Primorskoe Gold-Silver Deposit

I. Anisimov[✉], A. Sagitova, O. Troshina, and I. Agapov

Technology Research Department, Polymetal Engineering JSC,
St-Petersburg, Russia
anisimovis@polymetal.ru

Abstract. Ore variability study at Primorskoe gold-silver deposit demonstrated wide variety of mineral composition and gold and silver recoveries with cyanadation. The ore consisted of quartz-feldspar veins, quartz-rhodonite, quartz-"pyrolusite", quartz-epidote-garnet and quartz-Mn-silicates/hydroxides. Todorokite, birnessite, rancieite were the most common among the last ones. Statistical analyses of chemical and mineral composition, parameters of cyanidation tests showed occurrence of three main ore types – feldspatic, manganese silicate and oxide. Gold recovery effected by locking in Fe-oxides. The highest silver recovery strongly correlated with feldspathic cluster and whiter sample color reflecting Ag mineral forms: acanthite and electrum. Ore with silicate Mn showed good recoveries of acanthite, electrum and iodargirite associated with Mn silicates. Main silver losses were connected with Mn-oxides content and dark ore coloration where Ag chemically bound in Mn-oxides. Sr and Ba content along with sample color were indications that could be used as a proxy for recovery in geometallurgical mapping and ore-sorting.

Keywords: Silver · Predictive recovery · Cyanidation ·
Manganese mineralization

1 Introduction

Primorskoe Au-Ag deposit is located in Omsukchan district of Magadan oblast, Russia. It formed by hydrothermal veins of various composition: Qu-Fsp-Chl with Ep, Qu-Rdn-MnO$_x$, Qu-MnO$_x$ and Qu-Gar ± Wol. Noble metal mineralization represented with acantite, iodargirite, aurorite, jalpaite, pyrargyrite, electrum, kustelite and native silver and gold. Abundance of specific manganese oxides effected gold and silver extraction by cyanidation. The mineral composition study was aimed to define causes of possible gold and silver losses with cyanidation tails and possible ways to recover refractory silver.

© The Author(s) 2019
S. Glagolev (Ed.): ICAM 2019, SPEES, pp. 19–23, 2019.
https://doi.org/10.1007/978-3-030-22974-0_5

2 Methods and Approaches

72 individual and 57 composite geometallurgical samples of Primorskoe deposit were studied for chemical and mineral composition and cyanidation tests were performed. Ag varied from 176 to 36450 ppm in individual samples and gold – from 0.01 to 113 ppm. Ag content range in the composite samples was 49–1509 ppm averaging 606 ppm and Au – 0.04–9.92 ppm averaging 1.56 ppm.

Chemical composition was studied at ALS labs, Moscow. Au content was assayed by fire assay with atomic absorption finish, Ag and other 35 elements analyzed by ICP-AES after four acid digestion.

Quantitative X-ray powder diffraction by Rietveld refinement (QXRD) was carried out using Bruker D8 Advanced diffractometer with Linxeye XE detector at Technology Research Department, Polymetal Engineering.

Statistical analysis was done on filtered data with Aitchison transformation using PCA, regression analyses and Pearson correlations in Cytoscape software.

3 Results and Discussion

Mineral composition of the sample significantly varied in quartz (up to 94%), feldspar (up to 63%), micas (illite and muscovite – up to 25%). Epidote and piemontite formed up to 14.5%, grossular and calderite – up to 8.9%. Bustamite (up to 22.8%), rhodonite (up to 14.6%) and wollastonite (up to 16.7%) were observed in some samples. Apatite-magnetite-titanite-vermiculite association was common in few samples. Manganese oxides contents ran up to 34.4%.

Mn-mineralization presented in 18 minerals: silicates (piemontite, calderite, bustamite, rhodonite), carbonate (rhodochrosite), sulfide (alabandite), oxi-hydroxides. Todorokite, birnessite (7Å, 14-Å and amorphous), rancieite well spread among the last ones. Manganite, pyrolusite, ramsdellite, jacobsite, bixbyite, pyrochroite, chalcophanite, coronadite, aurorite were less common.

Data population for multivariate statistical analyses included 103 following parameters of chemical and mineral composition, color (RGB, brightness – BRT, and darkness - DRN), material fineness ($\gamma + 100$), silicate and oxide Mn ratio (Mnsil/Mnox), 14 parameters of cyanidation conditions (NaCN concentration in the pregnant solution – NaCNps; NaCNc and CaO consumption; final test pH – pHf; Au and Ag contents in pregnant solution and cake – AuPS, AgPS, AuK, AgK; Au and Ag recovery – εAu, εAg and losses – −εAu, −εAg (Fig. 1). Pearson correlations revealed occurrence of two main geochemical and mineral clusters – whiter feldspathic (rock-forming Ab-Olg-Chl) and darker manganese. Mgt-Tit-Ap-Vrm sub-cluster occurred in the feldspathic one showing accessory syenite mineralization. "Skarn" association (wollastonite, diopside, andradite, calcite, bustamite, rhodonite) and quartz vein (Qu, SiO_2, Cr and Iron – grinding contaminants) were distinguished. Ag recovery and NaCN content in pregnant solution tied with color cluster.

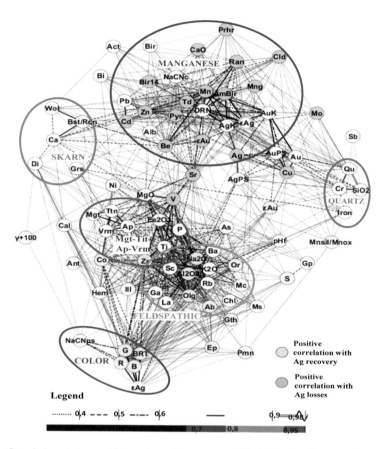

Fig. 1. Correlations between 88 transformed parameters of individual and composite samples of geometallurgical sampling campaign of Primorskoe deposit and proposed mineral association clusters. Legend shows Pearson correlation coefficient values

Reagents consumptions (CaO, NaCN) correlated with Mn-cluster, proving that ion exchange might take place during leaching. Zn, Cd, Pb tended to Mn-cluster without connection to a specific Mn-oxide. Sr related to both main clusters equally reflecting isomorphic distribution in feldspathic gangue as well as in MnO_x.

Ag recovery demonstrated positive correlation with feldspathic cluster and sample white color (BRT). Ag losses had strong connection with MnO_x. Au recovery tied together with color and quartz, Au losses correlated with MnO_x and locking.

PCA analysis showed same regularities. 6 principal components explained 54.78% of the total variance and described mineral composition (Fig. 2), Mn- and accessory mineralization, noble metal contents and recovery. PCA highlighted strong connection between Ag loses, todorokite and amorphous birnessite.

Regression Eq. (1) showed strong relation ($R^2 = 0.66$) between silver recovery, color (BRT) and some elements contents (in ppm). Variables listed in the order of their significance:

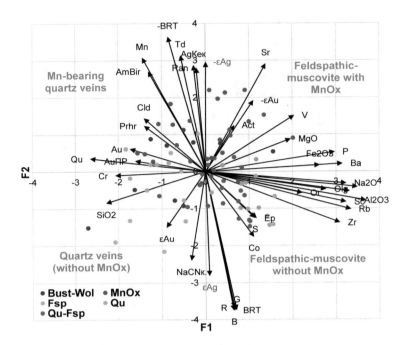

Fig. 2. Factor loadings and factor scores for factors 1 and 2 with their interpretation

$$\varepsilon Ag = 18.24 + 0.47BRT + 1.39Al2O3(wt\%) - 0.12V + 0.05Sr - 3.45Sc - 0.01Ag - 1.21Mn(wt\%) + 0.45Ca(wt\%) + 0.003Cu - 0.13Rb + 0.01S \quad (1)$$

4   Conclusions

Individual and composite samples of Primorskoe Au-Ag deposit demonstrated variety of mineral composition from high silica and feldspar to Mn-skarn association. Mn mineralization presented in silicates, oxides and hydroxides, carbonate, sulfide had great impact on Ag recovery. Wide range of vein gangue, ore and noble metal mineral associations were similar to ones characteristic for Dukat deposit, the largest silver deposit in Russia.

Ag mineral forms affected recovery rate from highest presented by acanthite, electrum in feldspatic association to the lowest locked in Mn-oxides, mainly in todorokite and birnessite. Ore sorting by color and element content can be used.

Acknowledgements. The authors are grateful to Sergey Kubyshkin for performance of cyanidation tests and Boris Milman for scientific advices.

Mineralogical-Geochemical Criteria for Geometallurgical Mapping of Levoberezhnoye Au Deposit (Khabarovsk Region, Russia)

I. Anisimov[✉], A. Sagitova, M. Kharitonova, A. Dolotova, and I. Agapov

Polymetal Engineering JSC, Saint-Petersburg, Russia
anisimovis@polymetal.ru

Abstract. Levoberezhnoye gold deposit is located in Khabarovsk region. It formed quartz-sulfide and quartz-adularia veining and fracture zones in argillic altered intermediate volcanic tuffs and lavas. 61 variability study samples were composed of quartz, feldspar, mica, kaolinite, chlorite with minor pyrite, arsenopyrite, jarosite and accessories. Multivariate statistics of mineral composition and multi-element assays distinguished following ore types: (1) primary quartz-feldspar sulfide-bearing breccia veins, (2) oxidized breccias with micas transformed to illite-smectite; (3) high sulfidation quartz-kaolinite. Gold leach recovery correlated with high sulfate content as well as mica and chlorite transformation to illite and smectite. Low sulfidation ores showed lower leaching recovery connected to gold encapsulation in pyrite. Thus, oxidized and sulfate ore types were amenable to cyanidation, while primary ore was recommended for sulfide flotation gold recovery. Molybdenum high content connected to Ag, Cu, Pb and As and supposed to be formed in a separate mineralization event from gold.

Keywords: Gold recovery · Cyanidation · Oxidation · Illite-smectite

1 Introduction

Levoberezhnoye deposit is located in Khabarovsky region in Estern Russia. It localized in intermediate volcanics and formed steeply dipping quartz-adularia Au-Ag breccia-vein system imbedded in rhyolites and extensively altered lake and flow tuffs and ignimbrite volcanics. The ore bearing rocks suffered multiple hydrothermal brecciation events with quartz±adularia-sulfide cement and fine sulfide dissemination in altered volcanics. The veining and rocks are fine grained and hard for visual mineral identification.

The samples characterized with drastic variations in gold recovery by cyanidation from 19 to 99%. The aim of the work was to determine compositional differences in ore types and ore characteristics effected gold recovery and cased metal losses.

© The Author(s) 2019
S. Glagolev (Ed.): ICAM 2019, SPEES, pp. 24–28, 2019.
https://doi.org/10.1007/978-3-030-22974-0_6

2 Methods and Approaches

61 composite drill core sample of geotechnical mapping of Levoberezhnoe were studied for cyanadation leaching and bulk mineral composition. Sample color was described with RGB-parameters.

Au was assayed with fire assay with atomic absorption finish, multi-element ICP-AES assays after four acid digestion of straight and diluted samples and XRF-analysis, sulfide and total S, total C estimated by LECO analysis.

Mineral phase identification and their quantification was done using Eva software and COD database. Quantitative X-ray powder diffraction with Rietveld refinement Topas software at Polymetal Engineering.

Multivariate statistical analysis was performed on filtered data with Aitchison transformation using Pearson correlations with Cytoscape software, PCA and regression analyses.

3 Results and Discussion

Three main ore types were distinguished based on mineral composition: 1 – quartz-albite with sulfide (arsenopyrite-pyrite) mineralization, 2 – kaolinite-dickite with sulfate, 3 – illite-smectite.

Wide structural and chemical variety of feldspars was observed: high and low microcline, orthoclase, albite and hyalophane. Balancing mineral content QXRD results for micas and K-felpdspars revealed significant potassium shortage and suggested high baddingtonite content in feldspar, which needs confirmation with assays. Observed anomalously high values of molybdenum were connected to Ag, Pb, Au and Cu.

Multivariate statistics analyses included 73 following parameters: chemical and mineral composition, color (RGB, brightness – BRT and darkness - DRN), material fineness (γ + 100), Au recovery – εAu and losses – $-\varepsilon$Au. Pearson positive correlations revealed occurrence of 5 geochemical and mineral clusters (Fig. 1):

1 – quartz-kaolinite-albite-sulfate with zeolites, Sr enriched (high sulfidation alteration); 2 – micas-chlorite with calcite group was linked to 3 - sulfide mineralization (beresite – low sulfidation); 4 – potassic feldspar-ankerite; 5 - oxidation cluster between kaolinite and sulfide/micas.

Qu-Kln-Ab cluster showed kaolinitization. K-Fsp and mica clusters tied together with K and Rb. Barium feldspar - hyalophane associated with K-Fsp. Mica cluster combined muscovite, illite, illite-smectite and chlorite. Sulfide cluster connected with mica cluster and gold losses. Au recovery connected with As, illite, illite-smectite indirectly through the minerals formed from sulfide oxidation: jarosite, goethite and scorodite. Mica transformation to illite and illite-smectite followed oxidation of sulfides

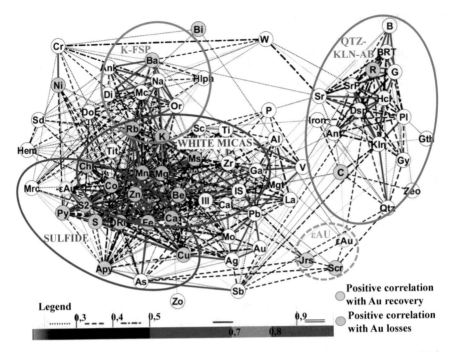

Fig. 1. Pearson correlations of 74 transformed chemical, mineral composition and cyanidation tests parameters of 51 small composite samples from Levoberezhnoye deposit

and, thus reflecting good Au leach results. Beresite group coupled with gold losses, the second ones indirectly correlated with gold recovery.

PCA analysis exposed 6 principal components, which explained 58.14% of the total variance. They described mineral composition (Fig. 2), oxidation rate, rare-metals and arsenic associations, color, grinding fineness. Results were similar to ones obtained with pair correlations: Au losses linked to sulfides, oxidation rate raised Au recovery (Fig. 2). Thus, flotation would be the best Au recovery solution from primary sulfide ore and tank cyanidation to oxide one.

Regression equation for gold recovery by cyanidation was calculated. It had relatively low $R^2 = 0.46$ (1) and connected color parameter (R/BRT) and elements contents (ppm and wt%):

$$\varepsilon AuCN = -96.91 + \frac{185.3R}{BRT} - 7.83Ti(\%) - 2.71Co - 0.73La - 1.15Pb + 0.53Sc - 0.004Zr + 1.81Ga - 13.09Ssfd(\%) + 10.90Fe(\%) - 0.54V + 0.34Ag - 1.94Ni - 75.18Ca(\%) + 0.01Ba + 0.01P$$

$$(1)$$

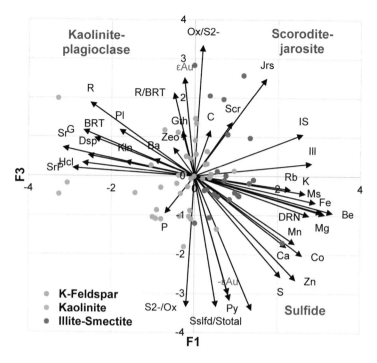

Fig. 2. Factor loadings and factor scores for factors 1 and 3 with interpretation

4 Conclusions

Mineral composition of the samples varied significantly from high quartz (up to 70%) and feldspar (up to 64%) to kaolinite (up to 45%) and illite-smectite (up to 40%). Au losses with cyanidation tails were bound to sulfides content. Sulfide oxidation to jarosite and scorodite and red component of sample color reflected increase in Au recovery. Absence of chlorite and transformation of muscovite to illite-illite-smectite also manifested Au cyanadation recovery improvement.

Revealed links between samples characteristics and regression equation, geometallurgical mapping and ore sorting can be performed based on sample color and chemical composition (Ti, Co, La, Pb, Sc, Zr, S, Fe, V, Ag, Ni, Ba).

Mineralogical Reasons of Au Recovery Variability from North-Western Pit of Varvara Au-Cu Mine (Kazakhstan) and Criteria for Geometallurgical Mapping

I. Anisimov[✉], A. Dolotova, A. Sagitova, M. Kharitonova,
and I. Agapov

Polymetal Engineering JSC, Saint-Petersburg, Russia
anisimovis@polymetal.ru

Abstract. Varvara Au-Cu mine deposit is located in Northern Kazakhstan. Mineralization is hosted in volcano-sedimentary, sedimentary rocks, metamorphosed and altered ultramafic and felsic rocks. Variability study was done on 58 composite samples represented five mineral ore types: serpentine-chlorite-talc; carbonate-chlorite-talc; quartz-sulfide; pyroxene-chlorite-prehnite $\pm$ garnet; quartz-feldspar $\pm$ pyroxene $\pm$ amphibole. Five processing ore types were defined: Au, Ni-As, pyrite, Cu and mixed. Mineralogy and geochemical studies revealed separate mineral associations carrying Cu, Ni-As and Au mineralization. Flotation and cyanidation tests were performed for each sample. Au losses with cyanidation cake occurred due to locking in sulfides. Floatation concentrate contamination with Mg-silicates (talc, serpentine) was connected to Au losses. Quartz-sulfide ore demonstrated better recovery by flotation. Cyanidation were most effective for pyroxene-chlorite-prehnite $\pm$ garnet and quartz-feldspar $\pm$ pyroxene ore compositions. Carbonate presence in the serpentine-chlorite-talc ore followed decrease in recovery by both extraction methods. Optimal viable ore treatment method can be chosen based on regression equations using ore chemical composition and color.

Keywords: Typification · Cyanidation · Flotation · Recovery estimation · Geometallurgy

1 Introduction

Varvarinskoe Au-Cu deposit is located in Kostanay region in Northern Kazakhstan and operated by several open pits. It is localized in volcanogenic, terrigenic, carbonate rocks (D_2–C_1) with ultrabasic and granodiorite intrusions (C_2–C_3). Silicification, argillic alteration, scarn processes were wide spread. Retrograde metamorphic changes were typical for ultramafic rocks.

This work was aimed at determining compositional differences in ore types and ore characteristics effected gold recovery by cyanidation and flotation.

© The Author(s) 2019
S. Glagolev (Ed.): ICAM 2019, SPEES, pp. 29–32, 2019.
https://doi.org/10.1007/978-3-030-22974-0_7

2   Methods and Approaches

58 small composite geometallurgical samples from Northwestern open pit were studied for bulk chemical and mineral composition, sample color with RGB-parameters. Cyanidation and sulfide flotation tests were done on the head samples.

Au was assayed with fire assay with atomic absorption finish, multi-element ICP-AES assays after four acid digestion of samples and XRF-analysis, sulfide and total S, total C estimated by LECO AES analysis.

Mineral phase identification and their quantification was done using Eva software and COD database. Quantitative X-ray powder diffraction with Rietveld refinement was done using Topas software at Polymetal Engineering.

Multivariate statistical analysis was performed on filtered data with Aitchison transformation using Pearson correlations with Cytoscape software, PCA and regression analyses. Regression analysis was carried out to predict gold extraction.

3   Results and Discussion

Mineral composition of the samples varied between serpentine-talc-chlorite $\pm$ carbonate, quartz-feldspatic and quartz-sulphide associations. Five main mineral ore types were distinguished based on bulk mineral composition: 1 – serpentine-chlorite-talc, 2 – carbonate-chlorite-talc, 3 – quartz-sulfide, 4 – pyroxene-chlorite-prehnite-garnet, 5 – quartz-feldspar-pyroxene-amphibole.

Flotation and leaching tests revealed five processing ore types: a - gold, b - copper, c - Ni-As, d - pyrite and e - mixed.

Multivariate statistics analyses were performed on all available samples parameters including mineral and chemical composition; color (RGB, brightness – lBRT and darkness - dBRT); material fineness (-71 μA); Au, Cu, As, Ni, sulfidic S, total S recovery to concentrate and losses to flotation tails (XF) (εAuKF, εCuKF, εAsKF, εNiKF, εSsKF, εSKF; εAuXF, εCuXF, εAsXF, εNiXF, εSsXF, εSXF) and cyanidation (εAuCN, εCuCN, −εAuCN, −εCuCN), parameters of cyanidation conditions (reagents uptake - NaCNi, CaCNi), cyano-soluble copper – αCuCNr, Au content in cake – AuXvA, characteristics of gravity con (mass pull - γgk-t, pyrite, arsenopyrite, Cr-spinel, chalcopyrite, Ni-minerals contents in gravity con: PyMrcGK, ApyGK, ChrGK, CpyGK, NiMGK).

Pearson correlations revealed occurrence of 3 geochemical and mineral clusters following bulk sample mineralogy (Fig. 1): quartz-sulfide, diorite (feldspatic) and serpentinite. Au losses with cyanidation tails connected with Au locked in sulfides, which was proved with SEM study. S and Cu contents and NaCN consumption were included in the cluster with Au losses with cyanidation. This fact may point to consumption of free CN^- by reaction with S^0 and Cu dissolution, thus leading to deficit of free CN^-. This reaction produced CuCN and rhodanates that could block surface gold particle impeding its dissolution. Concentrate contamination with self-floating Mg-silicates (talc, serpentine) was connected to Au losses with flotation tails.

Principal component analysis exposed 7 principal components, which explained 72.1% of the total variance and described mineral composition, geochemical

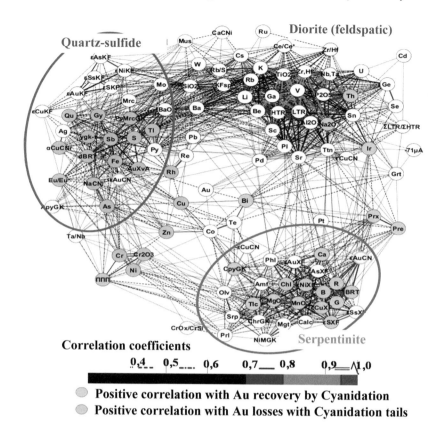

Fig. 1. Positive correlations between compositional and metallurgical parameters of samples from Varvara

associations, Au and Cu recovery by flotation and cyanidation (Fig. 2). 5 ore types were well separated in the coordinates of factors 2 and 3, 2 and 4. Quartz-sulfide ore samples demonstrated better recovery by flotation, while cyanidation was more effective for samples with pyroxene-chlorite-prehnite-garnet and quartz-feldspar-pyroxene-amphibole compositions. Carbonate occurrence in the serpentine-chlorite-talc ore reduced recovery by both concentrating methods.

Regression equations for Au recovery by flotation ($R^2 = 0.80$) and by cyanidation ($R^2 = 0.88$) included color parameter (BRT) and elements contents (ppm or wt%):

$$\varepsilon AuFl = 127.32 + 0.17Co - 0.01Cu - 4.78Fe(\%) + 5.19K(\%) + 0.39Mo + 0.39Th + 0.01Sb - 0.63BRT + 0.10Zn + 0.43Mg(\%) + 0.84SiO2(\%) - 0.81Li - 0.40Rb - 1.97Te + 5.03Cs$$

$$\varepsilon AuCN = 226.60 - 2.86Ca(\%) + 0.30Co - 7.00Fe(\%) - 2.08Sc - 0.16Sr + 0.13V - 0.24MgO(\%) - 3.37Al2O3(\%) - 0.55SiO2(\%) + 1.98Ga - 0.42BRT + 0.27LOI(\%) + 0.004Cu + 0.02Ni - 0.01As$$

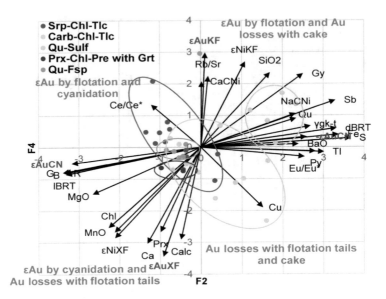

Fig. 2. Factor loadings and factor scores for factors 2 and 4 with interpretation

4   Conclusions

Mineral composition of the samples varied significantly from serpentine-talc-chlorite ±
carbonate to quartz-feldspar, skarn and quartz-sulfide associations. Au losses with
cyanidation tails connected with Au locked in sulfides. Au losses with flotation tails
connected to concentrate dilution with self-floating Mg-silicates (talc and serpentine).
Ore sorting can be done based on regression equations of element composition and ore
color to direct the ore to more economically viable process.

Acknowledgements. The authors are grateful to Nicolay Rylov for scientific advices, Lims-lab
and Nati-R for collaboration.

Mineralogical Breakthrough into Nanoworld: Results and Challenges

A. Askhabov[(✉)]

Institute of Geology Komi SC UB RAS, Syktyvkar, Russia
askhabov@geo.komisc.ru

Abstract. A bit more than a hundred years ago Wolfgang Ostwald, Professor at the University of Leipzig, published a book titled "The World of Neglected Dimensions" (Ostwald 1923). He announced a program of a research breakthrough into the world of microscopic particles, which was imminent by that time. A new stage of the intervention into "world of neglected dimensions" began near the end of the 20th century. The main objects at this stage were nanosized particles. The agenda raised issue of development of new sciences, including nanomineralogy. The subsequent mineralogical intervention into the "world of neglected dimensions" proved to be quite successful. We associate challenges of the next breakthrough with the study of objects and processes in the range from individual atoms and molecules to the first mineral individuals (nanoindividuals). The protomineral world is today a new "world of neglected dimensions".

Keywords: "world of neglected dimensions" · Nanoparticles · Quatarone concept · Protomineral world

1 Introduction

The unprecedented interest in nanoscale objects, which we witness in recent years, I called a new stage of intervention into the "world of neglected dimensions". The first stage of the intervention (research breakthrough) occurred at the beginning of the 20th century and was associated with W. Ostwald, who actually referred to the region of microscopic particles as "the world of neglected dimensions" (Ostwald 1923). Intensive researches in this area, which began then, resulted in formation of a new science - colloid chemistry. Its creators R. Zigmondi ("he opened access to the world of inaccessible sizes"), T. Svedberg and J. Perrin (for "a breakthrough into the world of discrete particles") were awarded the Nobel Prizes in 1925 and 1926. What is happening in our time is a secondary discovery of the "world of neglected dimensions", now not only at micro, but also at nanoscale.

Mineralogical intervention into the micro and nanoworld began long ago. There is generally nothing revolutionary in what is happening nowadays. Nano-mineralogy is a normal and inevitable stage in the development of mineralogical science. Moreover the role of mineralogy (crystallography) in the study of nanoscale objects is quite comparable with the role of physics, chemistry or biology. Suffice it to recall that structural mineralogy always operated on nanoscale elements, and nucleation and growth of

S. Glagolev (Ed.): ICAM 2019, SPEES, pp. 33–36, 2019.
https://doi.org/10.1007/978-3-030-22974-0_8

crystals is a typical nanoprocess, a crystalline nucleus is a nanocrystal, and opal is a nanostructured natural material. Nanotechnology is often a repetition of natural processes or nature-like technologies.

The very first results of the mineralogical intervention in the nanoworld were very impressive (Nanomineralogy… 2005). Among the most important achievements of the past years:

– - discovery of a new type of structurally and morphologically ordered objects – nanoindividuals. The likely morphological diversity of nanoindividuals is enormous and not limited by laws of classical crystallography;
– - significant expansion of boundaries of mineral world due to solid amorphous substances previously attributed to mineraloids. A discovery of a new class of structurally ordered mineral structures;
– - finding of the lower limit of mineral objects, beyond which matter is in a different, non-mineral (protomineral, quatarone, cluster) state;
– - identification of common laws of self-organization at nanoscale in the mineral and living worlds. Mineral and living matter, as is known, are formed at nanoscale. Both do not exist outside the lower limit of the nanolevel;
– - substantiation of fundamental role of forms of existence and properties of nanoscale particles in minerals and ores for the development of new technologies for deep and complex processing of mineral raw.

The research of the mineral nanoworld is just beginning. The study of mineral nanostructures, natural clusters, nanostructured natural materials, organo-mineral nanoobjects, nanodispersed phases in ores, development of new technologies for their extraction, modifying properties of minerals and mineral nanoparticles is an incomplete list of the nearest tasks of general and applied nanomineralogy. At that we will have to reconsider a number of fundamental concepts, significantly expand them or introduce new ones. Another argument for the expansion of mineralogical objects to nanoscale is that on this basis a great challenge for mineralogical science can be formed, which is now missing. It is not reasonable to reject a breakthrough promising unprecedented discoveries and a deeper understanding of nature of the mineral matter and its sources. It is hard to deny the intellectual appeal and charm of the nanomineralogical project. The opening door to the nanoworld should not be closed, even if there is a threat of erosion of foundations of classical mineralogy.

Significant progress in understanding properties of the nanoworld was made by the quatarone concept of cluster self-organization of matter developed by us (Askhabov 2011). Within this concept new models of crystal nucleation (Askhabov 2016), formation of various types of nanoparticles (including fullerenes (Askhabov 2005a)) and solid amorphous materials (Askhabov 2005b) are proposed. At that quatarons themselves are new nanoobjects without analogues in the macroworld. They are not small pieces, cut from a large piece of matter or obtained by successive division. They cannot be identified with ordinary clusters—equilibrium structures optimized by geometry or energy.

The quataron concept solved the fundamental problem of deciphering the mechanism of crystal growth in an amazingly simple way (Askhabov 2016). Quatarons proved to be ideal building units for crystal growth. Due to the dynamism of structure, their inclusion into the crystal lattice occurs with virtually no kinetic resistance and deformation of the crystal lattice. The quataron model of crystal growth acts as an alternative to the known models of crystal growth by attaching individual atoms or ready crystalline blocks.

The development of ideas of the quataron concept opened a window to the protomineral world. And this world is today regarded as a new "world of neglected dimensions" (Askhabov 2017). As a result, the program of W. Ostwald receives a new impetus and focuses on the study of objects of the protomineral world. We must answer not only the questions of how minerals are formed, but also why they are formed, why minerals are exactly as they are. The ontogenesis of minerals should begin not with the origin of the mineral, but with the protomineral state of mineral matter.

Thus, the world of minerals is preceded by the world of specific pre-mineral objects —the protomineral world (Fig. 1). This world requires interdisciplinary approaches and new instrumental methods with spatial angstrom-nanometer and temporal femto-nanosecond resolution. In connection with the emergence of the protomineral world in the current agenda of mineralogical science, it is highly desirable to draw up a program of top-priority experiments for implementation with the European free electron laser - a device that is potentially able to satisfy requests of mineralogical science and study in detail the process at the source of mineral matter.

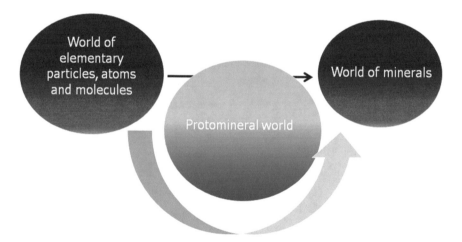

Fig. 1. Way from the world of separate atoms and molecules toward the world of mineral individuals comes through the protomineral world.

Acknowledgements. The work was accomplished with partial financial support by UB RAS program No. 18-5-5-45, and RFBR No. 19-05-00460a.

References

Ostwald W (1923) World of neglected dimensions. Introduction to modern colloid chemistry with an overview of its applications. Mir, Moscow, 228 p (in Russian)

Yushkin NP (2005) Ultra- and microdispersed state of the mineral matter. In: Askhabov AM, Rakin VI (eds) Nanomineralogy. Nauka, St. Petersburg, 581 p (in Russian)

Askhabov AM (2011) Quataron concept: main ideas, some applications. In: Proceedings of komi science center UB RAS.- №3, no 7, pp 70–77 (in Russian)

Askhabov AM (2016) Quataron crystal nucleation and growth models. In: Proceedings of RMS.- Ch. CXLV, no 5, pp 17–24 (in Russian)

Askhabov AM (2005a) The kvataron model of fullerene formation. In: Physics of solid state, vol 47, no 6, pp 1186–1190

Askhabov AM (2005b) Aggregation of quatarons as a formation mechanism of amorphous spherical particles. In: Doklady earth sciences, vol 400, no 1, pp 937–940

Askhabov AM (2017) New stage of mineralogical invasion into the "world of neglected dimensions": discovery of the protomineral world. In: Proceedings of anniversary congress of the Russian mineralogical society "200th anniversary of the Russian mineralogical society", St. Petersburg, vol 2, pp 3–5 (in Russian)

Case Study: Geochemistry and Mineralogy of Copper Mine Tailings in Northern Central-Chile

K. Berkh[✉], D. Rammlmair, M. Drobe, and J. Meima

Federal Institute for Geosciences and Natural Resources, Hanover, Germany
Khulan.berkh@bgr.de

Abstract. Selected mine tailings in northern-central Chile were geochemically and mineralogically studied for their economic potential and environmental impact. High bulk Co content up to 1500 ppm and Cu content up to 9100 ppm are caused by Co-bearing pyrite, chalcopyrite, and their secondary products such as malachite and Co-Cu-carrying Fe-hydroxides. Due to high amount of sulfide minerals acid mine drainage (AMD) is forming in the oxidized upper part of the tailing, which makes a retreatment in dispensable to reduce the environmental impact.

Keywords: Mine tailings · Geochemistry · Mineralogy · Cobalt · Copper · Reprocessing

1 Introduction

Since Chile is the largest Cu producing country with the biggest reserves, Chilean mining industry generates huge quantities of mining residues, amongst others in the form of tailings dumps. The fact that some of them were generated many decades ago, where process technology was inadequate in comparison to today, makes some of them economically interesting. Due to advanced weathering, the tailings, which are potentially acid producing, bear an environmental hazard. Therefore, our aim is to investigate the geochemistry and mineralogy of the tailings in order to determine their economic potential and environmental impact.

The studied mine tailings dump is located in the region of Coquimbo, where an arid Mediterranean climate with mean annual temperature of 14.6 °C, precipitation of 132 mm, and evaporation of about 1702 mm prevails (Mora et al. 2007). Geologically, the region is characterized by Early Cretaceous Chilean Manto-type (volcanic-hosted stratiform) Cu deposits (Kojima et al. 2003) that are the most probable source of the studied tailings.

2 Methods and Approaches

Eight drill cores with lengths of seven meters were taken from the tailings dump (Fig. 1). Bulk geochemistry of homogenized material for each meter was investigated using a standard WDXRF. Mineralogy was studied on representative grain

© The Author(s) 2019
S. Glagolev (Ed.): ICAM 2019, SPEES, pp. 37–40, 2019.
https://doi.org/10.1007/978-3-030-22974-0_9

concentrates obtained by gravity separation. Abundance of the minerals was analyzed by MLA and composition of the minerals was examined by EPMA.

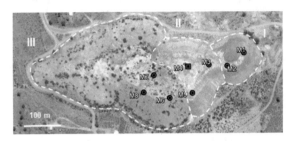

Fig. 1. Drill cores positions (M1–M8) in three heaps (I-III)

3 Results and Discussion

The tailings material consists of alternating layers of sand, silt and clay. According to geochemical pattern, the tailings dump can be subdivided into three groups, as shown in Fig. 1. **Heap I and II** are enriched in Fe and Co (Fig. 2). Special feature of heap I is a depletion in S and Ca but a high percentage of loss on ignition (LOI). It may point to a weathered part of the tailings dump, where oxidation of pyrite results in dissolution of calcite and accumulation of water bearing secondary clay minerals. **Heap III** contains high amounts of host rock and therefore elevated contents of Si, Al, Mg, Na, K and P can be seen. It is strongly enriched in Cu but only at the near surface level (Fig. 2).

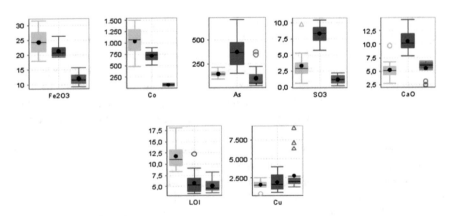

Fig. 2. Bulk content of relevant elements in three heaps (■ – heap I, ■ – heap II, ■ – heap III)

Primary gangue minerals in **heap I and II** are epidote, quartz, andradite, and albite pointing towards skarn mineralization. The only difference between the two heaps is the absence of calcite in the heap I confirming the bulk geochemistry. The most common primary ore mineral in both heaps is pyrite, which occurs as liberated grains. Three types of pyrite were identified. The first type is pure pyrite. The second type has an As-rich rim with up to 8.7 wt% As (Fig. 3a) and is predominantly present in the heap II resulting in an elevated bulk content of As (Fig. 1). The third type contains a significant amount of Co. Hereby, the Co concentration increases from rim to core and can reach up to 3.8 wt %. Additional trace elements are e.g. up to 0.5 wt% Ni, 0.1 wt% Cu, and 0.2 wt% Cd. Remaining primary ore minerals are magnetite, hematite and trace amounts of chalcopyrite. They also occur as well liberated grains and are occasionally inter grown with each other. These minerals do not contain significant amounts of trace elements.

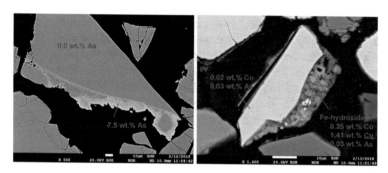

Fig. 3. BSE images of the pyrite grains: a As-rich rim on pyrite and b pyrite rimmed by Fe-hydroxide with significant amounts of Co and Cu

Secondary phases are gypsum and Fe-hydroxides as weathering products of calcite and pyrite (Fig. 3b) and preferentially occur in heap I. Such in-situ precipitation of Fe-hydroxides can only take place at near neutral pH proving an initial occurrence of calcite as a buffering agent in the strongly weathered heap I. The Fe-hydroxides contain high concentrations of Co (up to 0.9 wt%), Cu (up to 2.5 wt%) and Ca (up to 4.3 wt%), pointing to a dissolution of pyrite, chalcopyrite and calcite and fixation of released metals in Fe-hydroxides. The analyzed saturated soil extraction had a pH of 2. Its Co and Cu contents were 103 and 136 mg/l respectively, whereas Hg, Cd, As, Zn and Pb contents remained under 2.4 mg/l. Especially, the Co and Cu values exceed the international guidelines (IFC 2007 and Gusek and Figueroa 2009) by more than a factor of 100.In fact, the potential for AMD is high in the oxidized parts of the heaps.

The **heap III** predominantly consists of gangue minerals. The most common mineral is quartz followed by albite, anorthite, micas, K-feldspar, amphibole, epidote, pyroxene, chloride, and calcite. A Minor amount of Fe-oxides and chalcopyrite exists, either as liberated grains or finely intergrown with gangue minerals. Chalcopyrite is strongly replaced by Fe-hydroxides and Cu-carbonates along grain boundaries and micro fractures.

4 Conclusion

Bulk geochemistry provides a good prediction of the mineralogy. Elevated amounts of Fe, Co, and S in the heaps I and II correspond to Co-bearing pyrite-rich waste material. In contrast, heap III consists of host rock forming elements resulted by non-sulfidic gangue mineral waste. From an economic point of view, Co is the only valuable metal in the heap I and II. For the extraction magnetic separation should be performed to eliminate high amounts of magnetite. Heap I contains extensive amounts of Co- and Cu-rich Fe-hydroxides that usually coat pyrite grains. Therefore, leaching can be directly applied to extract acid soluble Co and Cu and also to liberate pyrite grains. Otherwise, pyrite cannot be floated. Afterwards, pyrite from both sulfidic heaps can be floated and Co can be extracted by bioleaching. In case of heap III, only the uppermost first meter, which is Cu-rich, should be treated. Cu occurs preferentially as acid soluble mineral such as malachite and Cu-rich Fe-hydroxides. However, calcite would have to be removed by gravity separation to reduce the consumption of sulfuric acid. Otherwise, itis economically not feasible. From an environmental point of view, heap III poses no environmental risk because it does not host AMD potential. Oxidation of minor chalcopyrite will be buffered by the carbonate content. In turn, the sulfidic tailings should be immediately treated, because the potential of AMD generation due to heavy rains is high. The hazardous potential of the impoundment should not be underestimated because of agricultural activities in the vicinity.

Acknowledgements. This investigation is supported by DERA the German Mineral Resources Agency of BGR.

References

Gusek JJ, Figueroa LA (2009) Mitigation of metal mining influenced water (Management Technologies for Metal Mining Influenced Water, vol 2), Soc Min Metall Explor. Littleton, Colorado

IFC (2007) Environmental, health and safety guidelines for mining. International Finance Corporation, World Bank Group, DC, 33 p

Kojima S, Astudillo J, Rojo J, Tristá D, Hayashi K (2003) Ore mineralogy, fluid inclusion, and stable isotopic characteristics of stratiform copper deposits in the coastal Cordillera of Northern Chile. Miner Deposita 38:208–216

Mora F, Tapia F, Scapim SA, Martins EN (2007) Vegetative growth and early production of six olive cultivars, in Southern Atacama Desert, Chile. J Cent Eur Agric 8(3):269–276

Application of Fluoride Technology for Processing of Off-Grade Aluminum Raw Materials

I. Burtsev, I. Perovskiy[⊠], and D. Kuzmin

Institute of Geology named after Academician N.P. Yushkin Komi Science Center of the Ural Branch of the Russian Academy of Sciences, Syktyvkar, Russia
igor-perovskij@yandex.ru

Abstract. We used the process of hydrofluoride desilication for off-grade aluminum raw materials widely distributed in the Komi Republic. The optimal ratios of fluorinating agent (NH_4HF_2) to the target component (SiO_2) were determined by the method of differential thermal analysis. At a fluorination temperature 300 °C and timing 30 min, we obtained concentrates containing up to 80% Al_2O_3 and up to 10% SiO_2.

Keywords: Bauxite · Aluminum raw materials · Desilication · Fluorination · Ammonium hydrofluoride

1 Introduction

One of the most important problems of Russian aluminum industry is the deficit of high-quality alumina raw materials, forcing to import it from other countries.

A major source of alumina raw materials is the Vezhayu-Vorykvinskoe and Verkhneshchugorskoe bauxite deposits, which are part of the Vorykvinskaya group of deposits in Middle Timan and developed by the United Company RUSAL. Mined bauxites are predominantly consumed for alumina production by the Bayer method.

The share of low-quality sintered bauxites is significant and amounts from 5 to 55% of balance reserves for Vezhayu-Vorykvinskoe and Verkhneshchugorskoe deposits. A sharp reduction in consumption of sintered bauxites makes an alternative processing a rather actual question. In addition, the prospected reserves contain a significant amount of off-balance ores, off-grade in quality: high-silica low-modulus (better M_{Si} 2.6–3.8) bauxites and associated high-aluminous allites.

Gravitational, flotation, magnetic, chemical, and other methods have been proposed to enrich the bauxites, but their use for enriching Middle Timan bauxite is limited due to a fine size of the minerals, a large proportion of amorphous phases, and a small efficiency of used processes in general (Burtsev et al. 2016).

Fluoride technologies are one of the most promising ways to process mineral raw materials (Dyachenko and Kraidenko 2007; Medkov et al. 2011). The purpose of the research – to assess the prospects for the use of fluorination to desilicate off-grade

S. Glagolev (Ed.): ICAM 2019, SPEES, pp. 41–44, 2019.
https://doi.org/10.1007/978-3-030-22974-0_10

aluminum raw materials (bauxite, allite, kaolinite clays), which are widespread in the Komi Republic.

2 Methods and Approaches

The objects of research - representative samples of bauxite, allite, kaolinite clays from the Vezhayu-Vorykvinskoe (V-V), Verkhneshchugorskoe (Vsch) deposits and occurrences of the Izhma area (IA). Samples were grinded in a disk eraser (ID-200) and separated by size classes 0.125–0.250 and 0.25–0.5 mm. Analytical work was carried out by the equipment of the Center for Collective Use GeoScience of the Institute of Geology Komi SC UB RAS using XRD, XRF, DTA. Fluorination was carried out in a tube furnace equipped with a gas-extraction system. Ammonium hydrodifluoride (NH_4HF_2) was used as a fluorinating component, which is an ecologically safe matter under standard conditions.

3 Results and Discussion

X-ray fluorescence analysis showed that the granulometric differentiation did not lead to a change in the chemical composition and silicon module of the samples (Table 1). We diagnosed reflexes of kaolinite with hematite admixture on diffraction patterns of samples No. 1, 2, and 3, and boehmite, kaolinite with anatase admixture on the patterns of sample No. 4.

Table 1. Chemical composition of off-grade bauxite ores

No.	Class, mm	Mass fraction, %						M_{Si}
		Al_2O_3	SiO_2	Fe_2O_3	TiO_2	MgO	K_2O	
1	0.25–0.5	43,82	50,83	2,93	1,66	0,31	0,10	0.86
IA	0.125–0.25	44,08	50,76	2,70	1,66	0,31	0,10	0.86
2	0.25–0.5	43,32	48,40	4,96	1,59	0,67	0,67	0.88
Vsch	0.125–0.25	43,63	48,67	3,92	1,78	0,79	0,71	0.90
3	0.25–0.5	35,94	39,83	19,84	1,37	1,76	0,68	0.90
V-V	0.125–0.25	36,01	39,99	19,75	1,27	1,77	0,65	0.90
4	0.25–0.5	59,85	23,91	12,98	1,12	1,03	0,61	2.50
Vsch	0.125–0.25	59,37	23,78	13,53	1,15	1,04	0,62	2.50

Admixture components Na_2O, CaO, MnO, P_2O_5 no more than 0.8%.

Thermodynamic calculations of fluorination of kaolinite NH_4HF_2 are given elsewhere (Rimkevich et al. 2016). However, the theoretical equation turns out to be complicated in stoichiometric terms, which leads to the overconsumption of the fluorination component. Our works on fluorination of titanium ores of the Yarega deposit in the Komi Republic showed the efficiency of applying the ratio calculation to the target component—silicon oxide (Perovskiy and Ignat'ev 2013; Perovskiy and Burtsev 2016). SiO_2 fluorination can be described by the following equation:

$$SiO_2 + 3NH_4HF_2 = (NH_4)_2SiF_6 + 2H_2O + NH_3 \uparrow$$

We studied the process of fluorination of off-grade bauxites at molar ratios SiO_2: NH_4HF_2 equal to 1: 1 and 1: 1.5 with the help of DTA. We established that ratio 1:1 was preferred. Excessive ammonium fluoride (1:1.5) did not result in a positive effect, but was accompanied by passivation of fluorination reactions and formation of a larger volume of gaseous products.

Taking into account DTA data, the temperature regime of fluorination consisted of sintering the sample with NH_4HF_2 at temperature 200 °C (30 min.) with subsequent sublimation of resulting salt $(NH_4)_2SiF_6$ at 300 °C (30 min.). Upon completion of the fluorination, we carried out water leaching, which allowed transferring the undecomposed fluoroammonium salts into the solution.

The results of fluorination, given in Table 2, show that SiO_2 content is significantly reduced in the samples. At the same time, the size of material does not affect the effectiveness of desilication.

Table 2. Chemical composition of samples after fluorination

No.	Class, mm	Mass fraction, %					
		Al_2O_3	SiO_2	Fe_2O_3	TiO_2	MgO	K_2O
1	0.25–0.5	80.77	7.75	6.75	3.60	0.33	0.25
IA	0.125–0.25	81.17	7.73	6.78	3.67	0.32	0.23
2	0.25–0.5	73.94	8.98	9.85	3.92	1.03	1.66
Vsch	0.125–0.25	71.81	11.24	10.07	3.67	1.00	1.58
3	0.25–0.5	55.07	7.28	30.20	2.38	2.72	1.40
V-V	0.125–0.25	52.87	9.17	30.61	2.29	2.74	1.36
4	0.25–0.5	68.00	9.99	17.55	1.45	1.46	1.01
Vsch	0.125–0.25	67.31	9.67	18.52	1.48	1.44	1.02

Admixture components Na_2O, CaO, MnO, P_2O_5 no more than 0.9%.

4   Conclusions

The process of hydrofluoride desilication has been applied to off-grade aluminum raw materials widespread in the Komi Republic. The optimal ratio of NH_4HF_2: SiO_2 equal to 1:1 was determined by DTA method. At the fluorination temperature 300 °C and timing 30 min, concentrates with Al_2O_3 content more than 70–80%, and SiO_2 - less than 10% were produced. This technology is an alternative to the acid and alkaline processing of high-siliceous aluminum raw materials. The use of fluoride technology allows not only to improve the quality of ores because of desilication, but also to obtain products that can be directly processed by electrothermal methods to silumin (Lepezin et al. 2014), ferroalloys (Bukin and Seregin 2014) and other products with high added value.

Acknowledgements. The work was carried out with the financial support of the State task "Scientific basis for effective subsoil use, development and exploration of mineral resource base, development and implementation of innovating technologies and economic zoning of the Timan-Nothern Ural region" No. AAAA- A17-117121270037-4.

References

Bukin AV, Seregin AN (2014) Development of technology for smelting ferrosilicoaluminium from sub-standart bauxite and aluminous waste of metallurgy and electrical power energetic. Probl Ferr Metall Mater Sci (2):31–36

Burtsev IN, Kotova OB, Kuzmin DV, Mashin DO et al (2016) Prototypes of new technologies for development of the mineral raw materials complex of the timan-north urals region. In: Proceedings of the Komi science centre of the Ural division of the Russian academy of sciences, vol 27, no 3, pp 79–88

Dyachenko AN, Kraidenko RI (2007) Separation of silicon-iron-copper-nickel concentrate by fluorammonium method into individual oxides. Bull Tomsk Polytech Univ 311(3):35–38

Lepezin GG, An'shakov AS, Faleev VA, Avvakumov EG, Vinokurova OB (2014) Plasma-chemical method of producing silumin and aluminium from the minerals of the sillimanite group. Doklady Chem 456(2):110–113

Medkov MA, Krysenko GF, Epov DG (2011) Ammonium bifluoride – the perspective reagent for complex processing of mineral raw materials. Vestn Far East Branch Russ Acad Sci. 159 (5):60–65

Perovskiy IA, Burtsev IN (2016) Influence of mechanical activation of leucoxene on efficiency of its processing by fluoride method. Perspektivnye Materialy (2):66–73

Perovskiy IA, Ignat'ev GV (2013) Ammonium fluoride method of desilication of leucoxene concentrate of Yarega deposit. In: Predictive estimate of technological properties of minerals by applied mineralogy methods. Proceedings of 7th Russian seminar of process mineralogy, Karelian Scientific Centre RAS, Petrozavodsk, pp 110–116

Rimkevich VS, Pushkin AA, Girenko IV, Leontyev MA (2016) Perspectives of complex processing the kaolin concentrates by hydrochemical method. Izv Samara Sci Cent Russ Acad Sci 18(2):186–190

Applied Mineralogy for Complex and Profound Mineral Processing

V. Chanturiya$^{(\boxtimes)}$ and T. Matveeva

Institute of Comprehensive Exploitation of Mineral Resources,
Russian Academy of Sciences, Moscow, Russia
vchan@mail.ru

Abstract. The main methodological parameters and the modes of investigation of samples containing micro- and nano-sized reagent phases are determined. Using scanning laser microscopy and an original method for analyzing the surface relief, a quantitative assessment of the adsorption layer of the collector reagent on Au-sulfide minerals was performed, the proportion of the molecular form of adsorption and the retention agent fixing strength was calculated. The theoretical basis for choosing the reagent mode for selective flotation of multicomponent ores was developed. The action mechanism has been revealed and the prospect of using novel selective collectors and environmentally friendly plant reagents for extracting non-ferrous and noble metals from complex sulfide ores has been substantiated.

Keywords: Reagent · Adsorption · Nanoparticles · Mineral · Microscopy · Flotation

1 Introduction

At present, modern mining and ore dressing plants have been facing a number of serious challenges in processing low-grade complex ores and technogenic resources, increasing demands of high quality metal concentrates and ecologically safe methods of beneficiation. In those conditions, the tasks of making mineral processing more complete and comprehensive and of creating highly effective technologies come to the fore. These should be based on the intensification of the existing methods and on creating new methods of mineral extraction from hard-to-enrich ores and from technogenic deposits by using the newest achievements of the fundamental sciences. The transition to the new strategy of primary processing is only possible on technological-mineralogical evaluation of raw minerals.

A modern complex of high-resolution physical methods allows to investigate the composition, structure and properties of geomaterials at the micro and nano-level, including:

- Identify micro and nano-sized particles of noble metals and surface natural and artificial nano-formations on minerals;
- Experimentally substantiate the structural, phase and chemical transformations of minerals under various energy methods of influence;

© The Author(s) 2019
S. Glagolev (Ed.): ICAM 2019, SPEES, pp. 45–48, 2019.
https://doi.org/10.1007/978-3-030-22974-0_11

- To substantiate the choice and mechanism of interaction of reagents with noble metals during flotation of complex ores of complex material composition;
- Investigate the structural, phase and chemical transformations of sulfides and rocks in heap leaching processes.

2 Methods and Approaches

Research methods are optical, confocal laser, analytical electronic, scanning probe microscopy, UV-spectrophotometry of reagent solutions, flotation of minerals. The KEYENCE scanning laser microscope with the surface analysis module VK-9700 enables making a non-contact measurement of the roughness of the surface of minerals and thus determining the height and size of the new formations obtained as a result of interaction with the reagents. The electronic microscope with energy-dispersive microanalyzer LEO-1420 VP INCA-350 allows determining the elemental composition of micro- and nanophases of reagents on the surface of minerals. The analysis of the surface of minerals before and after a contact with reagent solutions was carried out on polished sections made in the form of polished plates $10 \times 10 \times 2$ mm in size.

3 Results and Discussion

In IPKON RAS within the framework of the scientific school of academician V. A. Chanturiya a complex of theoretical and experimental studies on the research and testing of new classes of complex-forming reagents – collectors and modifiers for flotation extraction of non-ferrous and noble metals from refractory gold-bearing oreshave been made. To analyze the processes of physicochemical effects of flotation reagents on micro and nano inclusions of gold and platinum on the surface of sulfides, the authors first developed and improved methods for obtaining samples of mineral complexes that mimic natural sulfides containing "invisible" or submicron gold or platinum. The new highly effective reagents developed in IPKON RAS are modified diethyldithiocarbamate DEDTC and diisobutyldithiophosphinate DIFM, dithiazine derivatives MTX and dialkylpyrylmethane DAM. Those reagents showed the effect for extracting micro- and nanoparticles of noble metals while enriching the mineral raw materials of complex composition owing to formation of poorly water soluble gold compounds and their selective adsorption on gold containing sulfides, improving the flotation selectivity compared to traditional xanthate collector (Chanturiya 2017; Chanturiya et al. 2016; Matveeva et al. 2017a; Matveeva et al. 2017b). Studies were carried out using a set of UV and IR spectroscopy methods (UV-1700 Shimadzu and Infralum FT-8), analytical scanning electron (ASEM) (LEO 1420VP with INCA Oxford micro 350) and laser (KEYENCE VK-9700) microscopy, X-ray phase analysis (X-ray diffractometry). The results made it possible to establish the conditions for the formation of an adsorption layer of new reagents on micro- and nanoparticles of noble metals and to ensure an increase in gold recovery during flotation (Fig. 1).

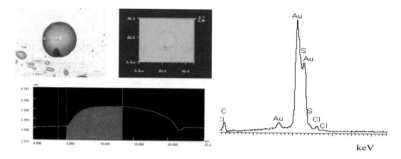

Fig. 1. Adsorbed layer of collector DEDTCm on the surface of Au-pyrite.

As the major losses, from 25% to 30% of valuable components (gold and minerals of the platinum group), are accounted for by the micro- and nano-size of the mineral particles whose concentration in the ore does not generally exceed 1,5–3,0 g/t, it was necessary to obtain samples imitating natural minerals. IPKON RAS has scientifically substantiated and offered procedures for artificially coating minerals with Au and Pt micro- and nano-particles, which made it possible to investigate the interaction mechanism for a new class of frothing agents with noble metals.

The introduction of the new agents both increased the extraction and improved the quality of the mineral concentrate in the mineral processing of rebellious ores of noble metals with a complex material composition, Table 1.

Table 1. Indicators of flotation of the low-grade sulfide ore of the Fyodorovo-Panskoye deposit when employing DEDTCm and ButX

Flotation products	βPd	βPt	εPd	εPt
ButX Conc.	3.88	1.02	73.97	72.33
Tails	0.21	0.06	26.03	27.67
Ore	0.7	0.19	100	100
DEDTCmConc.	10.14	4.24	82.23	85.31
Tails	10.18	0.06	17.77	14.69
Ore	0.93	0.38	100	100

The implementation of new technologies at the mining-enrichment works in Russia will make it possible to increase metal extraction by 10–15%, to obtain high-grade finished products that are competitive in the world market, to involve unpayable ores and technogenic raw materials into processing and considerably ameliorate negative effects on the environment in the mining industry regions.

4   Conclusions

The development of the mining sciences should be based on modern achievements of applied mineralogy and innovative development of comprehensive, economic and effective exploitation of mineral resources. The novel methods ensure both the higher level of mineral extraction, the higher grade of obtained valuable components and high level of ecological safety.

Acknowledgements. The authors are grateful to the Grant of the President of RF "Scientific School" of acad. V.A. Chanturiya and Program of Presidium of RAS P39.

References

Chanturiya VA (2017) Scientific substantiation and development innovative complex processing of mineral resources. Gornyi J 11:7–13

Chanturiya VA, Matveeva TN, Ivanova TA, Getman VV (2016) Mechanism of interaction of cloud point polymers with platinum and gold in flotation of finely disseminated precious metal ores. Miner Process Extr Metall Rev 37(3):187–195

Matveeva TN, Gromova NK, Minaev VA, Lantsiova LB (2017a) Modification of sulfide minerals surface and cassiterite by stable complexes Me–dibutylditiocarbamate. Obogashchenierud 5:15–20. https://doi.org/10.17580/or.2017.05.03

Matveeva TN, Ivanova TA, Getman VV, Gromova NK (2017b) New flotation reagents for extraction of micro- and nano-particles of noble metals from refractory ores. Gornyi J 11:89–93

Mineralogical and Technological Features of Tin Minerals at Pravourmiysky Deposit (Khabarovsk Region)

T. Chikisheva[1,2,3(✉)], S. Prokopyev[1,2,3], E. Kolesov[4], V. Kilin[4],
A. Karpova[1,3], E. Prokopyev[1,2], and V. Tukuser[1,3]

[1] LCC PC «Spirit», Irkutsk, Russia
chikishevatatyana@mail.ru
[2] Institute of the Earth Crust SB RAS, Irkutsk, Russia
[3] Irkutsk State University, Irkutsk, Russia
[4] PJSC «RUSOLOVO», Moscow, Russia

Abstract. The paper presents the data obtained in the process of mineralogical studies of technological samples of tin ore from the Pravourmiysky deposit. The authors studied in detail the mineralogical features of ores and tin-containing minerals and their significance for the enrichment technology. During of the study, the information on the main technological properties of the ore of the deposit was clarified and supplemented. As a result of the study of the mineral composition, the mineralogical features of the ore were identified, allowing to select the methods of ore enrichment and predict the quality of the concentrates and products obtained. The causes of loss of tin with tailings were established. The obtained data on the mineral composition, properties of ore minerals, and textural and structural features of the ores will be applied when modernizing their enrichment technology at the processing plant.

Keywords: Mineralogical research · Tin ores · Tin minerals

1 Introduction

In the 21st century, tin is very in demand in the global economy due to its use in new industries, the introduction of innovative technologies and the environmental friendliness of metal. At the same time, the tin market is volatile, depending on sharp fluctuations in the price situation and annual changes between supply and demand [1].

Russian industry consumes about 6.5–7 thousand tons of tin per year. About 90% of the mined tin is imported. Tin deposits in Russia are among the richest in the world. Pravourmiyskoe tin deposit is one of the promising ore deposits of tin. In addition, together with tin, tungsten can be mined from ore deposits. To achieve a more complete extraction of valuable components from the ore, a detailed study of the mineral composition of ores, the textural and structural features of the ore, the physical properties of minerals and the degree of their contrast is necessary. Mineralogical analysis is the basis for the study of the material composition, structure and texture, choice of directions and methods of preparing raw materials for processing, enrichment technologies and

© The Author(s) 2019
S. Glagolev (Ed.): ICAM 2019, SPEES, pp. 49–52, 2019.
https://doi.org/10.1007/978-3-030-22974-0_12

metallurgy [2]. the paper is devoted to the study of the mineralogical and technological features of the Pravourmiysky deposit tin ores in the Far East.

2 Methods and Approaches

The object of the research was technological samples of tin ore from the Pravourmiisky deposit. The mineral composition of the ore and the quantitative assessment of the contents of each mineral were determined using the methods of optical microscopy and using x-ray methods. Mineralogical analysis of ore crushed to a particle size of less than 2 mm was carried out according to the methods of the Scientific Council on methods of mineralogical research. The sequence of operations to determine the mineral composition of the ore samples consisted in dividing the initial material of the samples into size classes, followed by gravitational fractionation and studying the distribution of minerals into fractions for each size class.

3 Results and Discussion

The main ore mineral of tin at the Pravourmiysky deposit is cassiterite. Stannin, mawsonite, stannoidite are present in small quantities. More than 98% of the tin in the ore is in cassiterite, 1–2% of the tin is in the sulphide minerals of tin.

Streaks, inclusions and clusters of grains of cassiterite, arsenopyrite and lellingite represent ore mineralization. The size of individual grains of ore minerals is up to 5.0 mm, the apparent thickness of clusters is up to 60 mm. Cassiterite is observed in the form of xenomorphic grains, their clusters; crystals of a prismatic, pyramidal-prismatic appearance are less often observed (Fig. 1). The grains size of cassiterite varies from 0.01 to 2.0 mm, with a predominance in the range of 0.1–0.5 mm, the thickness of clusters of 3–5 mm, sometimes can reach up to 10.0 mm.

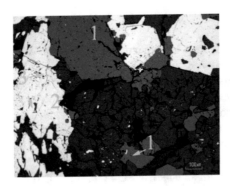

Fig. 1. Micrograph of cassiterite (1) in the intergrowth with arsenopyrite (2), lellingite (3).

According to the mineralogical analysis, crushed ore mainly consists of fragments of greisen, quartz, topaz, tourmaline, feldspar and mica. Ore minerals (arsenopyrite, lellingite, bornite and chalcopyrite) in crushed ore are in the form of fragments of crystals and grains of irregular shape. Sometimes they are found in intergrowths with each other and with rock-forming minerals.

Cassiterite is observed in the form of irregular shape grains, as well as in intergrowths with rock-forming minerals, less often with arsenopyrite or lellingite. The results of the study of the disclosure of cassiterite grains are shown in Table 1.

Table 1. Disclosure of cassiterite grains in crushed ore

Size class, mm	Cassiterite grains and cassiterite – rich intergrowth, %	Cassiterite-containing intergrowth, %	Total
−2+1	96,9	3,1	100,0
−1+0,5	98,9	1,1	100,0
−0,5+0,315	98,2	1,8	100,0
−0,315 +0,25	99,8	0,2	100,0
−0,25+0,125	99,8	0,2	100,0
−0,125+0,071	99,9	0,1	100,0
−0,071 +0,040	100,0	0,0	100,0
−0,040+0,0	100,0	0,0	100,0

According to Table 1, in ore crushed to a particle size of less than 2 mm, cassiterite grains are mainly in the form of free grains. The maximum number of intergrowths (3.1%) is noted in the size class −2 + 1 mm. The complete disengagement of cassiterite grains from intergrowths is achieved in size less than 0.071 mm.

Tin sulphide minerals (stannin, mawsonite, stannoidite) are associated with later copper mineralization. These minerals are in close intergrowth with chalcopyrite and bornite. Their size usually does not exceed 0.1 mm.

4   Conclusions

Tin minerals, which are valuable components of the studied samples, have a significantly higher degree of contrast of gravitational properties, which can be used for their primary concentration Together with cassiterite and stannine, wolframite, arsenopyrite with lellingite and copper sulphides will be extracted into the primary gravity concentrate. To obtain tin concentrate that meets the requirements for raw materials, arsenopyrite and copper sulfides can be extracted from the rough concentrate using flotation methods.

During the flotation process, the sulfide minerals of tin (stannin, mawsonite, stannoidite) are extracted into sulfide products along with the minerals of copper and arsenic. To extract tin sulfides minerals, selective flotation with preliminary fine grinding of sulfide products will be required.

Differences in the magnetic susceptibility of minerals, with the use of magnetic separation, wolframite, biotite and tourmaline can be distinguished.

Studying the contrast of the physical properties of ore minerals provided an idea of the mineral composition of the products of primary gravity enrichment, which will make it possible to develop an optimal scheme for finishing operations.

Acknowledgements. The Irkutsk State University, individual research grant № 091-18-231, supported this work.

References

1. Bashlykova TV (2005) Tekhnologicheskie aspekty racional'nogo nedropol'zovaniya: Rol' tekhnologicheskoj ocenki v razvitii i upravlenii mineral'no-syr'evoj bazoj strany [Technological aspects of rational subsoil use: The role of technological assessment in the development and management of the country's mineral resource base]. Moscow^ MiSIS (in Russian)
2. Danilov UG, Grigoryev VP (2017) Problemy i perspektivy razvitiya olovyannoj promyshlennosti Rossii. Gornaya promyshlennost' 5(135):83–87 (in Russian)

Ore Mineralogy of High Sulfidation Çorak-Taç Epimesothermal Gold Deposit (Yusufeli-Artvin-Turkey)

K. Diarra[1], E. Sangu[1], and E. Çiftçi[2(✉)]

[1] Department of Geological Engineering, Faculty of Engineering,
KOU, 41380 Kocaeli, Turkey
[2] Department of Geological Engineering, Faculty of Mines,
ITU, 33469 Maslak, Istanbul, Turkey
eciftci@itu.edu.tr

Abstract. Çorak and Taç, two nearby mineralizations, are located in the eastern black sea region, which is one of the most productive metallogenic belts of Turkey. It is characterized by a great number of Kuroko-type volcanogenic massive sulfide deposits as well as vein-type polymetallic deposits, porphyry and epithermal precious metal deposits. Subject neighboring deposits are hosted within the voluminous Cretaceous-Eocene granitoids and interbedded volcanic rocks and carbonates. Mineralogy of altered host rocks include quartz veins, carbonates, sericite, chlorite, chalcedony, and disseminated sulfides - mainly pyrite, sphalerite, galena, and chalcopyrite. The main texture encountered in the host rocks is hyalo-porphyry. Due to hydrothermal alterations primary minerals are mostly altered in which the ferromagnesian minerals are chloritized and calcified, while feldspars are altered into sericite, calcite, and albite. Silicification and argillic alteration (medium, moderate, high) are widely spread however; XRD analysis carried on 33 core samples from Çorak has also revealed local propylitic alteration, limoniti-zation and hematitization as well. The minerals assemblages that accompanied the different alterations include jarosite and alunite suggesting high sulfidation hydrothermal mineralization. Through the ore microscopic studies, pyrite, chal-copyrite galena, sphalerite, and a lesser amount of sulfosalts (tennantite-tetrahedrite and pyrargyrite) were determined. Quartz and calcite account for the main gangue minerals. While the Taç mineralization is pyrite, chalcopyrite and sphalerite dominating, the Çorak mineralization contains relatively less chal-copyrite and galena becomes prevalent with sphalerite. Gold in both sites may reach up to 10 ppm, on average 3 ppm. Silver occurrence is insignificant.

Keywords: Yusufeli · Artvin · Turkey · High sulfidation · Gold · Epimesotherma

1 Introduction

The study area is located in Yusufeli area in the eastern Pontides' northeastern most tip within a larger metal rich tectonic corridor stretching from southern Georgia and northern Armenia to Bulgaria and Romania and it is a host to numerous volcanogenic massive sulfide (VMS) and vein-type deposits dominantly of Late Cretaceous age.

© The Author(s) 2019
S. Glagolev (Ed.): ICAM 2019, SPEES, pp. 53–58, 2019.
https://doi.org/10.1007/978-3-030-22974-0_13

Taç and Çorak deposits occur in an area about 10 km × 3 km in size, oriented NW-SE and SW-NW, respectively. It is limited to the north by series of outcrop of large unaltered basaltic basement with some andesite enclave. They are covered by more recent tuffs, sandstone and old alluvial fan. The south is more dominated by andesites and tuff outcrops together with relatively recent sedimentary covers.

Mineralization and alteration are represented by extensive strata-bound argillic alteration cut by gold and base metal bearing quartz veins within a massive pyroclastic volcanic host-rock below a bedded volcanogenic succession, with additional receptive massive volcanic units occurring higher up in the sedimentary succession. Porosity is believed to be the most likely control on the stratiform nature of alteration. However, proximity to major structures along or close to the structural axis appears to be an important influence on the presence of alteration suggesting structural control on mineralization.

Some typical structures are associated with the Cu-Pb-Zn mineralization: The most obvious one is the hydrothermal alteration with its colored intense argillic and propylitic alterations. They are always associated with a local fault. Afterward, the large intrusive rocks are crosscutting the volcanic, the pyroclastic and the volcano sedimentary rocks.

In Çorak SW-NE section view, the high grade of base metal representing the mineralization is localized in volcano sedimentary rock (tuff) as well as in andesite and feldspar porphyry andesitic, the three sets are separated by set of faults.

2   Methods and Approaches

For mineralogical and ore microscopy studies, 35 thin and polish sections from core drill and outcrop, have been prepared and scrutinized under reflect and transmitted light microscope of the laboratories of Kocaeli university (KOÜ) and Istanbul technical university (ITU) in Turkey. 6 samples have been taken from alteration zone have been analysis in order to compute their CIA: ($[Al_2O_3/(Al_2O_3 + CaO + K_2O + Na_2O)]$ 100) and ICV: (ICV: $CaO + K_2O + Na_2O + Fe_2O_3$ (t) $+ MgO + MnO + TiO_2)/Al_2O_3$). The target was to determine their depositional environment as well as the alteration styles. 33 samples from different alteration zones from Çorak have been submitted to the lithological and XRD analysis using Rietveld methods.

15 polished thin sections have been subjected to Cathodoluminescence (CL) microscopy study to describe and to distinguish gangue minerals and minerals generations. Samples containing mostly sphalerite, quartz and carbonates were chosen, since they are able to produce CL emissions. The CL study was performed at the ITU Advanced Microscope lab using an optical cathodoluminescence system (CITL MK5 system).

3 Results and Discussion

The andesite and andesitic tuffs are the main lithologies hosting the mineralization in the study area. The mineralogy of the altered hostrock enclosed quartz veins, carbonate, sericite, chlorite, and chalcedony and disseminated sulfides. Their details microscopic studies have revealed the following common textures: trachytic; hyalo microlithic porphyry; vesicular microlithic porphyry; hyalo micro granular porphyry and quasi-oolithic. The hydrothermal alteration manifest itself in the form of silicification, epi-dotization, chloritization, seritization, and finally a calcification as a last stage. Colored argillic alteration is widespread in the area along with silicic and propylitic alteration. Drill holes intercept sulfide ore zones at different depth. The sulfide mineralizations may range from 50 cm to 1 m in thickness. Weakly mineralized zones of 1 to 2% with disseminated sulfides are intercalated between the massive or semi massive sulfide lenses. The altered interval shows 3 types of minerals assemblages: Quartz + sericite + calcite, quartz + epidote and quartz + chlorite + calcite: (I) **quartz + chlorite alteration**: In this facies, the ferromagnesian (Amphibole) are completely chloritized and calcified. Chlorites are filling cavity as well. They wrap also the fewer amount of chalcedony present in the groundmass. The opaque minerals represent less than 1–2% of the groundmass. The texture is vesicular hyalo-microlithic porphyry; (II) **quartz + epidote alteration:** In this facies, the light colored plagioclases are turned into epidote. The calcite as a last phase is filling fracture. (propylitic alteration); (III) **quartz + sericite alteration:** The sericite is not only replacing the ferromagne-sian and the plagioclase but also the matrix. Plagioclases turned into sericite. Euhedral quartz is present in the groundmass. The most prevalent ore mineral in Çorak is pyrite it is present in all the samples. It is followed by chalcopyrite, sphalerite, and galena. Compare to sulfides, sulfosalts namely tennantite-tetrahedrite are present in lesser amounts. Covellite and bornite are rare. Gangue minerals are mainly composed of quartz and calcite. Two main mineral's assemblages are prevailing whether it is from stockwork (disseminate and veinlet zone) or from the high grade sulfide zone. In the stockwork zone pyrite has euhedral to subhedral shape. Chalcopyrite is anhedral and replaces very often pyrite. In the massive ore zone, pyrite and chalcopyrite are accompanied by large sphalerite and galena grain. Galena always crosscut sphalerite, suggesting that it is later than sphalerite. Tennantite and tetrahedrite assemblage replace very often pyrite and chalcopyrite in this zone. The pyrargyrite is present in very small amount. The tennantite is dominant in Çorak's polymetallic sulfide zone. It has been also observed galena replacing gangue minerals quartz suggesting symplectic inter-growths of ore minerals with silicate. Colloform texture is observed but they are rare. Except the pyrite, all the minerals are anhedral. Minerals size span from 50 to 300 micron. The sulfosalts are the smallest ones (Figs. 1a–f).

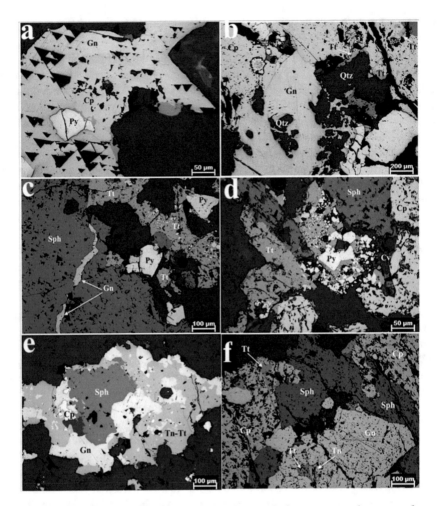

Fig. 1. (a) galena replacing chalcopyrite; (b) galena replacing quartz, replacement of gang minerals by ore minerals (symplectic intergrowths) and tennantite replacing chalcopyrite; (c) galena veins sphalerite, while pyrite replaced by tennantite; (d) tennantite replacing pyrite and chalcopyrite; (e) tetrahedrite veining galena and tetrahedrite-tennantite replacing sphalerite (Py: galena; Cp: chalcopyrite; Sph: sphalerite; Tn-Tt: Tennantite-Tetrahedrite; Qtz quartz)

The following paragenesis is deducted from the ore microscopy study: **Pyrite – Chalcopyrite (I) – Sphalerite – Galena – Sulfosalts- Chalcopyrite (II) - Bornite – Covellite**.

The cathodoluminescence study on sphalerite has revealed heterogeneity in the composition of the sphalerite (Fig. 2). These states that those sphalerite grains are originated from more than one generation phase and/or from fluids with changing composition.

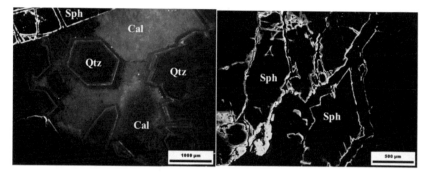

Fig. 2. Optical cathodoluminescence microscopy images (OCLMI) of sphalerite and quartz grains with infilling late calcite, showing compositionally heterogeneity sphalerites

4   Conclusions

Taç and Çorak deposits occur in north easternmost tip of the eastern Pontides tectonic belt of Turkey. They are two separate ore bodies within the same system. Hydrothermal alteration is intense and extensive in the form of silicification, epidotization, chloritization, seritization, and finally a calcification as the final stage. Three main alterations zone are succeeding from the proximal, intermediate to distal zone: Quartz + Chlorite alteration, quartz + epidote alteration, quartz + sericite alteration. Two main mineral's assemblages are prevailing whether it is from stockwork (disseminated and veinlet) or from the high grade sulfide zone. They are respectively: pervasive pyrite, chalcopyrite with sphalerite and galena.

Argillic, propylitic and silicic alteration induced by acidic hydrothermal fluids have led to the formation of quartz, muscovite, orthoclase, gypsum, dolomite, kaolinite, pyrite, sphalerite, galena and the pair jarosite-alunite assemblage. The presence of alunite and jarosite minerals indicate that Çorak and Taç are high sulfidation epithermal deposit with mesothermal signatures.

Acknowledgements. This study was financially supported by BAP Project unit (No: 2017/80) of Kocaeli University in Turkey. We are thankful to the university and to Santral Mining Co. and their employees in Yusufeli for their courteous support for the field work.

References

1. Bogdanov B (1980) Massive sulphide and porphyry copper deposits in the Panagjurishte district, Bulgaria. In: Jankovic S, Sillitoe Richard H (eds) European copper deposits; Proceedings of an international symposium. Springer, Berlin-Heidelberg-New York, pp 50–58
2. Ciftci E (2000) Mineralogy, paragenetic sequence, geochemistry, and genesis of the gold and silver bearing upper cretaceous mineral deposits, Northeastern Turkey. PhD dissertation, University of Missouri-Rolla, Missouri, USA, p 251

3. Kouzmanov K, Moritz R (2009) Late Cretaceous Porphyry Cu and epithermal Cu-Au association in the Southern Panagyurishte District, Bulgaria: The paired Vlaykov Vruh and Elshitsa deposits
4. Muntean JL, Einaudi MT (2000) Porphyry gold deposits of the Refugio district, Maricunga belt, Northern Chile. Econ Geol 95:1445–1472

Mineralogical and Technological Aspects of Phosphate Ore Processing

A. Elbendari[✉], V. Potemkin, T. Aleksandrova, and N. Nikolaeva

Mineral Processing Department, Saint Petersburg Mining University,
Saint Petersburg, Russia
abdullah_elbendary@yahoo.com

Abstract. The article studies the mineralogical features of phosphate ores. In the conditions of declining industrial reserves of apatite-containing ores, issues of a more comprehensive and in-depth study of the mineral and material composition, as well as the improvement of existing technologies for the processing of this type of raw material, become topical. Using optical methods of analysis, electron microscopy with automated mineralogical analysis (MLA), mineral and elemental composition of apatite was obtained. Taking into account the studied mineralogical and material composition, experiments on grinding and flotation were carried out. Based on these data, it was concluded that the optimal scheme for the processing of phosphate ores is a flotation scheme with preliminary selective disintegration.

Keywords: Mineralogical composition · Grinding · Beneficiation · Phosphorus-bearing minerals

1 Introduction

Phosphates are one of the most important minerals on Earth, as they are used as fertilizers for agriculture and as a necessary raw material for the chemical industry (Brylyakov 2004; Abouzeid 2007). In addition, phosphates are the source of rare-earth elements. They are used in many commercial and industrial products, such as: detergents, toothpastes and fireproof materials. Worldwide consumption of P_2O_5 in all of the areas above is projected to grow gradually from 44.5 million tons in 2016 to 48.9 million tons in 2020 (Jasinski 2017).

In the conditions of declining industrial high-quality reserves of phosphorus-containing ores, issues of a more comprehensive and in-depth study of the mineral and material composition, as well as improvement of existing technologies for processing this type of raw material, become urgent. The study of the influence of the mineral raw materials composition on the features of the beneficiation schemes construction is given in the works of many authors (Aleksandrova et al. 2012; Evdokimova et al. 2012; Gerasimova et al. 2018; Litvintsev et al. 2006; Mitrofanova et al. 2017).

© The Author(s) 2019
S. Glagolev (Ed.): ICAM 2019, SPEES, pp. 59–65, 2019.
https://doi.org/10.1007/978-3-030-22974-0_14

2   Methods and Approaches

The object of the study was apatite-nepheline ore of the Khibiny deposits group (Russia) and phosphate ore of the Abu-Tartur deposit (Egypt). For the development of beneficiation schemes and modes, complex studies were carried out on mineral and elemental composition, including optical methods of analysis, electron microscopy using automated mineralogical analysis (MLA), etc. As a result of the work, the mineral composition of apatite-nepheline ore (ANO) and phosphate ore (PO) was studied taking into account the data of optical and electron microscopic studies, spot X-ray spectral and chemical analyzes, atomic emission spectrometry, automated mineralogical analysis. The chemical composition of ANO and PO is given in the Table 1.

Table 1. The chemical composition of ANO and PO

Apatite-nephelineores (Russia)		Phosphoriteore (Egypt)	
Component	Content, %	Component	Content, %
SiO_2	32.50	SiO_2	19.4
TiO_2	2.20	TiO_2	0.06
Al_2O_3	15.52	Al_2O_3	2.6
Fe	5.45	Fe	5.3
CaO	18.28	CaO	37.6
MgO	0.92	MgO	0.8
MnO	0.16	MnO	0.05
K_2O	4.20	K_2O	0.59
Na_2O	5.56	Na_2O	0.57
P_2O_5	12.50	P_2O_5	21.8
SO_3	0.04	SO_3	1.6
CO_2	0.05	CO_2	1.2
LOI	2.62	LOI	8.43
Total	**100.00**	Total	**100.00**

To study the possibility of increasing efficiency of the flotation process, studies were carried out on the selection of the optimal grinding mode and a series of flotation experiments.

3   Results and Discussion

According to the mineralogical analysis, the main primary minerals of ANO are apatite and nepheline, the contents of which are respectively 30.67 and 30.88%; minor quantities contain pyroxenes, mica, feldspars, as well as natrolite and kaolinite – secondary minerals formed due to the destruction of the primary mineral phases.

Phosphorus-containing minerals of the sample are apatite, eschynite, phosphates of rare-earth elements and lomonosovite, with a distribution to these minerals of 99.94, 0.01, 0.02 and 0.03% phosphorus respectively. Valuable minerals of the sample are apatite – the main mineral concentrating phosphorus and nepheline – the main mineral concentrating aluminum.

Apatite is the main mineral concentrator of phosphorus in the ore; it forms disseminated prismatic crystals, vein clusters of crystals, less often massive clusters of fractured xenomorphic grains, often included in the grains of other minerals - pyroxenes, mica, sphene, nepheline (Fig. 1).

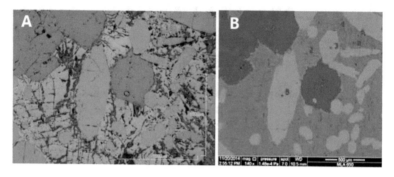

Fig. 1. Apatite crystals in the aegirine-augite matrix. Spectra in (b): 1, 7 - nepheline; 2, 4 - aegirine-augit; 3, 8 - apatite; 5, 6 - arfvedsonit

Image: (a) - in reflected light; (b) - in backscattered electrons. Apatite is characterized by a pronounced idiomorphism of grains that have clear crystallographic outlines; the shape of apatite grains is columnar, prismatic, acicular, which causes a weak connection between them in aggregates. The crystalline form of apatite, the natural brittleness of the mineral will contribute to the primary destruction of the mineral during ore grinding and the concentration of apatite in smaller grades.

According to MLA data, 32.78% of the mineral in ore is distributed into free particles, 22.29% into binary and 44.92% into polymineral intergrowth.

Phosphate minerals in PO are represented by carbonate-fluoroapatite, fluoroapatite, hydroxylapatite, and francolite. The most common clay minerals in the studied sediments are smectites. The amount of kaolinite and illite in general is insignificant, although their content is also high enough. Among the non-phosphate components of the PO, detrital quartz, as well as ankerite and pyrite cement in unaltered phosphates, are predominant. Pyrite and ankerite in many cases replace partially or completely phosphate grains. Elements such as Ba, Cr, Ni, Sr, Y, and Zr are found in relatively high concentrations, while Co, Nb, Pb, Rb, Th, and U are found in relatively low concentrations (Abdel-Moghny and Zhabin 2011; Baioumy 2013) (Fig. 2).

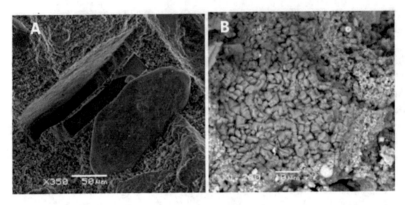

Fig. 2. Electron microscopic images of PO (Baioumy 2013). (a) – rounded silica inclusions; (b) – kaolinite plates

To determine the optimal processing scheme for ANO and PO, samples with a grain size of less than 2 mm and a mass of 550 g were ground in a ball mill in accordance with the specifications given in Table 2:

Table 2. Mill characteristics and experiment conditions

Mill	Inner diameter (D), mm			125			
	Length, mm			170			
	Volume, cm^3			2085			
	Critical speed, rpm			120			
Media (Balls)	Material			Alloy steel			
	d, mm	19	22	25	27	28	29	30
	Number	1	2	2	4	4	3	1
	d, mm	32	34	35	36	39	40	41
	Number	2	1	2	1	1	2	1
	Specific gravity			7.8			
	Mass of balls, g			3388.5			
Material	Igneous & sedimentary phosphate ore						
	Specific gravity			3.1 & 2.7			
	Powder weight, g			550 gm			

The grinding results are shown in Fig. 3.

As a result of the grinding process study, it was found that with an increase in the grinding time, the particle size sharply decreases, and pulp grinding with a solid content of 50% gives the best results. Studies have also been conducted for the process of grinding apatite ore with the addition of tributyl phosphate in amount of 500 and

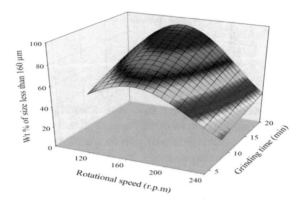

Fig. 3. 3D relationship between the rotational speed and grinding time

1000 ml/ton. Studies have shown that the addition of surface-active substances (tributyl phosphate) during grinding of ANO and PO does not only increase the efficiency of grinding, but also partially convert rare-earth metals into soluble form with their subsequent extraction.

Beneficiation of ANO and PO samples was carried out according to the flowchart shown in Fig. 4.

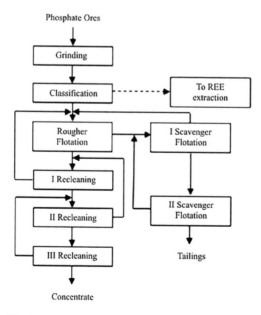

Fig. 4. Block diagram of phosphate ore beneficiation

Based on the studies, it was found that this scheme is optimal for both samples (sedimentary and volcanic).

4 Conclusions

In the conditions of declining quality of industrial reserves of phosphate ores, issues of a more comprehensive and in-depth study of the mineral and material composition, as well as the improvement of existing technologies for the processing of this type of raw material become topical. Achieving this goal is complicated by the constant decline in the quality of ores involved in processing and requirement of 100% recycled water supply implementation. Based on the mineralogical, chemical, material composition, as well as technological research on the possibility of processing phosphate ores, it was concluded that the optimal scheme for the extraction of apatite is a flotation circuit with preliminary selective disintegration. At the same time, the possibility of extracting rare earth metals has been established.

Acknowledgements. The work is carried out under financial support of the Ministry of Education and Science of the Russian Federation, the project RFMEFI57417X0168.

References

Abdel-Moghny MW, Zhabin AV (2011) Mineralogical features of phosphate rocks Egypt. Bull Voronezh State Univ Ser Geol 2011(1):37–49

Abouzeid AM (2007) Upgrading of phosphate ores – a review. Powder Handling Process 19:92–109

Aleksandrova TN, Litvinova NM, Gurman MA, Aleksandrov AV (2012) Comprehensive utilization of the far eastern apatite-containing raw materials. J Min Sci 48(6):1047–1053

Baioumy HM (2013) Effect of the depositional environment on the compositional variations among the phosphorite deposits in Egypt. Geol Geophys 54(4):589–600

Brylyakov YE (2004) The development of the theory and practice of complex enrichment of apatite-nepheline ores of the Khibinsky deposits

Evdokimova GA, Gershenkop ASh, Fokina NV (2012) The impact of bacteria of circulating water on apatite-nepheline ore flotation. J Environ Sci Health - Part A Toxic/Hazardous Subst Environ Eng 47(3):398–404

Gerasimova LG, Nikolaev AI, Maslova MV, Shchukina ES, Samburov GO, Yakovenchuk VN, Ivanyuk GY (2018) Titanite ores of the khibiny apatite-nepheline-deposits: selective mining, processing and application for titanosilicate synthesis. Minerals 8(10):446

Jasinski SM (2017) Phosphate Rock, U.S. Geological Survey, Mineral Commodity Summaries

Litvintsev VS, Melnikova TN, Yatlukova NG, Litvinova NM (2006) Mechanoactivation in the processes of ore preparation. Mt J (6):95–96

Mitrofanova GV, Ivanova VA, Artemiev AV (2017) Use of reagents-flocculants in water-preparation processes during phosphorous-containing ore processing. In: International multidisciplinary scientific GeoConference surveying geology and mining ecology management, SGEM 2017, vol 17, no 11, pp 1143–1150

Properties and Processing of Ores Containing Layered Silicates

A. Gerasimov[✉], V. Arsentyev, and V. Lazareva

REC Mekhanobr-tekhnica, St-Petersburg, Russia
gerasimov_am@npk-mt.spb.ru

Abstract. The solid mineral raw materials of the sediment deposits containing the argillaceous variations such as the coal and the potash ores are the most multi-tonnage solid minerals mined and processed in the territory of the Russian Federation. Both types of the deposits have relatively simple geology aspects with large and medium-sized solids and as a rule, the most often, with the regular bedding which are characterized by the fitchery thickness and the internal structure. Such deposits are very difficult for mining and processing. The thermal modification of this of ores allows changing the physico-chemical properties of its components in such a way that their further processing - grinding, separation, transportation and storage can be carried out without the use of process water or significantly reduces its consumption.

Keywords: Argillaceous minerals · Hydrochemical modification · Thermochemical modification · Water saving

1 Introduction

Many deposits of minerals are associated with the presence of layered silicates such as the clay, mudstones, argillaceous slate - highly hydrophilic mineral varieties that have the ability to swell in the water. This ability to self-disperse creates the great difficulties both for the extraction and for the processing of such ores. Since the processing of most minerals is associated with the use of wet processes, where in contact with water the swelling layered silicates are dispersed to a particle size of less than 50 μm, but due to their high hydrophylic property they form bonds that promote the formation of so-called structured suspensions having increased viscosity. In this case, the structured suspensions make it difficult to perform the separation operations as the classification, the gravity and magnetic separation, the flotation and thickening and filtration operations (Arsentyev et al. 2014).

In order to eliminate the influence of swelling silicates on the treatment processes it is necessary to use the diluted suspensions with the solids content of 10–20% while the usual solids content in processing suspensions is 35–50%. This causes the increase in 3–4 times the consumption of water during the processing of ores containing the swelling sheet silicates (Zhdanovich et al. 2011).

The most large-tonnage processed commercial mineral containing the swelling silicates are the coal and the potassium salts. Considering these minerals from the point

© The Author(s) 2019
S. Glagolev (Ed.): ICAM 2019, SPEES, pp. 66–69, 2019.
https://doi.org/10.1007/978-3-030-22974-0_15

of view of their advanced processing, it is important to note the similar processing characteristics:

- comparatively high content of the useful component which is 60–80% for the bituminous coal and 20–30% for potassium ores;
- polydispersity of the impregnations of the useful component is from 0.5 to 10 mm;
- relatively low density of the constituent rocks is from 1.4 to 2.3 g/cm^3;
- the presence of argillaceous material such as the layered silicates swelling in the water.

2 Methods and Approaches

Coal, sylvinite and kaolin were selected for the dry processing study.

The presented coals are dark gray, dense, laminated and fractured with the matte surface in the place of spalling. The samples of the coal are brittle and under a slight stress are destroyed along the microfracture (Gerasimov et al. 2016).

The sylvinite sample is a rock salt-sylvinite-carnalite formation with the varying ratio of the main rock-forming minerals represented mainly by readily soluble congeries and individuals of the chloride class (Arsentyev et al. 2017a, b).

The kaolin ore is represented by the kaolin argillites mainly gray, dark gray, greenish-gray, less often yellowish-ochreous and sorrel, occasionally milky-gray colors cured during the compaction, dehydration and cementation processes. The structure is massive often the breccia structure and splintery, sometimes the oolitic structure, here and there the indistinctly-banded. The structure is the pelitic, aleuropelitic, aleuro-psammito-pelitic, occasionally oolitic (Arsentyev et al. 2017a, b).

3 Results and Discussion

The thermal treatment of the high-ash coals (low-temperature pyrolysis reaction) modefies the physicochemical properties of the minerals that make up the ash reducing their ability to swell in water and increasing the magnetic susceptibility which makes it possible to significantly intensify the processes of their magnetic-flotation processing to produce the high-quality solid fuels. In the coal, under the influence of temperatures in the process of the coal pyrolysis, the initial fracture structure is transformed into the cracked-pore structure with the increase in the number and dimensions of the poriness, vesicles and cracks. The sample of the coal subjected to medium-temperature pyrolysis reaction compared to the initial coal has 20% lower apparent density and the specific energy consumption of crushing is 20–30% lower.

The specific ash content per unit of the heat of combustion indicates that the combined dry processing of coal allows reducing the specific ash content by 1.3–1.6 times i.e. the formation of the ash and slag during the combustion is reduced.

It has been established that the thermal treatment of the sylvinite ore; containing the argllous-carbonate insoluble minerals in the range of 100–500 °C practically does not affect the structure of salting minerals of the halite and the silvinite but it significantly modifies the structure of minerals entering into the insoluble fraction which has the positive effect on the flotation of these ores and the process of sedimentation of the slurry fractions in the flotation tails.

For reducing the energy costs for the heat treatment of sylvinite ores it has been proposed to use microwave treatment.

The hydrothermal treatment of kaolin raw material in the autoclave can significantly reduce the viscosity of kaolin suspensions and further efficient processing to produce the concentrates for the production of alumina by using the wet screening process for the suspended matters with the relatively high density.

4 Conclusions

An analysis of international practice has shown that fresh water is becoming the most scarce resource for the mining and processing industry, which leads to a significant increase in the cost of water supply and circulation systems, with their growth rates being four to five times higher than the growth rates of the ore mining output.

Studies of the problems of reducing water consumption in mineral processing should be combined with energy consumption assessments due to the close relationship between water and energy consumption values within a single processing system of any plant. Most often, a technology that ensures water saving requires higher energy consumption and vice versa.

The presence of layered silicates with mobile crystal lattices in the mined ore significantly complicates both its wet and dry processing.

A promising solution, aimed at efficient use of resources in processing of ores containing layered silicates, implies the inclusion of thermal and hydrothermal modification of such ores as the primary stages of the process chain. These modifications would ensure a reduction or elimination of the negative impact of such materials currently experienced in ore preparation, separation and dewatering and in the storage of processing products.

The solution requires additional energy inputs, but ensures significant savings in other resources.

Acknowledgements. Financial support was provided by the Russian Scientific Fund (project 18-17-00169).

References

Arsentyev VA, Vaisberg LA, Ustinov ID (2014) Directions of creation of low-water technologies and devices for the processing of finely divided mineral raw materials. Obogashenie Rud (5):3–9

Arsentyev VA, Gerasimov AM, Kotova EL (2017a) Thermochemical modification of sylvinite ore using the microwave heating. Obogashenie Rud (6):3–8

Arsentyev VA, Vaisberg LA, Ustinov ID, Gerasimov AM (2016) Prospects for reducing water use in coal-processing. Mining J (5):97–100

Arsentyev VA, Gerasimov AM, Mezenin AO (2017b) Study of technology of kaolin processing using hydrothermal modification. Obogashenie Rud (2):3–9

Gerasimov AM, Dmitriev SV (2016) Combined technology of dry coal processing. Obogashenie Rud (6):9–13

Lyskova MY (2016) Geoecology in the modern construction of enterprises for the extraction and enrichment of potassium-magnesium salts, News of Tula State University. Geosciences (4):39–49

Mozheiko FF, Potkin TN (2008) Physico-chemical basis of processing of high-arcilla off-balance sylvinite ores. Vesti, National Academy of Sciences of Belarus, Series of Chemical Sciences, no 4, pp 25–32

Titkov SN et al (2013) The technology of dry crushing of potash ore to flotation size. In: IX Congress of enrichers of the CIS countries. Collection of materials, vol 2. Digest MISA. M. 2013, pp 578–583

Zhdanovich IB et al (2011) Influence of heat treatment of saliferous arcilla on the structural-geological properties of their dispersions. Vesti, National Academy of Sciences of Belarus, Series of Chemical Sciences, no 3, pp 113–117

Applied Mineralogy of Anthropogenic Accessory Minerals

A. Gerasimov[1]([✉]), E. Kotova[2], and I. Ustinov[1]

[1] REC «Mekhanobr-Tekhnika», St-Petersburg, Russia
gornyi@mtspb.com
[2] Mining Museum, St-Petersburg Mining University, St-Petersburg, Russia

Abstract. The term "accessory minerals" is applicable to an expanding range of mineral resources, including rare minerals of anthropogenic deposits and mineral metallurgical wastes. Typical accessory minerals of sedimentary rocks, metallurgical slags and calcines include sulphide minerals of iron, copper, and zinc. This work covers certain chemical transformations of accessory sulphides contained in the wastes of the metallurgical industry. Note that similar behaviors of sulphides may also be observed in the thermal processing of coal. It is shown that, in terms of thermodynamics, thermal processing of products containing precious metals enables application of various methods for generating artificial covellite with silver and gold content through the use of accessory sulphide minerals.

Keywords: Accessory minerals · Precious metals · Thermal processing · Thermodynamics

1 Introduction

Accessory minerals, in the classical sense, are minerals contained in rocks in small quantities (less than 1%). It may be assumed that the term is also applicable to an expanding range of mineral resources, including rare minerals of anthropogenic deposits and mineral metallurgical wastes. Typical accessory minerals of sedimentary rocks, metallurgical slags and calcines include sulphide minerals of iron, copper, and zinc, often containing precious metals.

This work covers certain chemical transformations of accessory sulphides contained in the wastes of the metallurgical industry. Note that similar behaviors of sulphides may also be observed in the thermal processing of coal. Process mineralogy (Bradsow 2014), recently expanded to include thermal processes (Rubin et al. 2014), is the critical tool when studying the behavior of accessory sulfide minerals.

2 Methods and Approaches

The traditional technology used for the processing of zinc concentrates obtained in flotation concentration of polymetallic sulphide ores envisages oxidizing roasting of the zinc concentrate, in which the main mineral, ZnS sphalerite, is converted into zinc

© The Author(s) 2019
S. Glagolev (Ed.): ICAM 2019, SPEES, pp. 70–74, 2019.
https://doi.org/10.1007/978-3-030-22974-0_16

sulphate $ZnSO_4$. The inevitable impurities of pyrite FeS_2 and chalcopyrite $CuFeS_2$, present in the zinc concentrates, are also oxidized into the corresponding simple sulfates and oxides at temperatures above 770 °K (Naboichenko et al. 1997). At present, fluidized bed furnaces are predominantly used for oxidizing roasting, with the oxidation process occurring in a hot gas flow in a suspended state.

Despite the obvious benefits, such as high yield and roasting rates, fluidized bed furnaces also have certain disadvantages, including partial underburning of sphalerite due to the rapid formation of a poorly gas-permeable zinc sulfate film on its surface. This inevitably leads to underburning of fine inclusions in sphalerite grains of other sulfide minerals, primarily chalcopyrite. Chalcopyrite is the most common impurity in sphalerite, which is due to the high similarity of the structures of their crystal lattices. It is known that chalcopyrite is most frequently found in the sphalerite structure in the form of very fine grains, down to emulsion shots (Betekhtin 2008). The behavior of chalcopyrite inclusions in sphalerite is significant for subsequent stages of processing of the calcined concentrate.

After the oxidizing roasting, the resulting zinc calcine is subjected to staged sulfuric acid and aqueous leaching of zinc sulfate and oxide-sulfate with the recovery of a zinc sulfate solution to be subsequently used for zinc metal precipitation by electrolysis. Acid and neutral zinc sulfate leaching is performed at temperatures of up to 80–100 °C, additionally enabling chemical transformations of insoluble calcine components on the grain surfaces. The insoluble cake obtained after leaching includes a certain amount of underburned sulphide minerals containing precious metals, primarily silver. It was previously demonstrated (Otrozhdennova et al. 1997; Geikhman et al. 2003) that sulphide minerals remaining in the cake may be flotation concentrated, yielding a froth product containing up to 18% of sulfide sulfur and up to several kilograms of silver per ton.

We have studied the composition of concentrates obtained by flotation concentration of the cake generated after zinc sulphate leaching at three metallurgical enterprises in different countries. The studies were performed using a Camscan scanning microscope, an Analysette laser microanalyzer, a Geigerflex X-ray phase spectrometer, optical microscopy tools, etc.

It was found that there was no fundamental difference in the material composition of such concentrates at different metallurgical plants. This is explained by the similarity of the ore bases and of the metallurgical treatment structures used at the mining enterprises analyzed. Sphalerite (70–80%) is the main mineral found in the composition of the concentrates, followed by such spinel-structure minerals as $MeFe_2O_4$, covellite, oxide compounds of zinc and iron (several percent), and traces of chalcosite.

In terms of their elementary chemical composition, the concentrates were represented by zinc 45–48%, total sulfur 20–22%, sulfur sulfide 17–19%, iron 11–14%, silica 2.5–4%, and copper 3–4%. The silver content was 0.2–0.4% (2–4 kg/t); the gold content was 10–15 g/t.

3   Results and Discussion

The most important finding is the presence of copper sulphide covellite CuS in the flotation concentrate, in which the bulk of silver is concentrated. The availability of silver-containing films of copper sulfides (without their identification) on the surfaces of sphalerite contained in the zinc cake flotation concentrate has been indicated previously (Otrozhdennova et al. 1997). We have established that covellite forms distinct substances on the sphalerite surface. No native silver was found in the samples. Gold is present in the form of fine grains containing silver and copper. The gold sample is 700–930. Copper-containing metals in the source zinc concentrate entering the roasting process are represented almost exclusively by chalcopyrite, including its aggregates with sphalerite.

Two main methods of covellite formation in the metallurgical processing of zinc may be considered in this regard. The first is the interaction of dissolved copper sulfate with the sphalerite surface in sulfuric acid leaching of zinc calcines, with the formation of covellite films in the course of the resulting exchange reaction and with the transition of zinc ions into the solution. The second possible source of covellite is the transformation of fine chalcopyrite inclusions contained in sphalerite grains with the migration of the resulting covellite to the sphalerite surface due to the strong differences in the structures of their crystal lattices. None of these methods has been subjected to special experimental verification. Given that even isomorphous substitution of copper with silver can occur in chalcopyrite, despite the difference in the electron spiral radii of silver of 1.26 Å and copper of 0.96 Å, it may be assumed that silver-containing covellite may be generated using these methods, especially considering the effects produced on such processes by associated minerals.

Let us consider the thermodynamic feasibility of the two hypothetical processes of formation of man-made covellite. In the below calculations of the free energy of the processes and their enthalpy and entropy increments, known reference data are used for reactions (I) and (II). These calculations show that reaction (I) can occur spontaneously, which is facilitated by the entropy effect, and reaction (II) requires external energy inputs, which does not contradict the conditions of respective production processes (Tables 1 and 2).

Table 1. Thermodynamic characteristics

Substance	ΔH^0, kJ/mol	S^0_{298}, J/mol K	C_P, kJ/mol K
ZnS	−206	57.7	46.0
CuSO4	−771.4	109.2	98.9
CuS	53.1	66.5	47.8
ZnSO4	−982.8	110.5	99.2
Реакция	−58.5	10.1	2.1

Table 2. Thermodynamic characteristics

Substance	ΔH^0, kJ/mol	S_{298}^0, kJ/mol K	C_P, kJ/mol K
$CuFeS_2$	−194.93	124.9	95.67
CuS	53.10	66.5	47.8
FeS	−100.00	60.3	50.5
Реакция	41.8	1.9	2.63

$$ZnS \ + \ CuSO_4 \ \rightarrow \ CuS \ + \ ZnSO_4 \qquad (1)$$

$$\Delta H^0_{798} = \Delta H^0_{298} + \int_{298}^{798} \Delta c_p dT = -58.5 \cdot 10^3 + 2.1 \cdot (798 - 298) = -57.45 \, \text{kJ/mol}$$

$$\Delta S^0_{798} = \Delta S^0_{298} + \int_{298}^{798} \frac{\Delta c_p}{T} = 10.1 + 2.1 \cdot \ln \frac{798}{298} = 12.17 \, \text{J/molK}$$

$$\Delta G^0_{798} = \Delta H^0_{798} - 798 \cdot \Delta S^0_{798} = -57450 - 798 \cdot 12.17 = -67.2 \, \text{kJ/mol}$$

$$CuFeS_2 \ \rightarrow \ CuS \ + \ FeS2 \qquad (2)$$

$$\Delta H^0_{798} = \Delta H^0_{298} + \int_{298}^{798} \Delta c_p dT = 43,15 \, \text{kJ/mol}$$

$$\Delta S^0_{798} = \Delta S^0_{298} + \int_{298}^{798} \frac{\Delta c_p}{T} = 4,4 \, \text{J/molK}$$

$$\Delta G^0_{798} = \Delta H^0_{798} - 798 \cdot \Delta S^0_{798} = 39,56 \, \text{kJ/mol}$$

4 Conclusions

In terms of thermodynamics, combined thermal processing of products containing precious metals enables application of various methods for generating artificial covellite with the participation of accessory sulphide minerals. Respective statements may be useful for both metallurgical processes and thermochemical processing of hard coals, which always contain impurities of sulphide minerals.

Acknowledgements. The work was carried out with the support of the Russian Science Foundation (project No. 17-17-30015).

References

Betekhtin AG (2008) Course of Mineralogy. Publishing House University, Moscow (2008)

Bradsow D (2014) The role of 'Process mineralogy' in improving the process performance of complex sulfide ores. In: Proceedings XXVII IMPC, C.14, pp 1–23

Geikhman VV, Kazanbaev LA, Kozlov PA (2003) Industrial tests of flotation of zinc cakes. Nonferrous Met (1):29–32

Otrozhdennova LA, Maksimov II, Khodov NV (1997) Combined technology of silver recovery from zinc cakes. Min J (4):39–40

Naboichenko SS (1997) Processes and devices of non-ferrous metallurgy. USMU, Ekaterinburg

Rubin S, Aksenov A, Senchenko A, Lagutina S (2014) Argentojaroside synthesis and research. In: Proceedings of the XXVII IMPC, C.19, pp 209–213

Crystal-Chemical and Technological Features of the KMA Natural Magnetites

T. Gzogyan$^{(\boxtimes)}$ and S. Gzogyan

Belgorod National Research University, Belgorod, Russia
mehanobr1@yandex.ru

Abstract. The results of researches of crystal-chemical features of magnetites obtained from the KMA ferruginous quartzites deposits are presented. It is shown that all magnetites are represented by cation-scarce mineral differences. The correlation dependences between the technological indexes (yield, extraction) and the magnetite concentration in quartzites are established.

Keywords: Crystal-chemical features · Ferruginous quartzites · Magnetite · Maghemite · Mossbauer spectra · Magnetic properties · Defectiveness

1 Introduction

The most important geochemical feature of iron is the presence of several oxidation degrees. By crystal-chemical properties of the Fe^{2+} ion is close to the Mg^{2+} and Ca^{2+} ions and to the other main elements. Due to crystal-chemical similarity Fe replaces Mg and partially Ca in many silicates. In quartzites, the main ore mineral is magnetite, the crystal-chemical structure characteristics of which determine the technological properties of quartzites, according to which quartzites are characterized by high variability of composition and properties, and the results of enrichment depend on their material composition. It is worth noting the diversity of mineral composition, texture and structural features of ores and, as a consequence, a wide range of physical, mechanical and technological properties (Gsogyan 2010).

2 Methods and Approaches

In this work the crystal-chemical features of the main ore minerals of unoxidized and oxidized quartzite KMA deposits are investigated using modern research methods (nuclear gamma-resonance spectroscopy (Mossbauer effect), X-ray analysis and X-ray fluorescence). Measurements of the dependence of magnetization on the applied magnetic field and saturation magnetization by the vibration method are carried out.

Magnetite is a typomorphic mineral composition and properties of which depends on the conditions of formation and has got a structural spinel. This is the ohFd3 m space group, where cations occupy 16 octahedral (B) and 8 tetrahedral (A) positions. Therefore, for ideal crystals of $Fe_1^{3+}[Fe_1^{3+}F_1^{2+}]O_4$ the reversed spinel, the ratio of the intensities of the leftmost lines of the mossbauer absorption spectra, taking into account

S. Glagolev (Ed.): ICAM 2019, SPEES, pp. 75–79, 2019.
https://doi.org/10.1007/978-3-030-22974-0_17

the difference in the probabilities of the effect for the octahedral and tetrahedral sub-lattices, should be slightly less than two.

3   Results and Discussion

The characteristic lines of the obtained Mössbauer absorption spectra of quartzites represent the superposition of a series of six-peak spectra. The ratio of the intensities corresponding to half of the ions in the A and B positions is practically not satisfied, despite the fact that the selected sample is closest to the monomineral difference. In addition, there is also a slight broadening of the leftmost and rightmost lines of the total absorption.

Mossbauer spectra indicate that the intensity ratio 1:2 is not satisfied. Violation of this ratio may be due either to an isomorphic substitution of iron ions in the B-position of $Fe_1^{3+}\left[Fe_{5/3-2/3X}^{3+}F_{X-Y}^{2+}Me_{y\square1/3-1/3X}^{2+}\right]O_4$ or to the phenomenon of non-stoichiometry and the presence of Fe_1^{2+} vacancies in the same position $Fe_1^{3+}\left[Fe_{5/3-2/3X}^{3+}F_{X-Y}^{2+}Me_{y\square1/3-1/3X}^{2+}\right]O_4$. Obviously, at x = 1, this is the stoichiometric magnetite $Fe_1^{3+}[Fe_1^{3+}F_1^{2+}]O_4$, and at $x = 0 - \gamma - Fe_2O_3\left(Fe_1^{3+}\left[Fe_{5/3\square2/3X}^{3+}\right]O_4\right)$. The appearance of isomorphically replacing ions in the octahedral position of the magnetite spinel structure leads to a partial removal of electron exchange in position B between Fe^{2+} and Fe^{3+} and, as a result, to a change in the ratio of the intensities of the lines corresponding to iron ions in positions A and B with simultaneous broadening of not only the extreme right groups of lines, but also the entire set of lines of the total resonance absorption spectrum (broadening is especially pronounced with an increase in the concentration of isomorphically substituting ions). The appearance of a cationic vacancy in position B of the spinel structure does not lead to experimentally observed broadening but only to a redistribution of the intensities of the lines corresponding to Fe^{3+} and $[Fe^{3+}, Fe^{2+}]$ in positions A and B due to increasing in the contribution of Fe^{3+} which does not participate in electronic exchange Fe^{2+} with position B due to vacancies. But in both cases the integral intensity remains constant. However, the presence of a cationic vacancy in the octahedral position in the spinel structure sig-nificantly affects the technological properties. The combination of experimental data obtained allows us to write the crystal-chemical formula of magnetites in the form $Fe_1^{3+}\left[Fe_{5/3-2/3X}^{3+}Fe_{x\square1/3-1/3X}^{2+}\right]O_4$, where $\square$ – is a cation vacancy (in the structure of magnetite as isomorphic substituting ions can be Mg^{+2}, Mn^{+2}, Ni^{+2}, Ti^{+2}, etc.), and x – is non-stoichiometric (the defectiveness is the Fe^{2+} deficiency widely varying in area of deposits). At x = 0 we have a six-peak absorption spectrum corresponding to maghemite. The condition of electro-neutrality is observed due to the formation of defects that is, electron holes. The relationship between Fe^{3+} and $[Fe^{3+}, Fe^{2+}]$ deter-mines the behavior of such magnetites in magnetic fields.

In addition, martite lines are present on the Mossbauer spectra of magnetites and this indicates that the crystal lattice of magnetites is coherently connected with the lattice of the "parent" magnetite. The presence of a coherently coupled

$Fe_1^0 \left[Fe_{5/3\square1/3}^{3+} \right] O_4$ structure significantly affects the degree of its extraction during the enrichment. Traces of hematite and various non-metallic particles are present with magnetite.

Thus, magnetite is a very sensitive indicator of the conditions of deposits formation and its "life", started in nature, is manifested and sometimes significantly in the kinetics of technology, because rapid processes intensively destroy its information structure and slowly transfer it to the products resulting from processing (Gsogyan 2010).

Features of heterogeneity of the composition and properties of magnetite affect the variability of its oxidation degree in the course of crushing process and, as a result, the technological indicators of enrichment (Gsogyan 2010). Quantitative XRF of quartzites revealed the presence in them along with magnetite, hematite and maghemite; moreover, the presence of maghemite is the characteristic of all mineral types of quartzite. XRF showed that the highest content of maghemite ($\sim 11\%$) is the characteristic of magnetite quartzite, somewhat lower ($\sim 8\%$) is observed in semi-oxidized and the lowest ($\sim 4.5\%$) in oxidized.

Morphological series observed in nature in ferruginous quartzites: Fe_4O_4 (wustite) $\rightarrow$ $Fe^{2+}Fe_2^{3+}O_4$ (magnetite) $\rightarrow$ γFe_2O_3 (maghemite) $\rightarrow$ αFe_2O_3 (hematite) $\rightarrow$ $Fe_2O_3 \cdot H_2O$ (goethite) $\rightarrow$ $\gamma FeO(OH)$ (lepidocrocite) $\rightarrow$ $Fe_2O_3 \cdot xH_2O$ (limonite /martite) emphasizes the complex relationships between their structures. With the transition from one structural type to another, some features of the original type are inherited. The edge of the unit cell, depending on the composition of the oxidation of wustite to magnetite and maghemite consistently decreases with the number of Fe^{2+} cations. Linear dependence of the lattice parameter on the composition emphasizes the similarity between the structures of these compounds. Reducing edges of the unit cell of magnetite in the transition to maghemite caused by the replacement of Fe^{2+} with an ionic radius of 0,80 Å on Fe^{3+} with an ionic radius of 0,67 Å with simultaneous removal of one third of the Fe^{2+} ions from the structure of magnetite. The magnetites associated with the incomplete processes of martitization are highly heterogeneous, which is associated with the manifestation of morphotropic transformations in the series indicated above, the content of Fe in which falls from 72.4% (magnetite) to 62.0% (limonite).

The average edge size of the unit cell varies from 8.371 Å (Lebedinsky) to 8.395 Å (Mikhailovsky) and this is somewhat less than that of a magnetite of stoichiometric composition (8.396 Å). The evaluation of magnetite deformations allows us to conclude that a change in the main interplanar distances of magnetite samples is more characteristic of Lebedinsky and Prioskolsky.

Microprobe analysis revealed iron deficiency in magnetite grains of various generations, the content of which varies from 66.89% (Mikhailovsky) to 72.3% (Stoilensky).

Thermomagnetic analysis in almost all magnetites at T = 340–350 °C shows a weak exothermic effect corresponding to maghemite and its presence is confirmed by electron diffraction patterns, which indicate the absence of periodicity in the arrangement of some reflexes and the presence of defects and dislocations in the magnetite lattice (Fig. 1).

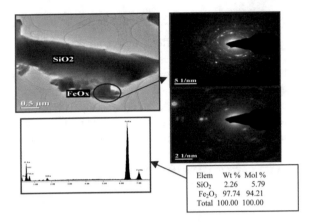

Fig. 1. The electron phase, taken from the magnetite particles at different angles of orientation of the device

Maghemite is marked from 11% (Mikhailovsky and Prioskolsky) to 6% (Korobkovsky). The data of high-temperature magnetometry confirm the lack of Fe^{+2} in the KMA magnetites and their difference from stoichiometric: the Curie point for magnetite averaged 571.8 Å (Stoilensky) to 578.3 Å (Prioskolsky). The change in the ratio of Fe^{3+} and Fe^{2+} causes the inextensive magnetic properties which are one of the most important factors affecting the enrichment. These studies suggest that, due to the lack of a sufficient amount of isomorphic impurities, we are dealing with alternative behavior in the two-phase system of $Fe_2O_3 \div Fe_3O_4$, that is, the formation of metastable transitional phases with a structure closer to the matrix than the equilibrium phase. This is confirmed by the results of determining the concentrations of isomorphically substituting impurities of Mg^{2+}, Mn^{2+}, Ti^{2+} and Ni^{2+}, which did not exceed 0.2% NiO; 0.25% MnO_2 and 0.4% MgO.

All Mossbauer absorption spectra contain maghemite lines. Consequently, the formation of solid solutions of $FeO:Fe_2O_3$ coherently coupled with γ-Fe_2O_3 and semicoherently with Fe_2O_3 is strictly observed in deposits. This led to the widespread development of quartzites heterogeneity in their composition and properties. The coherent phases formed in this way can be clusters of dissolved atoms in the lattice of the parent material that is often observed. The presence of "non-stoichiometric" cation-deficient differences of magnetites led to the need to find the value that determines their defectiveness. The crystal-chemical formula of such magnetites is: $Fe_1^{3+}\left[Fe_{5/3-2/3X}^{3+}Fe_x^{2+}Fe_{1/3-1/3X}^{2+}\right]O_4$ where x is defectiveness (Fe^{2+} deficiency in the octahedral position of the spinel structure) (for $x = 1$ is ideal magnetite $Fe_1^{3+}\left[Fe_1^{3+}Fe^{2+}\right]O_4$ (content of 72.4%), and at $x \rightarrow 0$ a complete transition up to γ-Fe_2O_3 (crystallographic symmetry) $Fe_1^{3+}\left[Fe_{5/3}^{3+}Fe_{1/3}^{2+}\right]O_4$ when a part of iron ions is recharged to the state of Fe^{3+} and the iron content drops to 69.9%).

The dependence of the magnetic properties on the composition and crystal-chemical features of magnetites is performed by the vibration method in a uniform magnetic field H up to 71.62 kA/m in the temperature range from 20 to 800 °C. The obtained dependence of the specific saturation magnetization has an inflection point which apparently is associated with the influence of a sufficiently large weight fraction of $Fe_1^{3+} \left[Fe_{5/3}^{3+} \ Fe_{1/3}^{2+} \right] O_4$ The unevenness of magnetic properties is due to the change in the ratio of Fe^{3+} and Fe^{2+}. Specific magnetic susceptibility indicates a different petrographic composition of ores and the size of magnetite impregnation and their significant heterogeneity which is a consequence of their genesis and is expressed in the difference in their technological properties. The saturation magnetization is a multiparameter function for ferruginous quartzites.

In this regard, the dependences of technological indicators of enrichment were obtained as a function of the reduced value (σ_s/σ_s^{id}), where σ_s are the obtained values of the specific magnetization of this sample; σ_s^{id} is the specific magnetization of the isolated magnetite fraction ($Fe_{tot} = 72.2\%$, SiO_2 less than 0.2%).

4 Conclusions

- magnetites from ferruginous quartzite deposits of the KMA are represented by cation-deficient mineral differences with a wide variation of the magnitude of defects (x) in the area of occurrence;
- the presence of coherently bound $\gamma\text{-}Fe_2O_3$ and semi-coherently bound $\alpha\text{-}Fe_2O_3$ is determined depending on the degree of oxidation which undergoes spinodal and intracrystalline decomposition up to hematite;
- technological indicators of enrichment depends on the crystal-chemical and magnetic properties of quartzites. These allow to find optimal conditions for concentrates obtaining, i.e. implement the "geological" management of the production.

Reference

Gzogyan TN (2010) To the question of the heterogeneity of the KMA magnetite deposits. Min Inf-Anal Bull (5):256–259

Practical Application of Technological Mineralogy on the Example of Studying of Sulphidization in the KMA Ferruginous Quartzites

S. Gzogyan[(⊠)] and T. Gzogyan

Belgorod National Research University, Belgorod, Russia
mehanobrl@yandex.ru

Abstract. The results of mineralogical-petrographic researches of relationship of iron sulphidic minerals with a magnetite and their influence on the technological properties of the KMA ferruginous quartzites are presented. A classification of ferruginous quartzites by sulphidic factor on the basis of textural and structural features of the divided mineral components is developed.

Keywords: Ferruginous quartzites · Pyrite · Pyrrhotine · Magnetite · Sulphidic factor

1 Introduction

Actual problems of processing iron ores and quartzites remain increasing in the metallurgical value of the concentrate, reducing metal losses and costs in the production of commercial products. In increasing the efficiency of processing ferruginous quartzites a special role belongs to the scientific direction – the technological mineralogy. The transition from the descriptive methodology to the methodology of genetic analysis allowed giving a scientific interpretation of many problems associated with the technology of processing ferruginous quartzites.

2 Methods and Approaches

The object of a research is more than 900 group geological and technological samples of the KMA ferruginous quartzites obtained during the operational exploration of deposits, which reflect the most representative texture-structural and mineralogical features of the KMA quartzite deposits.

The study of the composition and properties of the samples taken was carried out using optical microscopy, NMR-spectroscopy, high-temperature magnetometry, microprobe analysis and X-ray diffractometry.

The physical properties are used to determine the microhardness of minerals and magnetic properties. The main attention is paid to relationship of iron oxides and sulphides and their distribution in the finished product. Researches were carried out on samples of polished sections and briquettes.

© The Author(s) 2019
S. Glagolev (Ed.): ICAM 2019, SPEES, pp. 80–83, 2019.
https://doi.org/10.1007/978-3-030-22974-0_18

3   Results and Discussion

The object of a research is more than 900 group geological and technological samples of the KMA ferruginous quartzites obtained during the operational exploration of deposits, which reflect the most representative texture-structural and mineralogical features of the KMA quartzite deposits.

The study of the composition and properties of the samples taken was carried out using optical microscopy, NMR-spectroscopy, high-temperature magnetometry, microprobe analysis and X-ray diffractometry.

The physical properties are used to determine the microhardness of minerals and magnetic properties. The main attention is paid to relationship of iron oxides and sulphides and their distribution in the finished product. Researches were carried out on samples of polished sections and briquettes.

The second category (medium enriched) includes ferruginous quartzites in which the structure of relationship between ore minerals and sulphidic minerals is more complex, that contributes to the transition to magnetite concentrate of both free grains of pyrrhotite and aggregates of magnetite with pyrite (Fig. 1b). From ferruginous quartzites of the second category (sulfur content 0.18), in the laboratory concentrate 0.09% remains (in the industrial one – 0.085%).

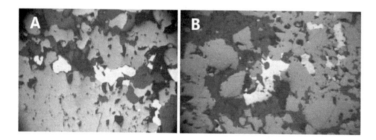

Fig. 1. Pyrite in magnetite aggregates (a) and internal and external intergrowths of pyrite with magnetite, poikilitic inclusions of pyrrhotite in magnetite (b), zoom 12.5 × 8 × 10.

The third category of quartzites (hardly enriched) is represented by ferruginous quartzites with close structural intergrowths of sulphidic minerals with magnetite, poikilitic inclusions of pyrite in magnetite and pyrrhotite and vice versa, and that leads to the transition and concentration of sulphides in the process of wet magnetic separation into magnetite concentrate (Fig. 2).

With a total sulfur content in the feedstock of 0.42–0.47 in the laboratory magnetite concentrate, its size reaches 0.42–0.78% (in the industrial one – 0.40–0.74%).

Thus, optical and mineralogical researches have shown that clogging of magnetite concentrate with sulfur occurs not only because of the ferromagnetic properties of the monoclinic pyrrhotite variety, but also due to the inclusion of pyrite grains in the magnetite and pyrrhotite grains and vice versa (Table 1).

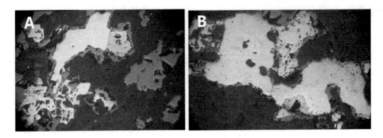

Fig. 2. Substitution structures of magnetite and pyrrhotite by pyrite, zoom 25 × 8 × 10 (a), 12.5 × 12.5 × 10 (b).

Optical and mineralogical researches of finished products have established that sulphidic minerals in concentrate are as in the form of intergrowths and inclusions in magnetite grains as in the form of separate grains (Fig. 3).

Table 1. Grain composition and distribution of chemical components by grain size in the magnetite concentrate

Grain size, mm	Yield of grain size,%	Content, %			
		Fe		S	
		Aggregate	Magnetite	Pyrite	Pyrrhotine
+0.071	1.20	29.4	27.53	0.046	0.040
−0.071 + 0.045	4.20	55.29	33.37	0.024	0.018
−0.045 + 0.032	27.33	66.92	64.88	0.028	0.019
−0.032 + 0.020	45.03	71.05	66.55	0.023	0.016
−0.020	22.24	70.47	62.77	0.021	0.014
Total	100.00	68.57	63.39	0.0242	0.0168

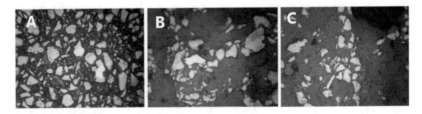

Fig. 3. Magnetite concentrate: aggregates of iron sulphides with magnetite and their separate grains (a, b) stand out against the background of free magnetite grains, poikilite inclusions in magnetite (c), polished briquette, zoom 132.

4 Conclusions

Thus, applying the methods of technological mineralogy and based on the studied nature of the variability of structural transformations, the classification of quartzites was developed according to enrichment by the sulphidic factor.

The next important achievement of technological mineralogy in the context of a particular sulphidic problem is the clearly shown migration of non-magnetic iron sulphides into magnetite concentrate using magnetic separation processes.

Ore Mineralogy of Kirazliyayla (Yenişehir-Bursa-Turkey) Mesothermal Zn-Pb-(±Cu) Deposit: Preliminary Results

F. Javid and E. Çiftçi[✉]

Department of Geological Engineering, Faculty of Mines, ITU, 33469 Maslak, Istanbul, Turkey
eciftci@itu.edu.tr

Abstract. The Kirazlıyayla deposit is one of Pb-Zn (±Cu) deposits associated with andesitic volcanism cutting through the metamophics of Karakaya complex. It is a structurally controlled, vein-stockwork style, mesothermal ore deposit. Replacement, brecciation, vein/veinlets, carries, sea-island, and dissemination textures were identified. Pyrite, sphalerite, chalcopyrite, galena, tennantite, and covellite constitute the ore mineral paragenesis. Quartz, calcite and dolomite with kaolinite account for the gangue minerals. Supergene stage is insignificant. The Kirazlıyayla mineralization is a Zn-Pb (±Cu) mineralization hosted by tectonically controlled andesitic volcanism within the Karakaya metamorphics and has common features with the other occurrences in the western Anatolia.

Keywords: Kirazlıyayla · Mesothermal · Lead · Zinc · Karakaya complex · Structurally controlled

1 Introduction

Turkey hosts noteworthy variety of mineral deposits owing to its geological evolution within the Alpine-Himalayan orogenic belt and its complex tectonic setting. Mineral deposits of Turkey are better understood through the understanding tectono-magmatic evolution of Turkey. Within this framework, there are a number of various ore deposits disseminated throughout the country including the Kuroko-type VMS deposits (strictly in the Eastern Pontides tectonic belt), Cyprus-type VMS deposits (along the Bitlis-Zagros suture zone in SE Anatolia and Küre-Kargı trend in the Central Pontides), Besshi-type VMS deposits (in the Central Pontides and along the Bitlis-Zagros suture zone), epithermal deposits of both LS and HS type (mostly in western Anatolia) and significant number of IS type with noteworthy lead and zinc presence, and carbonate-hosted sulfidic and nonsulfidic Pb-Zn deposits (along the Tauride Belt). In addition, skarn-type (mainly Fe producing) deposits (disseminated throughout the country), porphyry Cu-Mo (disseminated throughout the country), mesothermal cupriferous Pb-Zn mineralizations (along the Eastern Pontide belt), and podiform chromite deposits (along the suture zones throughout the country) are also important part of the metallogeny of Turkey.

© The Author(s) 2019
S. Glagolev (Ed.): ICAM 2019, SPEES, pp. 84–89, 2019.
https://doi.org/10.1007/978-3-030-22974-0_19

Although Turkey produces zinc and lead concentrates, due to the lack of smelters, about 1.5 billion dollars each year paid for the import of those metals. In the recent years, the government is in an attempt to revive the sector through various incentives. The western Anatolia is host to great number of mineral deposits. As for the lead-zinc occurrences, it can be summarized in 3 types in terms of their genesis in the region: (I) distal skarn occurrences associated with the carbonates of the metamorphic basement; (II) structurally controlled veins within the basement metamorphics and the overlying volcanics or along their contacts; and (III) replacements along the mafic dykes.

The study area is located in one of the tectonic elements of Turkey – the Sakarya terrane, which is an elongate crustal ribbon extending from the Aegean in the west to the Eastern Pontides in the east. It is consisted of sandstones of Lower Jurassic age, which sits on a fairly complex metamorphic basement that contains a high-grade Variscan basement metamorphics of Carboniferous age (Topuz et al. 2004, 2007; Okay et al. 2006), Paleozoic granitoids (Delaloye and Bingöl 2000; Okay et al. 2002, 2006; Topuz et al. 2007), and a low grade metamorphic complex - the Lower Karakaya Complex constituted by Permo-Triassic metabasite with lesser amounts of marble and phyllite. The Lower Karakaya Complex represents the Permo-Triassic subduction-accretion complex of the Paleo-Tethys as indicated by the presence Late Triassic blueschists and eclogites (Okay and Monié 1997; Okay et al. 2002), accreted to the margin of Laurussia during the Late Permian to Triassic. The complex is overlain by a thick series of strongly deformed clastic and volcanic rocks with exotic blocks of Carboniferous and Permian limestone and radiolarian chert. This complex basement was overlain unconformably in the Early Jurassic by a sedimentary and volcanic succession. The Early Jurassic is represented by fluvial to shallow marine sandstone, shale and conglomerate in the western part of the Sakarya Zone. The metamorphic basement is cut by Eocene volcanisms.

The Kirazliyayla Zn-Pb ore deposit is spatially and temporally related with Eocene intermediate extrusive rocks - andesite and trachyandesite with NE-SW extension and covered by clastic and carbonate rocks. The main purpose of the investigation is to determine the genesis of Kirazliyayla Zn-Pb ore deposit and its place in the metallogenic evolution of the region to contribute the understand of the metallogeny of Turkey. In that, geochemical characteristics, mineralogy of both host rocks and ore minerals, its tectonic setting are the main issues are covered.

2   Methods and Approaches

A total of 30 samples representing the ore deposit from open pit – main production step and from the boreholes were taken, from which polished sections were prepared and Nikon Eclipse LV100 reflected light microscopy integrated with a CITL MK5 Cathodoluminescence system was employed for mineralogical examination. The ore minerals and the paragenesis were identified on the basis of their petrographical features and their textural relationships, respectively. Electron Probe Micro Analysis (EPMA) and Secondary Electron Microscopy-Energy Dispersive Spectroscopy (SEM-EDS) were routinely used for confirming the minerals and chemistry of sulfide minerals when needed.

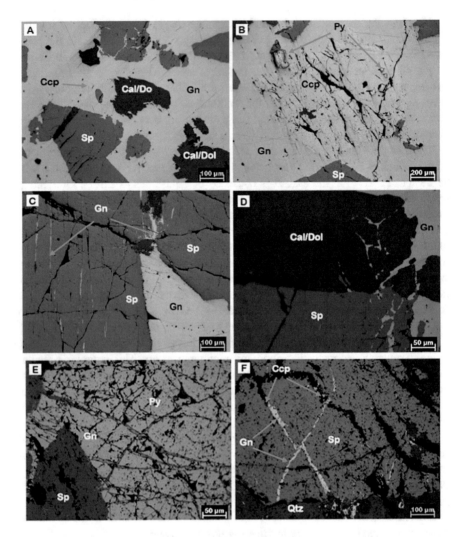

Fig. 1. (A) sphalerites and carbonates are partially replaced by late stage galena; (B) chalcopyrite and pyrite are partially replaced by late stage galena; (C) galena replaces coarse sphalerite; (D) galena replaces coarse sphalerite and also carbonate gangue; (E) large cataclastic pyrite veined by sphalerite and late stage galena; (F) large cataclastic sphalerite with replacing galena and chalcopyrite, all in silicic matrix; (G) Large sphalerite grains with late stage galena, chalcopyrite and tennantite veins; (H) large pyrite and chalcopyrite veined by late stage galena; (I) Euhedral quartz, large sphalerite grains with late stage galena; (J) sphalerite veined by late stage chalcopyrite and pyrite; (K-L) Late carbonates veining sphalerite (Py: pyrite; Ccp: chalcopyrite; Sp: sphalerite; Gn: galena; Tn: tennantite; Cal/Dol: calcite/dolomite)

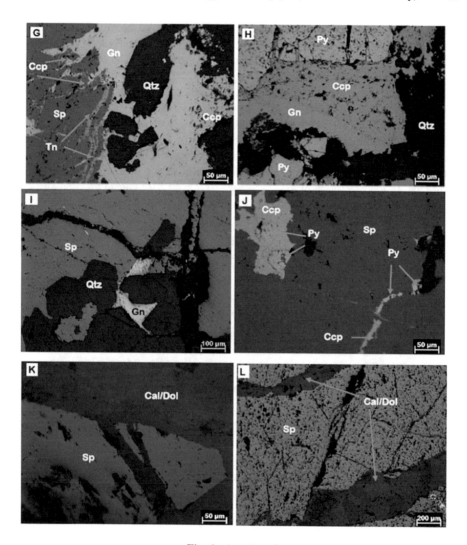

Fig. 1. (*continued*)

3   Results and Discussion

Field observations indicated that kaolinitization with silicification is pervasive alterations indicating low pH hydrothermal activity resulted in the mineralization. The mineralization is associated with andesitic volcanics in that vein/veinlets, replacement in places, stockwork, and breccia ore structures are prevalent. Thickness of veins may reach to 1 m in rare occasions. Samples were collected from the production steps in the open pit and from the boreholes cutting the mineralized zones.

Based on the studies of the ore samples, the major ore minerals are sphalerite and galena as zinc and lead carrier. Pyrite is a ubiquitous. Chalcopyrite is a minor to trace sulfide phase along with trace tennantite (Fig. 1K–L). Most of the samples indicate that the mineralization is medium to high grade. Sphalerite and galena occur as large grains (sometimes up to mm size) in most of the samples, indicative of being precipitated out of supersaturated fluids within narrow spaces. Both are also fairly inclusion-free (clean). When not, galena occurs with tennantite, sphalerite with pyrite and chalcopyrite. Galena has also chalcopyrite encapsulation in places.

Interpretation of the intergrowth ore textures suggests that mineralization started with pyrite crystallization then a brief precipitation of first generation of chalcopyrite, then followed by major sphalerite crystallization, which is followed by a major carbonatization, followed by a brief tennantite and second generation of chalcopyrite formation. In the final stage of the ore mineralization a major galena precipitation took place. Figure 2 summarizes the mineralization event. Ore deposit experienced very weak supergene stage in which only traces of covellite formed.

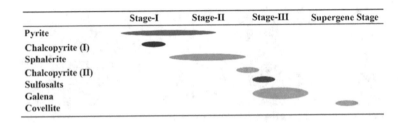

Fig. 2. Proposed ore mineral paragenetic sequence for the Kirazlıyayla Zn-Pb mineralization

4   Conclusions

Major ore minerals include sphalerite and galena. Chalcopyrite is minor while tennantite is trace. Pyrite is always present, but not as much as sphalerite and/or galena. Tennantite is the only fahlerz that occurs in some ore zones, overall in trace quantities. Sphalerite and galena occur as large grains in most of the samples. Both are also fairly inclusion-free (clean). When not, galena occurs with tennantite, sphalerite with pyrite and chalcopyrite. Galena has also chalcopyrite encapsulation in places. Paragenetic succession for the ore minerals appear to be (from early to late):

Pyrite-Chalcopyrite (I)-Sphalerite-Chalcopyrite (II)-Tennantite-Galena

Calcite/dolomite and quartz account for the gangue minerals. Kaolinite is the major clay mineral. Sericite also locally become significant.

Coarse nature of the major minerals (majority of the galena and sphalerite is larger than 100 microns) suggests high liberation (>90%). Small size occurrences (<10%) are mostly in the form of veinlets within each other, so during milling, most of that size range may be liberated.

Sulfosalts or fahlore are represented only by tennantite ($Cu_6[Cu_4(Fe,Zn)_2]As_4S_{13}$).

Preliminary results show that the Kirazlıyayla mineralization is a Zn-Pb ($\pm$Cu) hosted by andesitic volcanics whose emplacement within the metamorphic complex should be tectonically controlled.

Acknowledgements. The authors would like to acknowledge support received from Meyra Mining company for their courteous support during the field work.

References

Delaloye M, Bingöl E (2000) Granitoids from western and northwestern anatolia: geochemistry and modeling of geodynamic evolution. Int Geol Rev 42:241–268

Okay AI, Tüysüz O, Satır M, Özkan-Altıner S, Altıner D, Sherlock S, Eren RH (2006) Cretaceous and Triassic subduction-accretion, HP/ LT metamorphism and continental growth in the Central Pontides, Turkey. Geol Soc Am Bull 118:1247–1269

Okay Aİ, Monié P (1997) Early Mesozoic subduction in the Eastern Mediterranean: evidence from Triassic eclogite in northwest Turkey. Geology 25:595–598

Okay Aİ, Monod O, Monié P (2002) Triassic blueschists and eclogites from northwest Turkey: vestiges of the Paleo-Tethyan subduction. Lithos 64:155–178

Topuz G, Altherr R, Schwarz WH, Dokuz A, Meyer HP (2007) Variscan amphibolite facies metamorphic rocks from the Kurtoğlu metamorphic complex (Gümüşhane area, Eastern Pontides, Turkey). Int J Earth Sci 96:861–873

Topuz G, Altherr R, Kalt A, Satır M, Werner O, Schwartz WH (2004) Aluminous granulites from the Pulur Complex, NE Turkey: a case of partial melting, efficient melt extraction and crystallisation. Lithos 72:183–207

Use of Borogypsum as Secondary Raw

A. Khatkova[1], L. Nikitina[2], and S. Pateyuk[2(✉)]

[1] Department of Mineral Technology, School of Geology,
Transbaikal State University, Chita, Russia
[2] Department of Geology, University of Kazan, Kazan, Russia
nesvvik@gmail.com

Abstract. We have considered problem of accumulation, storage, utilization and recycling of wastes of various industries. We chose borogypsum as the object of study, which contains gypsum and silicon dioxide, which can be used in various industries. We proposed a new flotation reagent for separation of silicon concentrate from wastes of boric acid production. Using methods of mathematical planning we conducted a multifactorial experiment, which allowed identifying optimal flotation mode. We developed a technology for processing of borogypsum.

Keywords: Secondary raw · Borogypsum · Flotation · Perlastane · White soot · Silicon concentrate

1 Introduction

Over the years the enterprises of mining and chemical industries in the Far East accumulated millions tons of industrial wastes that are currently not being recycled. Thus, the total amount of wastes from the production of boric acid - borogypsum - in the Far Eastern region is more than 25 million tons. Borogypsum contains gypsum and silicon dioxide, which can be used in various industries. In this regard, the problem of complex processing of these wastes to obtain various functional materials is a very urgent task (Gordienko et al. 2014).

Highly dispersed amorphous silica is referred to as white soot. White soot of all kinds, alongside with black ones, is used to strengthen rubber, being absolutely indispensable for silicone rubbers. It is also used as fillers in the production of rubber linoleum, as well as in plastics, paints, lubricants and other materials to give them valuable properties.

In addition to silicon dioxide, another component of borogypsum, gypsum and anhydrite, can be used in industry. All possible products (component for cement production, dry building mixtures, finished building and architectural products, gypsum boards) have positive market prospects in Russian markets and in some neighboring foreign markets.

In Russia there are projects for the production of white soot with a high silica content by separating from datolite tailings. However, such technologies are not implemented at industrial scale, because they are quite expensive due to the low content of silicon dioxide in the original product.

© The Author(s) 2019
S. Glagolev (Ed.): ICAM 2019, SPEES, pp. 90–93, 2019.
https://doi.org/10.1007/978-3-030-22974-0_20

To reduce the consumption of acids and the cost of production of white soot, it is necessary to increase the content of silicon dioxide in the original product by other cheaper methods such as flotation.

The table below shows the dependence of the need for concentrate on the content of silicon dioxide (Table 1).

Table 1. The dependence of the need for concentrate on the content of SiO2

SiO_2 content in concentrate	Need for concentrate, t/hour	Need for sludge t/hour, with a concentrate yield 15%	Need for pulp volume, m^3/h, at 20% solids
65	1,35	9	40,1
60	1,47	9,8	43,7
55	1,60	10,7	47,7
50	1,76	11,7	52,1

According to the table, it is necessary to ensure the content of silicon dioxide from 50 to 65% with a product yield 15% and above using flotation.

2   Methods and Approaches

A sample of waste of boric acid production, borogypsum, obtained from Dalnegorsky MPP was selected as the object of study.

For the flotation of gypsum-containing raw, fatty acid reagents are commonly used (Bulut et al. 2008; Matsuno et al. 1958). We proposed a new flotation reagent perlastan ON 60 to isolate the silicon concentrate by reverse flotation. It was efficiently used at the flotation of non-metallic fluorite ores (Dolgikh 2012), and is a promising flotation reagent for borogypsum.

To determine optimal conditions for flotation, we used the method of rational planning of a multifactorial experiment (Malyshev 1977).

3   Results and Discussion

After conducting a multifactorial experiment, the following regularities were identified.

Increasing temperature of flotation leads to decreasing extraction, but at the same time, the content of silicon dioxide increases. It was also found that the temperature affects the floatability of gypsum, however, almost no effect on the floatability of silicon concentrate. Consequently, by changing the temperature, it is possible to regulate the process, depending on the task. Changes in flotation pH slightly affect the content and recovery, therefore, pH control is not effective, which in turn is an advantage of this method, for it does not require consumption of additional reagents. Increasing concentration of perlastane increases the content of silicon concentrate in the chamber product, however, the extraction drops significantly. The agitation time has a linear effect on the extraction, at the same time we didn't observe a clear dependence of the agitation time on the content. Therefore it is necessary to agitate perlastane with minerals for at least 3 min. As the flotation time increases, the content slightly decreases due to the fact that alongside with calcium sulphate, silica dioxide begins to shift to the foam product.

On the basis of the revealed regularities we determined optimal conditions of flotation. However, the content of silicon concentrate in the chamber product remains insufficient.

From the literature it was found that sodium sulfide or liquid glass was often used to suppress flotation of quartz. Experiments with these reagents showed that they practically did not suppress flotation of amorphous silicon dioxide, and could also suppress flotation of gypsum, therefore it would be not advisable to use them for borogypsum flotation.

We determined that it would be much more effective to increase the concentration of perlastane at the main flotation, increasing its time than to feed the reagent fractionally in different operations. Increasing the concentration of perlastane up to 1 kg/t, as well as flotation time up to 9 min, allows obtaining a product with a content of 53.6%, with a yield 32.5% and a recovery 64%.

4   Conclusions

We suggested a new technology for processing wastes of boric acid production according to the scheme presented in Fig. 1. The proposed scheme has its advantages in comparison to previously known technologies. Thus it does not require regulation of pH of the medium, consists of a single operation, which significantly reduces flotation time and also allows to increase the yield of the chamber product significantly.

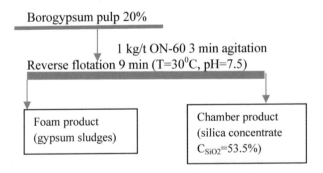

Fig. 1. Diagram of borogypsum flotation

References

Bulut G, Atak S, Tuncer E (2008) Celestite-gypsum separation by flotation. Can J Metall Mater Sci 47(2):119–126

Dolgikh OL (2012) The use of perlastane reagent as an alternative to oleic acid during fluorite flotation. Bull ZabGU 9(88):20–26

Gordienko PS, Kozin AV, Yarusova, SB, Zgibly IG (2014) Integrated processing of waste from the production of boric acid to produce materials for the construction industry. Min Inf Anal Bull (S4-9):60–66

Malyshev VP (1977) Mathematical planning of metallurgical and chemical experiments. Alma-Ata: Sci 35

Matsuno T, Kadota M, Ishiguro Y (1958) Separation of gypsum by the flotation process. Bull Soc Salt Sci 12(2):73–78

New Approaches in X-ray Phase Analysis
of Gypsum Raw Material of Diverse Genesis

V. Klimenko[✉], V. Pavlenko, and T. Klimenko

Belgorod State Technological University named after V G Shukhov,
Belgorod, Russia
klimenko3497@yandex.ru

Abstract. Modern software was used to conduct X-ray phase analysis of gypsum raw material of different genesis: gypsum from several deposits, citrogypsum, phosphogypsum, vitamin gypsum, chemically pure gypsum, hydration products from different gypsum and anhydrite cementing and composite materials on their bases. Two criteria for ranging gypsum raw material with the account of genesis and structural characteristics and predicting properties of gypsum bindings and materials based on them were suggested: structure sophistication criteria (K_g) and identity criteria (K_i). They were calculated by the results of X-ray phase analysis of calcium sulphate.

Keywords: Gypsum · X-ray phase analysis · Structure sophistication criteria · Identity criteria · Gypsum texture and structure

1   Introduction

There are convincing theoretical and experimental data about physical-chemical processes of gypsum dehydration, their structural defects dominating role, an origin and transformation mechanism of these defects during calcinations, at the same time there are no data about influence of the gypsum raw material formation on these processes. Thermodynamic characteristics of gypsum with different structure are rather close (Kelly et al. 1948; Reference... 2004) and it is difficult to study gypsum raw material deformations according to them. There are data (Gordashevski 1963) about application of differential thermal and x-ray analysis methods for these purposes. To characterize gypsum from different deposits P.F. Gordashevskiy (Gordashevski and Sakhno 1963) suggested identity criteria – the value received by division of reflexes intensivities difference at $2\Theta = 20°$ and $23°$ by their sum. On the other hand, gypsum crystal lattice defects can be characterized by crystallinity degree, which is determined by reflex with d = 2.81 Å in the X-ray patterns (Boldyrev 1983). Crystallinity degree diagnostics is possible in relation to diffraction reflection doublet intensivity and by crystallinity index determination. Some authors suggest studying crystal lattice defects by studying X-ray diffraction line broadening (Ginje 1961). Literature analysis shows wide researches in this sphere as well as absence of practical standards. In connection with this the purpose of this work was to find new criteria allowing ranging gypsum raw material with the account of genesis and structural characteristics at designing composite materials.

© The Author(s) 2019
S. Glagolev (Ed.): ICAM 2019, SPEES, pp. 94–98, 2019.
https://doi.org/10.1007/978-3-030-22974-0_21

2   Materials and Methods

As raw materials we studied gypsum of Shedok, Baskunchak, Novomoskovsk, and Peshelansk Deposits; man-induced gypsum – chemically pure gypsum, citrogypsum, vitamin gypsum, Voskresensk phosphogypsum, synthetical gypsum; gypsum thermal treatment and rehydration products, composite materials based on gypsum binders, multiphase gypsum binders (MGB). To receive composite materials, we used crashed glass withdrawals (CGW) and iron-ore concentrate of Lebedinsk mining and processing enterprise (OMC). X-ray phase analysis of the studied species was done with X-ray diffraction meter DRON-4 by powder pattern method. Grain-size analysis of powdery material was done by laser granulometry method with a MicroSizer 201 installation.

3   Results and Discussions

The X-ray phase analysis of the initial raw material showed that intensity and area of the main reflexes depend on its genesis. Intensity (L) and reflex area (S) at interplanar spacing (d) 7,628 Å are strongly dependent on gypsum raw material. Gypsum raw material genesis influences somewhat lower the intensity and reflex area at d = 4.291 and 3.069 Å. Reflex intensity at d = 3.809 Å almost independent of the gypsum raw material nature. The received data evidence that the studied gypsum samples have equal crystal lattice dimensions.

To determine the gypsum raw material genesis influence on its structure and crystal lattice defects several studies of the initial gypsum raw material crystallinity degree and its products of the heat processing have been studied. Here indexes were suggested calculated by reflex area difference division by their sum. The calculations were done for reflexes with the most intensity such as: 7.628 Å; 4.291 Å; 3.809 Å; 3.069 Å; 2.880 Å; 2.687 Å.

We found that most indexes slightly depended on raw material genesis. Certain dependencies were observed in changes of two indexes chosen for further work. The first index was structure sophistication index (K_g), and the second was identity index (K_i). K_g, was determined by X-ray phase analysis results, as ratio of reflex area difference division at $2\Theta = 29.09°$ (d = 3.069 Å) and $2\Theta = 31.08°$ (d = 2.876 Å) by their sum, a K_i as a ration of reflex area difference division at $2\Theta = 20.68°$ (d = 4.291 Å) and $2\Theta = 31.08°$ (d = 2.876 Å) by their sum.

$$Kg = \frac{S_{29.09} - S_{31.08}}{S_{29.09} + S_{31.08}} \qquad K_i = \frac{S_{20.68} - S_{31.08}}{S_{20.68} + S_{31.08}}$$

For natural gypsum K_g = 0.42–0.46, and for production induced gypsum was 0.560–0.903 (Table 1). The greater the value K_g, the more sophisticated crystal structure was and gypsum raw material was less stable.

We think that K_i value depends on gypsum raw material micro-assembly dimensions. Powdery-material grain-size composition determined by laser granulometry analysis proves a finely crystalline structure of phosphogypsum and citrogypsum.

Table 1. X-ray phase analysis of gypsum of different genesis

№	Gypsum genesis	Index value	
		Ki	Kg
1	Chemically pure gypsum	0.612	0.660
2	Shedok deposit gypsum	0.362	0.440
3	Baskunchak deposit gypsum	0.448	0.420
4	Novomoskovsk deposit gypsum	0.448	0.450
5	Peshelansk deposit gypsum	0.215	0.450
6	Vitamin gypsum	0.746	0.903
7	Voscresensk phosphogypsum	0.404	0.560
8	Citrogypsum 1	0.312	0.640
9	$CaCl_2$ and $(NH_4)_2SO_4$ gypsum	-	0.749
10	Anhydrite cement (1,5% $(NH_4)_2SO_4$ + 0,5%$CuSO_4$)	0.277	0.169
11	Composite material based on MGB and CGW remains (Klimenko et al. 2013)	0.286	0.156
12	Composite material: 70% $CuSO_4$ II + 30% LC	0.370	-
	70% Г-4 + 30% LC (Klimenko et al. 2018)	0.270	-

Crystal size of production induced chemically pure gypsum and vitamin gypsum is higher. They can be referred to large grained gypsum. By value K_i Shedok natural gypsum can be referred to finely crystalline raw material (finely tessellated oriented structure). Baskunchak natural gypsum has medium tessellated chaotic structure, and according to value K_i it takes intermediate position.

Hence, for large grained gypsum $K_i = 0.612–0.746$, for finely crystalline gypsum $K_i = 0.312–0.488$. Granulometry of produced gypsum changes unevenly. There are fractions, which number depends greater on genesis, and there are fractions, which number depends less on raw material genesis. To analyze the influence of gypsum heat treatment parameters and amount of residue hydrate water on value K_g and K_i we used Baskunchak natural gypsum. The results show that K_g of calcium sulphate hydration products can be both greater and less K_g of natural gypsum. Gypsum produced during heat treatment products hydration with hydrate water 11.00–14.42 mass%, has sophistication structure index value greater than that of natural gypsum. The smallest value Kg, irrespective of heat treatment parameters is of calcium sulphate hydration products with hydrate water amount 3–4 mass%. Calcium sulphate hydration products having value K_g close to K_g of the initial gypsum, have maximum strength. The amount of residue hydrate water of these products is equal 1.0–1.5 mas. %. By value K_i (0.39–0.45) the produced gypsum can be classified as fine- or medium crystalline structure.

Analysis of identity index value suggests that calcium sulphate hydration products with the amount of hydration water 10–15 mass% have bigger crystal micro cluster size and hydration products β-$CaSO_4$·$0.5H_2O$. β-centrifuged hemihydrates of calcium sulphate and soluble anhydrous plaster (β-$CaSO_4$·III) have smaller size of gypsum crystal micro clusters ($K_i = 0.35–0.43$).

4 Conclusion

A method and parameters (structure sophistication index (K_g) have been suggested and identity index (K_i)) allowing ranging gypsum raw material with the account of genesis and structural characteristics and forecast properties of gypsum cementing components and materials on their basis. It has been determined that for natural gypsum K_g = 0.42–0.46 and for produced gypsum K_g = 0.560–0.903. Value K_i depends on micro clusters sizes of gypsum raw material structure. For large grained gypsum K_i = 0.612–0.746. for finely crystalline gypsums – 0.312–0.488. At producing gypsum cementing agents it is necessary to observe that their K_g is closer to value K_g of natural gypsum and K_i is in the range 0.38–0.41.

Acknowledgements. The work is realized within the framework of the Program of flagship university development on the base of the Belgorod State Technological University named after V.G. Shukhov. using equipment of High Technology Center at BSTU named after V.G. Shukhov.

References

Boldyrev VV (1983) Experimental methods in mechanochemistry of solid inorganic matters. Science, Novosibirsk, 36 p

Ginje A (1961) Crystal X-ray filming. State Publishing House of Physic-Mathematical Literature, Moscow, 188 p

Gordashevskiy PF, Sakhno ZA (1963) About some properties of gypsum raw material of different crystalline structure, no 26. Collection of works/SRILCM, Moscow, pp 25–27

Gordashevskiy PF (1963) Thermal and X-ray phase gypsum analyses results. Constr Mater (12):28–30

Kelly N, Suttard D, Anderson K (1948) Thermodynamic properties of gypsum. BTI MCMPI, Moscow, pp 38–42

Klimenko VG, Kashin GA, Prikaznova TA (2018) Plaster-based magnetite composite materials in construction. In: IOP Conference Series: Materials Science and Engineering, vol 327, p 032029. https://doi.org/10.1088/1757-899x/327/3/032029

Klimenko VG, Pavlenko VI, Gasanov SK (2013) The role of pH medium in forming binding substauces on base of calcium sulphate. Middle-East J Sci Res 17(8):1169–1175

Reference book. Gypsum materials and articles (production and application) (2004) Under general editorship of professor, doctor of sciences A.V. Ferronskaya. Construction Universities Association Publishing House, Moscow, 485 p

Nanotechnologies in Mineral-Geochemical Methods for Assessing the Forms of Finding of Gold, Related Elements, Technological Properties of Industrial Ores and Their Tails

R. Koneev[1(✉)], R. Khalmatov[2], O. Tursunkulov[2], A. Krivosheeva[1],
N. Iskandarov[2], and A. Sigida[1]

[1] National University of Uzbekistan, Tashkent, Uzbekistan
ri.koneev@gmail.com
[2] Centre for Advanced Technology, Tashkent, Uzbekistan

Abstract. Seven mineralogical-geochemical types of ores have been identified in the gold orogenic deposits of Uzbekistan: /Au-W/Au-Bi-Te/Au-As/Au-Ag-Te/Au-Ag-Se/Au-Sb/Au-Hg/. Non-industrial are Au-W and Au-Hg. For each industrial type, certain gold compounds are characteristic: maldonite, Au-arsenopyrite and Au-pyrite, petzite, physhesserite, petrovskaite, aurostibite, which form regular micro- nanoensembles with the corresponding minerals Bi, Te, Se, S, As, Sb. They are direct indicators of the type and technological properties of ores. Pyrite and arsenopyrite are preserved in the processing wastes of Au-Bi-Te and Au-As ores, in which gold (901‰), maldonite (Au_2Bi), headleyite (Bi_7Te_3) and others were detected, 300–700 nm in size. Waste is suitable for the secondary extraction of gold.

Keywords: Nanomineralogy · Gold · Ores · Tails · Properties · Technologies

1 Introduction

The Republic of Uzbekistan is one of the leading gold mining countries in the world. Such large deposits as Muruntau, Myutenbay, Kokpatas, Zarmitan and others (Kyzylkum-Nurata ore region) are known. As a result of many years of production, the mining and processing industry faces the problem of transition from oxidized ores with large "free" gold to the processing of refractory ores with "invisible" gold. One of the effective approaches to the assessment of such ores is the use of nanotechnology, nanomineralogy and nanogeochemistry (Hochella 2002; Koneev et al. 2010).

2 Methods and Approaches

The section of mineralogy that studies the conditions of formation, the physicochemical properties of natural compounds, the size of which, at least in one dimension, enters the nanoscale (10^{-6}–10^{-9} m) is called nanomineralogy. There are nanominerals that are formed by the "down-top" technology, from atoms to chemical compounds, and

© The Author(s) 2019
S. Glagolev (Ed.): ICAM 2019, SPEES, pp. 99–102, 2019.
https://doi.org/10.1007/978-3-030-22974-0_22

nanoparticles that are formed by the "top-down" technology during dispersion. The anomalous properties of nanominerals and nanoparticles are determined by a significant increase in the specific surface energy. As a result of "size effects", nano-objects acquire high chemical, catalytic, sorption activity and other properties. "Noble", chemically inert gold in the nanostate becomes active and is found in ores not only in the form of native, but also in the form of a compound with various elements up to sulfur and oxygen. When studying ores, the main focus is on ore minerals. Non-metallic minerals, carbonaceous material rather negatively affect the enrichment processes, making it difficult to crush, float or adsorb gold. The study of mineralogy and nanomineralogy is important due to the fact that minerals of different composition and size differ in their ability to crush, float, or cyanate. The studies used an electron probe microanalyzer Superprobe JXA-8800R, Carl Zeise Oxford instrument (SEM-EDX); ICP MS 7500 Agilent Technologies.

3 Results and Discussion

The gold ore deposits of the Kuzylkum-Nurata region of Uzbekistan are confined to the South Tien Shan orogenic belt. Placed in the "black" shale, terrigenous, carbonate, volcanogenic and intrusive rocks. Tails of ore enrichment stored in long-term facilities were also studied. It has been established that in the orogenic deposits of the Kyzylkum-Nurata region, seven mineral-geochemical ore types are developed: /Au-W/Au-Bi-Te/Au-As/Au-Ag-Te/Au-Ag-Se/Au-Sb/Au-Hg/. Each type differs in the ratio of non-metallic and ore (sulphide) minerals, but primarily in the composition of micro-nanoensembles of minerals, compounds and gold fineness. In the studied deposits, the leading industrial ones are Au-Bi-Te, Au-As, Au-Ag-Se, Au-Sb (Table 1). In addition to native gold, electrum and kustelite, maldonite, Au-arsenopyrite, Au-pyrite, petzite,

Table 1. Types, composition of ores of gold and gold-silver deposits of Kyzylkum-Nurata region

Deposits	Main ores types		Leading mineral of ore	Micro-nanomineral ensemble	Gold compounds
Gold-quartz type Muruntau Myutenbay Triada	Au-Bi-Te, telluride; Au-As, arsenopyrite;	bismuth- pyrite-	Quartz, albite, arsenopyrite, pyrite, scheelite	Pilsenite, hedleyite, ingodite, tsumoite, joseite, tetradymite, kobellite, bismuth	Native, maldonite (Au₂Bi), Au-arsenopyrite Au-pyrite
Gold-sulfide-quartz type Zarmitan Urtalik Gujumsay	Au-Bi-Te, telluride; Au-As, arsenopyrite; Au-Sb, sulfoantimony;	bismuth- pyrite- antimony-	Quartz, carbonate, albite, arsenopyrite, pyrite, scheelite	Pilsenite, хедлейит, tsumoite, joseite, kobellite, bismuth, boulangerite, jamesonite, tetradymite	Native, electrum, maldonite (Au₂Bi), aurostibite (AuSb₂), Au- arsenopyrite Au- pyrite
Gold-sulfide type Amantaytau Daugyztau Kokpatas	Au-As, arsenopyrite; Au-Sb, sulfoantimony;	pyrite- antimony-	Quartz, carbonate, pyrite, arsenopyrite, antimonite	Jamesonite, bournonite, boulangerite, zinkenite, tetradymite	Native, electrum, aurostibite (AuSb₂), Au- pyrite
Gold-silver type Kosmanachi Visokovol'tnoe Adzhibugut	Au-As, pyrite-arsenopyrite; Au-Ag-Se, sulfosal-selenide;		Quartz, carbonate, pyrite, arsenopyrite, galena	Silver, freibergite, pyrargyrite, polybasite, naumannite, aguilarite	Electrum, kustelite, petrovskaite (AuAgS), fishesserite (Ag₃AuSe₂)

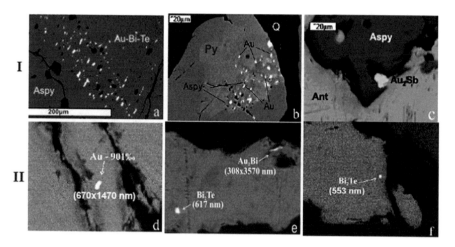

Fig. 1. Micro-nanomineral forms of gold release and its compounds in various types of ores (I) and tails (II). a - maldonite, Bi-tellurides in arsenopyrite; b - gold, arsenopyrite in arsenic pyrite; c - aurostibite with arsenopyrite in antimonite; d - nanogold in pyrite; e - maldonite and hedleyite and arsenopyrite; f - hedleyite in arsenopyrite.

calavertite, fishesserite, petrovskaite, aurostibite were detected in ores. Gold release forms and its micro- nanoanesmble in ores in Fig. 1, row I. The study of Au-Bi-Te and Au-As ore tails showed that some of the sulfide minerals still remain, in which the nanoforms of gold, its compounds and associated bismuth tellurides are preserved 300–700 nm in size (Fig. 1, row II).nts.

4   Conclusions

As a result of the research the following were established: - Seven mineral-geochemical types of ores were identified in the orogenic deposits of the Kyzylkum-Nurata region of Uzbekistan: /Au-W/Au-Bi-Te/Au-As/Au-Ag-Te/Au-Ag-Se/Au-Sb/Au-Hg/. Industrial types are /Au-Bi-Te/Au-As/Au-Ag-Se/Au-Sb/ ores possessing different technological properties; - Gold is represented by its native type, electrum, cyustelite, maldonite, petzite, aurostibite, fishesserite, petrovskaite, Au-arsenopyrite, Au-pyrite, forming micro- nanoensembles with the Bi, Te, Sb, Se, As, S minerals; - Pyrite and arsenopyrite remain in the enrichment tails of Au-Bi-Te and Au-As ores, in which nanoinclusions of gold, maldonite and bismuth tellurides are found. Tails are recyclable; - Nanotechnological methods allow to increase the efficiency and unambiguity of exploration, assess the prospects for hidden and unconventional deposits, as well as assess the technological properties of ores and enrichment tails for their complete processing.

Acknowledgements. The authors are grateful to the Navoi Mining and Metallurgical Combinate for their assistance in research.

References

Hochella M (2002) There's plenty of room at the bottom: nano-scince in geochemistry. Geochem Cosmochem Acta 66(5):735–743

Koneev RI, Khalmatov RA, Mun YS (2010) Nanomineralogy and nanogeochemistry of ores from gold deposits of Uzbekistan. Geol Ore Deposits 52(8):755–766

Applied Mineralogy of Mining Industrial Wastes

O. Kotova[1([⊠])] and E. Ozhogina[2]

[1] Institute of Geology Komi SC UB RAS, Syktyvkar, Russia
kotova@geo.komisc.ru
[2] FSBE "All-Russian Institute of Mineral Raw Materials" (VIMS),
Moscow, Russia

Abstract. Recycling and disposal of mining wastes has been very important and considered as an independent task. Features of composition and structure of mining wastes, identified by a set of mineralogical analysis methods, allowed predicting their possible involvement in the secondary processing. This was illustrated by the example of metallurgical iron slags.

Keywords: Mining wastes · Technological mineralogy · Iron slags

1 Introduction

The intense development of the mineral resource complex inevitably leads to accumulation of significant amounts of waste, which negatively affect natural ecosystems. Therefore, the disposal and recycling of wastes is important at the national level and considered within framework of the priority direction of development of science and technology in the Russian Federation – "Rational environmental management". Today we can state the fact that recycling is an independent major task for our industry.

Mining and processing wastes are very diverse. They include overburden, host rocks, dry raw processing, off-balance sheet and non-standard ores, which composition and properties are not only close to their natural analogues, but usually used the same way. Processing wastes of metallurgical, chemical, heat power industries are more abundant. Slags, muds, ash slags and oil muds, burnt rocks, pyrite cinders, clinkers, dusts are significantly different from natural ores and rocks. They are characterized by variable granular composition, often a high dispersion, presence of amorphous formations, complex interrelationship of mineral and (or) technogenic phases, including presence of eutectic colonies or structures of decomposition of solid solutions, a small amount of one or more minerals, polymineral (polyphase) aggregates, presence of isomorphic minerals and polytypic modifications, secondary changes associated mainly with hypergenesis.

© The Author(s) 2019
S. Glagolev (Ed.): ICAM 2019, SPEES, pp. 103–106, 2019.
https://doi.org/10.1007/978-3-030-22974-0_23

2 Methods and Approaches

The study of the composition, structure and technological properties of mining wastes is based on modern scientific, methodological, technical and instrumental support for the researches of technogenic raw and allows predicting its possible involvement in secondary processing, including elimination of environmental consequences of industrial processing (Ozhogina et al. 2018; Ozhogina et al. 2017; Chanturia et al. 2016; Yakushina et al. 2015; Ozhogna and Kotova 2019; Burtsev et al. 2018). Necessary and sufficient mineralogical information about the object is possible to get by a complex of mineralogical and analytical methods (optical and electron microscopy, X-ray, X-ray tomography, micro X-ray spectral analysis). For different types of wastes, an individual set of mineralogical analysis methods is used, which allows to obtain complete and reliable information, including information on the phase composition of technogenic formations, as well as the form of finding useful elements, granular composition, morphometric parameters, nature of localization of specific phases (Ozhogina et al. 2017).

Mineralogical study of mining wastes is carried out mainly according to the methodological documents developed for the analysis of natural mineral raw. Special methods of mineralogical analysis of industrial wastes do not exist. The study of such objects is of an interdisciplinary nature because of a reasonable combination of methods from various fields of science and adapted to solving mineralogical, technological and environmental problems.

3 Results and Discussion

The bulk of the processing wastes are slags. For example, iron-containing metallurgical slag is a loose material with dense lumpy aggregates and few octahedral crystals. More than 80% of the slag is represented by a thin material with a particle size of less than 0.2 mm. The main beneficial elements are iron (42.5%) and chromium (3.15%), which form their own mineral phases. Nickel (0.4%), associated with trevorite, and cobalt (0.08%), with unknown occurrence form, can possess industrial value.

A significant part of the slags (more than 75%) is formed by black magnetic material prone to artificial segregation, represented by spinelides forming a continuous isomorphic series spinel-magnetite-chromite. The main mineral is magnetite, which isolations are always non-uniform and contain numerous inclusions of non-metallic phases represented by olivine, pyroxene, mica, corundum, feldspar and glass (Fig. 1a). Sometimes the grains are surrounded by a fairly flat border of iron hydroxides and contain veinlets that do not extend beyond their boundaries (Fig. 1b). Two varieties of magnetite are noted; eutectic colonies are evidence of their simultaneous presence. This may be due not only to the closest intergrowth of the phases, or the oxidation process of a magnet and its partial transition to maghemite, but also continuous isomorphic substitutions of the ferrospinels, including heterogeneous -2 and 3 valence cations.

(a)   **(b)**

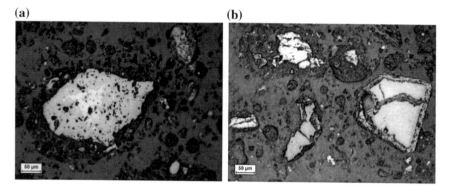

Fig. 1. A. Magnetite (light gray) with inclusions of non-metallic minerals, in the center of the grain is maghemite (white). B. Magnetite (light gray) with veins and border of iron hydroxides. Polished thin section, nicols are parallel

The features of the mineral composition and structure of iron-containing slags (phase composition, distribution of ore phases and aggregates, their morphostructural characteristics, heterogeneous structure of magnetite), identified by a complex of mineralogical analysis methods, suggest the expediency of chemical processing methods at recycling (Yakushina et al. 2015).

4   Conclusions

Mining wastes, being main types of technogenic raw, possess mineralogical characteristics (mineral and (or) phase composition, useful component occurrence form, morphostructural features and distribution pattern, real composition and structure) that determine the strategy and tactics of its secondary use:

(1)   as initial raw without processing, for example, for extraction of valuable metals;
(2)   as initial raw after additional processing to obtain material resources in the industry.
(3)   as object of disposal.

References

Burtsev IN, Kotova OB, Kuzmin DV, Mashin DO, Perovsky IA, Ponaryadov AV, Shushkov DA (2018) The role of technological research in the development of the mineral resource complex of the Timan-North Ural region. Explor Prot Miner Resour 5:38–47 (in Russian)

Chanturia VA, Ozhogina EG, Shadrunova IV (2016) Tasks of ecological mineralogy during the development of the Earth's interior. Phys Tech Probl Dev Miner Resour (SB RAS) 5:193–196 (in Russian)

Ozhogina EG, Kotova OB, Yakushina OA (2018) Mining wastes: mineralogical features. Vestnik of IG Komi SC UB RAS, pp 43–49 (in Russian)

Ozhogina EG, Shadrunova IV, Chekushina TV (2017) The role of mineralogical studies in solving environmental problems of mining areas. Min J 11:105–110 (in Russian)

Ozhogina E, Kotova O (2019) New methods of mineral processing and technology for the progress of sustainability in complex ore treatment. In: 29th International Mineral Processing Congress IMPC 2018, Canadian Institute of Mining, Metallurgy and Petroleum, 2-s2.0-85059377649

Yakushina OA, Ozhogina EG, Khozyainov MS (2015) Microtomography of technogenic mineral raw materials. IG Komi Science Center, Ural Branch of the Russian Academy of Sciences. vol 11, pp 38–43 (in Russian)

High-Tech Elements in Minerals of Massive Sulfide Deposits: LA-ICP-MS Data

V. Maslennikov[1]([⊠]), S. Maslennikova[1], N. Aupova[1], A. Tseluyko[1], R. Large[2], L. Danyushevsky[2], and U. Yatimov[1]

[1] Institute of Mineralogy, Ural Branch of RAS, Miass, Russia
mas@mineralogy.ru
[2] CODES ARC Centre of Excellence in Ore Deposits, University of Tasmania, Hobart, Australia

Abstract. LA-ICP-MS data on trace element zonation reflecting a local variation of physicochemical conditions of mineralization in black, grey and clear smokers from the Pacific and Atlantic oceans are used for comparison with the ancient chimneys of the Urals, Rudniy Altai, Pontides and Hokuroko massive sulfide deposits. Host rocks also influence on high-tech trace element assemblages in chalcopyrite: ultramafic (high Se, Sn, Co, Ni, Ag and Au) → mafic (high Co, Se, Mo and low Bi, Au and Pb) → bimodal mafic (high Te, Au, Ag, Bi, Pb, Co, moderate Se, and variable As and Sb) → bimodal felsic (high As, Sb, Mo, Pb, moderate Bi, and low Co, Te and Se). In sphalerite of the same range, the contents of Bi, Ga, Pb, In, Ag, Au and Sb increase versus Fe, Se, Sn and Co. The variations in pyrite coincide with these changes. Diagenetic evolution of high-tech elements is recognized in sulfide nodules.

Keywords: LAICP-MS · Trace elements · Chimneys · Massive sulfide deposits

1 Introduction

The bulk minor and trace element analyses indicate significant influence of host rocks on the metal inventory of the modern and ancient massive sulfide deposits. However, these general data can't provide dramatic improvement of the extraction of high-tech elements from massive sulphide ores. In contrast, the results of the quantitative laser ablation inductively coupled plasma mass spectrometry (LA-ICP-MS) show concentration of most trace metals in pyrite, chalcopyrite and sphalerite varieties as a function of physicochemical parameters of the fluid as well as host rock composition. Local LA-ICP-MS analyses of the sulphides provides a new insight into the processes of ore treatment and recovery of high-tech elements. The chimney fragments, sulfide turbidites, and their ore diagenites are suitable subject of this style of the research. The study is a key basement for fither improvement of mining and metallurgical processes.

© The Author(s) 2019
S. Glagolev (Ed.): ICAM 2019, SPEES, pp. 107–110, 2019.
https://doi.org/10.1007/978-3-030-22974-0_24

2 Methods and Approaches

In this paper, we use new LA-ICP-MS data on minor and trace elements in sulfides from Rudniy Altai, Pontides and Hokuroko massive sulfide deposits. Original LA-ICP-MS data on smokers from the Pacific and Atlantic oceans are used for comparison with the ancient chimney material. The sulfide nodules are the other important subject of the research reflecting chemical evolution of diagenetic and metamorphic processes.

Quantitative LA-ICP-MS analysis of sulfides from chimneys for a wide range of major and trace elements (Fe, Cu, Zn, Co, Ni, Au, Ag, Bi, Pb, Tl, Cd, As, Te, Se, Mo, Sn, V, Ti and Mn) was carried out using a New Wave 213 nm solid-state laser microprobes coupled to an Agilent 4500 quadrupole ICP-MS housed at the CODES LA-ICP-MS analytical facility, University of Tasmania and Institute of Mineralogy UB RAS. For this study, quantitative analyses were performed by ablating spots ranging in diameter from 40 to 60 microns. Data reduction was undertaken according to standard methods (Danyushevsky et al. 2011). The LA-ICP-MS was also used for detection of rare mineral microinclusions in the sulfides studied and for analysis of trace elements in telluride phases. Each zone of the chimney and nodules are characterized by LA-ICP-MS-images (Large et al. 2009).

3 Results and Discussion

A variety of very well preserved modern black, grey, white and clear smokers (Pacific and Atlantic oceans) and fossil vent chimneys from the Urals, Rudniy Altai, Pontides and Hokuroko massive sulfide deposit show systematic trace element distribution patterns across chimneys. Chalcopyrite is enriched in Se, Sn, Bi, Co, Mo, and Te. Sphalerite in the conduits and the outer chimney wall contains elevated Sb, As, Pb, Co, Mn, U, and V. He highest concentrations of most trace elements are found in colloform pyrite within the outer wall of the chimneys, and likely result from rapid precipitation in high temperature-gradient conditions. The trace element concentration in the outer wall colloform pyrite decrease in the following order, from the outer wall inwards; Tl > Ag > Ni > Mn > Co > As > Mo > Pb > Ba > V > Te > Sb > U > Au > Se > Sn > Bi, governed by the strong temperature gradient. In contrast, pyrite in the high- to mid-temperature central conduits exhibit concentration of Se, Sn, Bi, Te, and Au. The zone between the inner conduit and outer wall is characterised by recrystallization of colloform pyrite to euhedral pyrite, which becomes depleted in all trace elements except Co, As and Se. The mineralogical and trace element variations between chimneys are likely due to increasing fO_2 and decreasing temperature caused by mixing of hydrothermal fluids with cold oxygenated seawater (Maslennikov et al. 2009).

Host rocks also influence on high-tech trace element assemblages in chalcopyrite: ultramafic (high Se, Sn, Co, Ni, Ag and Au) → mafic (high Co, Se, Mo and low Bi, Au and Pb) → bimodal mafic (high Te, Au, Ag, Bi, Pb, Co, moderate Se, and variable As and Sb) → bimodal felsic (high As, Sb, Mo, Pb, moderate Bi, and low Co, Te and Se). In sphalerite of the same range, the contents of Bi, Ga, Pb, In, Ag, Au and Sb increase versus Fe, Se, Sn and Co. The variations in pyrite coincide with these changes (Maslennikov et al. 2017).

The next stage of mineral and chemical differentiation is halmyrolysis of clastic sulfide sediments followed by diagenesis, anadiagenesis and metamorphism. Halmyrolysis as a leaching process of clastic sulfide sediments give way to redeposition of high-tech element such as Se, Te, Au, Ag, Sn, and In as own minerals in enrichment submarine supergene zones. The inclusions of authigenic mineral such as Pb, Ag, Au-tellurides, Ag and Pb-selenides, In, Cu, Ag, Sn, Ge-sulfides, and Sn-oxide are increased in sizes in the enrichment submarine supergene zones of massive sulfide deposits.

In many massive sulfide deposits, sulphide nodules have a zonal structure: the nucleus of the poikilite pyrite (zone A) is successively surrounded by the fringes of metacrystalline pyrite (zone B) and marcasite (zone C). Each zone is characterized by own mineralogical features, which are reflected in the results of LA-ICP-MS-images ("microtopochemistry") of the surface of the nodule cut. It is assumed that at the first stage, the source of the substance for nodule growth was the products of the halmyrolysis of hyaloclastites containing an admixture of sulphide material. In the diagenetic nucleus of nodule, the chemical elements typical of illite (Si, Al, K, Mg, V, Cr,), rutile (Ti, W), apatite (Ca, Mn, U), galena (Pb, Bi, Sb, Ag), bismuth sulphosalts (Bi, Cu, Pb), bornite (Cu, Bi, Ag), tetrahedrite-tennantite (Cu, As, Sb), chalcopyrite (Se, Te, Cu), native gold (Au, Ag, Hg) and barite poikilites. In the pyrite, the high contents of Co and Ni are suggested to be substituted for Fe^{2+}. Only signs of tellurides of bismuth are guessed. Zone B, probably formed in the early anadiagenetic stage, is depleted of most trace elements, with the exception of Cu, Pb, and Ag. Formation of latest zone C was accompanied by of Cd, In, Tl, As, Sb, Mo, and Ni saturation.

4   Conclusions

The mineralogical and trace element variations between chimneys are likely due to increasing of O_2 and decreasing temperature caused by mixing of hydrothermal fluids with cold oxygenated seawater. The next stage of trace element differentiation is halmyrolysis of sulfide sediments followed by diagenesis, anadiagenesis and metamorphism. Halmyrolysis as a leaching process of clastic sulfide sediments gives a way to redeposition of high-tech element such as Se, Te, Au, Ag, Sn, and In as own minerals in enrichment submarine supergene zones. The enrichment in the core of sulfide nodules is changed to trace element depletion in later epigenetic stages.

Local LA-ICP-MS analyses of the sulphides provides new insight into the processes of selective ore treatment and recovery of high-tech elements from different genetic ore varieties in implication to diverse ore formational types of massive sulfide deposits.

Aknowledgements. LA-ICP-MS analyses were carried out during visiting programs (2005, 2009, 2012, 2013, 2015) sponsored by the ARC Centre of Excellence grant to CODES. The research of sulphide nodules was supported by the Russian Foundation for Basic Research (project no. 17-05-00854) in the Institute of Mineralogy UB RAS.

References

Danyushevsky LV, Robinson R, Gilbert S, Norman M, Large R, McGoldrick P, Shelley JMG (2011) Routine quantitative multi-element analysis of sulfide minerals by laser ablation ICP-MS: standard development and consideration of matrix effects. Geochim Explor Environ Anal 11:51–60

Large RR, Danyushevsky L, Hillit H, Maslennikov V, Meffere S, Gilbert S, Bull S, Scott R, Emsbo P, Thomas H, Singh B, Foster J (2009) Gold and trace element zonation in pyrite using a laser imaging technique: implications for the timing of gold in orogenic and Carlin-style sediment-hosted deposits. Econ Geol 104:635–668

Maslennikov VV, Maslennikova SP, Large RR, Danyushevsky LV (2009) Study of trace element zonation in vent chimneys from the Silurian Yaman-Kasy VHMS (the Southern Urals, Russia) using laser ablation inductively coupled plasma mass spectrometry (LA-ICP MS). Econ Geol 104:1111–1141

Maslennikov VV, Maslennikova SP, Ayupova NR, Zaykov VV, Tseluyko AS, Melekestseva IY, Large RR, Danyushevsky LV, Herrington RJ, Lein AT, Tessalina SG (2017) Chimneys in Paleozoic massive sulfide mounds of the Urals VMS deposits: mineral and trace element comparison with modern black, grey, white and clear smokers. Ore Geol Rev 85:64–106

Absolutely Pure Gold with High Fineness 1000‰

Z. Nikiforova[✉]

Diamond and Precious Metal Geology Institute SB RAS, Yakutsk, Russia
znikiforova@yandex.ru

Abstract. It is identified for the first time that, during process of complicated deformation in eolian conditions - mechanical transformation of flaky gold into toroidal form and then into globular-hollow form, gold is cleaned up to absolutely pure metal with fineness 1000‰. Note that, fineness of this eolian gold is higher than fineness of the reference object, shown by the detecting device (JXA-5OA micro-analyzer). In this connection, identified natural process of gold cleaning in eolian conditions can be successfully used in gold metallurgy to obtain absolutely pure gold.

Keywords: Gold · Fineness · Trace elements · Structures · Eolian process

1 Introduction

Native gold in exogenetic conditions, depending on its environment, undergoes gradual changes, in morphology and material composition. It is known that, gold cleaning in exogenetic conditions occurs mainly under chemical and physical-chemical influence in weathering crusts, and also as a result of simple strain in geodynamic conditions. This report will be focused on gold cleaning as a result of impact of mechanogenic processes in eolian conditions.

2 Methods and Approaches

Chemical composition of eolian gold of the Lena-Viluy interfluve (east Siberian platform) was studied by atomic-absorption spectrography (30 objects), spectral quantitative analysis (50 objects), and at JXA-5OA micro-analyzer (30 objects and 200 identifications).

3 Results and Discussion

During mechanical impact of sand grains on gold in eolian conditions, not only the form is transformed, but also inner structure and chemical composition are regularly changed. It is identified that, process of gold cleaning is more intensive in eolian conditions, than in hydrodynamic conditions (Nikiforova 1999).

© The Author(s) 2019
S. Glagolev (Ed.): ICAM 2019, SPEES, pp. 111–114, 2019.
https://doi.org/10.1007/978-3-030-22974-0_25

According to the analysis of eolian gold by atomic-absorption method, increase of gold fineness during transformation of flaky forms into globular-hollow forms from 810 to 970‰ is identified. Flaky gold with scarcely noticeable elevation at the periphery, has fineness range 810–970‰, with average gold fineness 890‰. In toroidal gold, interval of gold fineness fluctuation is identified - 920–970‰, with average index 940‰. Globular-hollow form is characterized by high gold fineness - 960–990‰, with average fineness 970‰.

Spectral quantitative analysis identified that, constant trace elements of flaky forms with characteristic features of eolian transformation are Fe- 0,1; Pb- 0,003; Sb- 0,002 Cu-0,017; Mn- 0,01; Pd- 0,002.; Ni-traces; Hg- traces. And some other. In toroidal gold, a smaller set of trace elements is identified Fe- 0,1; Cu-0,02; Mn- 0,03; Ni- traces; Hg-traces, and in globular-hollow body only these trace elements are identified – Fe- 0,1; Cu-0,05; Mn- 0,001.

Study of fineness of different areas of gold particles (Table 1 and Fig. 1) allowed identifying that, a flake with medium-grained structure (grain C-9a) has fineness from 747 t0 780‰, and its more high-standard shell – from 950 to 988‰. Flake P-138 with partially recrystallized rim has gold fineness 900–970‰, and its central part 814–860‰. Fully recrystallized flake showed maximum gold fineness – 990–1000‰. Fineness of globular-hollow gold within one sample is not just high, but has an absolute value - 1000‰. For example, 17 identifications in the grain 60a found insignificant fluctuations of fineness within interval 992–1000‰, and 13 identifications in the grain 60b showed the highest gold fineness – 1000‰.

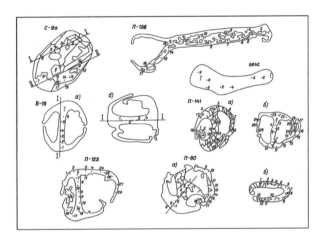

Fig. 1. Section of gold particles and points of fineness identification

Table 1. Fineness of particular areas of gold particles, ‰

| Points of fineness measurement | Morphologic type of eolian gold | | | | | | | | | |
| | Flaky | | | Globular recrystallized | | | | | | |
Index of the sample	C-9a	P-138	8642	Б-19a	B-196	P-141a	P-1416	P-123	P-60a	P-606
1	759	973	998	903	860	990	997	1000	996	1000
2	747	941	996	912	867	Not identified	992	955	996	1000
3	759	933	998	907	869		983	997	999	1000
4	756	Not identified	984	904	919	990	991	999	1000	1000
5	757	908	994	912		988	985	996	999	1000
6	755	920	997	902		996	996	995	997	1000
7	758	860	996	901		980	987	1000	992	1000
8	769	892	997			999	984	938	992	1000
9	766	978	1000			980	970	930	999	1000
10	953	814				989	995	993	994	1000
11	783	896				994	985	997	1000	1000
12	979	806				993	995	995	1000	1000
13	986	956				990	987	not identified.	1000	1000
14	970	937				920	995		1000	
15	974	973				932	1000	903	1000	
16	988	974				994	988	912	1000	
17	970	900				998	987	910	1000	
18						987	1000	1000	1000	
19						998		999		
20						984		Not identified		
21						991				
22						1000				
23						983				
24						996		1000		
25						998		1000		
26						999		1000		
27								1000		
28								1000		

It should be emphasized that, shell of globular-hollow forms constantly shows absolutely high fineness 1000‰. Just in several samples, primary gold preserved in partition, is characterized by lower fineness within a range from 860 to 919‰, for example grain 19b.

Increase of gold fineness in eolian conditions is explained by the fact that, as a result of complex deformation of flaky gold, very thin films of gold (fraction of mcm) are formed, being overlapped on each other, generate a shell of globular forms. In addition, surface for active chemical interaction of metal with environmental elements is increased, that contributed to the maximum removal of silver and trace elements from primary gold particles.

It should be emphasized that, shell of globular-hollow forms constantly shows absolutely high fineness 1000‰. Just in several samples, primary gold preserved in partition, is characterized by lower fineness within a range from 860 to 919‰, for example grain 19b.

Increase of gold fineness in eolian conditions is explained by the fact that, as a result of complex deformation of flaky gold, very thin films of gold (fraction of mcm) are formed, being overlapped on each other, generate a shell of globular forms. In addition, surface for active chemical interaction of metal with environmental elements is increased, that contributed to the maximum removal of silver and trace elements from primary gold particles.

The process of mechanically gold cleaning was proved by (Lechtman 1979) experimentally. According to the results of his experiments, clean layer of gold appeared after multiple alteration of forging of gold and copper alloy, with its processing in weak ammonia solution, where initial gold content did not exceed 12%.

4 Conclusions

It is identified for the first time, that during complex deformation, when flaky gold is mechanically transformed into toroidal form and then into globular-hollow form, metal is cleaned to absolutely pure gold with fineness 1000‰. Identified natural process of gold cleaning in eolian conditionsв can be successfully used in gold metallurgy to obtain absolutely pure gold.

Acknowledgements. The work is implemented within scientific-research projects of Diamond and Precious Metal Geology Institute, Russian Academy of Sciences, project № 0381-2016-0004.

References

Lechtman H (1979) A pre-Columbian technique for electrochemical replacement plating of gold and silver on copper objects. J Metals 31(12):154–160

Nikiforova ZS (1999) Typomorphic features of eolian gold. ZVMO. N5, pp 79–83

New Data on Microhardness of Placer Gold

Z. Nikiforova[(⊠)]

Diamond and Precious Metal Geology Institute SB RAS, Yakutsk, Russia
znikiforova@yandex.ru

Abstract. Microhardness of eolian gold – new morphologic type of placer gold, is studied. Flaky gold particles with a elevation at the periphery, as well as gold of toroidal and globular-hollow forms belong to eolian gold. Genesis of eolian gold is related to mechanical transformation of flaky gold particles into toroidal, and then into globular-hollow form in eolian conditions, that is experimentally proven. Previous studies determined changes of microhardness, mainly from 47 kg/mm^2 to 100 kg/mm^2, lower limit – 40 kg/mm^2. But, low microhardness was identified in globular-hollow gold for the first time, which stood at 21 kg/mm^2. This is due to the fact that, as a result of transformation of flaky gold in eolian conditions, under mechanical and chemical processes, silver and trace elements were removed, that led to fineness increase up to 1000‰, and to decompaction of inner structure of gold, that influenced microhardness indices. Identified patterns in nature, microhardness changes under mechano-genic and chemical process impact in eolian conditions, can be successfully used in gold metallurgy to produce gold alloys with very low microhardness.

Keywords: Gold · Microhardness · Fineness · Trace elements · Morphology · Process

1 Introduction

Microhardness of gold was earlier studied by Lebedeva (1963), Badalova et al. (1968), Petrovskaya (1968), Popenko (1982). Previous researchers identified that, microhardness changes on the average from 40 to 100 kg/mm^2 and depends on chemical composition of gold (percentage of silver and trace elements) and its inner structure. It is known that, Ag and Cu trace elements significantly increase hardness of gold alloys. Trace element presence (Pt, Sn, Al) also causes sharp increase of gold hardness. Gold of low fineness (550–650‰) has the biggest microhardness (Popenko 1982).

2 Methods and Approaches

Microhardness of globular-hollow eolian gold was studied by microhardness tester PMT-3. In order to detect microhardness, globular-hollow forms of gold were mounted in epoxy specimen. In such a manner, optimum section of gold particles was obtained, necessary for further study; it is possible to identify microhardness indices in its central part, end and shell. 55 identifications were performed at 12 sections of globular-hollow gold particles (Fig. 1 and Table 1). Since, gold fineness is increased in eolian

S. Glagolev (Ed.): ICAM 2019, SPEES, pp. 115–118, 2019.
https://doi.org/10.1007/978-3-030-22974-0_26

conditions, high-standard shell is formed in end parts, and central parts remain unchanged; microhardness was measured in end parts of the globular-hollow gold, and in partition of hollow ball, being a relic of the flake.

3   Results and Discussion

Microhardness of native gold, being in eolian conditions, is studied for the first time. It is identified that, in four gold particles (1, 2, 3, 4), increase of microhardness from partition center to the end is clearly observed. Microhardness between individual grains of gold particles (1, 2) is 2 times different. Uniform microhardness is typical for different parts of the globular-hollow forms (6, 7, 9). Sample 12 yields increase of microhardness from outer part of the end to partition center. Other four gold particles do not have any clear regularity concerning microhardness distribution.

When microhardness of eolian gold was studied, it was found that, microhardness depends on degree of inner structure changes and chemical composition of gold in exogenetic conditions. When flaky gold is transformed in a complicated way into toroidal form, and then into globular-hollow form in eolian conditions, not only gold morphology is changed, but also chemical composition is changed, in particular gold fineness is increased and amount of trace elements is decreased (Nikiforova 1999). It is explained by the fact that, as a result of complicated transformation of flaky gold in eolian conditions, very thin gold films are stretched from the flake end and overlap each other, generating globular-hollow form. In this connection, surface for chemical impact is increased in exogenetic conditions, that contributed to gradual removal of silver and trace elements, and sharp decrease of microhardness indices of eolian gold.

In addition, earlier unknown lower limit of microhardness of the globular-hollow gold is identified - 21 kg/mm^2.

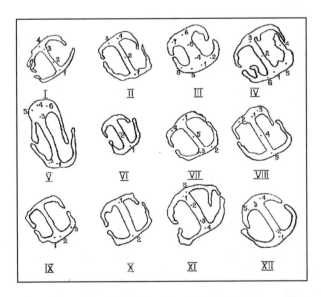

Fig. 1. Location of the points of identification of globular-hollow gold microhardness

Table 1. Microhardness of the elements of globular gold section, kg/mm^2

Sample №	Elements of section of globular form					
	End 1	Partition	End 2	Base of the		Shell
				End 1	End 2	
I	$\frac{64,5^*}{4}$	$\frac{41,2}{2,3}$	$\frac{55}{1}$			
II	$\frac{25,7}{3,4}$	$\frac{21}{2}$	$\frac{28,6}{1}$			
III	$\frac{41,2}{6}$	$\frac{25,7}{5,4}$	$\frac{32,1}{1}$	$\frac{55}{7}$	$\frac{41,2}{2}\,\frac{25,7}{3}$	$\frac{23,2}{8}$
IV	$\frac{46,2}{1}$	$\frac{42,0}{2,3}$		$\frac{64,2}{6}$		$\frac{57,2}{5}\,\frac{46,7}{4}$
V	$\frac{47,3}{4}$	$\frac{32,1}{3}\,\frac{25,7}{2}$	$\frac{25,7}{1}$	$\frac{55}{5}\,\frac{23,2}{6}$		
VI	$\frac{43,0}{1}$	$\frac{41,2}{2}$				
VII	$\frac{64,5}{4}$	$\frac{64,5}{5}$	$\frac{64,5}{2}$	$\frac{64,5}{3}$	$\frac{41,2}{1}$	
VIII	$\frac{47,3}{1}$	$\frac{47,3}{4}$	$\frac{32,1}{5}$	$\frac{32,1}{2,3}$		
IX	$\frac{28,6}{1}$			$28,6$		$\frac{28,6}{3}$
X	$\frac{47,3}{1}$	$\frac{55,0}{2}$				
XI	$\frac{36,2}{1}$	$\frac{47,3}{2}\,\frac{36,2}{3}$	$\frac{36,2}{4}$	$\frac{41,2}{5}$		
XII	$\frac{47,3}{3}$	$\frac{55}{2}$	$\frac{41,2}{1}$	$\frac{36,2}{4}$	$\frac{32,1}{5}$	

*Above line – microhardness, below line – point of microhardness measurement (Fig. 1).

In general, it is true that microhardness of eolian gold relatively low in comparison with native gold from other exogenetic conditions. It may be that, low microhardness is due to decompaction of some parts of the shell and the end of globular-hollow gold. Petrovskaya (1973) identified that, structures of recrystallization cause decompaction, stress relief, removal of silver and trace elements, that lead to increase of gold fineness up to 1000‰. Structures of decompaction, high fineness and paucity of trace elements are observed in the studied gold, that is why microhardness indices of globular-hollow gold reached such low limits.

In addition, earlier unknown lower limit of microhardness of the globular-hollow gold is identified - 21 kg/mm^2.

In general, it is true that microhardness of eolian gold relatively low in comparison with native gold from other exogenetic conditions. It may be that, low microhardness is due to decompaction of some parts of the shell and the end of globular-hollow gold. Petrovskaya (1973) identified that, structures of recrystallization cause decompaction, stress relief, removal of silver and trace elements, that lead to increase of gold fineness up to 1000‰. Structures of decompaction, high fineness and paucity of trace elements are observed in the studied gold, that is why microhardness indices of globular-hollow gold reached such low limits.

4   Conclusions

Thus, low microhardness – 21 kg/mm^2 is identified in globular-hollow gold is identified for the first time. Identified patterns in nature, microhardness changes under mechanogenic and chemical process impact in eolian conditions, can be successfully used in gold metallurgy to produce gold alloys with very low microhardness.

Acknowledgements. The work is implemented within scientific-research projects of Diamond and Precious Metal Geology Institute, Russian Academy of Sciences, project № 0381-2016-0004.

References

Badalova RP, Nikolaeva EP, Tolkacheva LF (1968) Study of microhardness of minerals of the silver-gold series from gold deposits of Uzbekistan. "Physical features of rare-metal minerals and methods of their study" collected articles, Moscow, Nauka

Filippov VE, Nikiforova ZS (1988) Transformation of native gold particles during process of eolian impact. AN SSSR Report, vol 299, № 5, pp 1229–1232

Lebedeva SI (1963) Identification of microhardness of minerals. AN SSSR Publishing House, Moscow

Nikiforova ZS (1999) Typomorphic features of eolian gold. ZVMO, N5, pp 79–83

Petrovskaya NV (1973) Native gold. Nedra, Moscow

Popenko GS (1982) Mineralogy of fold from the Quaternary deposits of Uzbekistan, Tashkent, FAN

Modern Methods of Technological Mineralogy in Assessing the Quality of Rare Metal Raw Materials

E. Ozhogina[✉], A. Rogozhin, O. Yakushina, Yu. Astakhova,
E. Likhnikevich, N. Sycheva, A. Iospa, and V. Zhukova

FSBE "All-Russian Institute of Mineral Raw Materials" (VIMS),
Moscow, Russia
vims@df.ru

Abstract. Modern technologies of mineralogical study and evaluation of rare metal raw quality are focused on its variety. Methods of the mineral processing, allowing to optimize monitoring of ore properties defining technological processes and quality of expected products are presented. Some examples of rare metal ores mineralogical study are given. The main challenging tasks in rare metal ores quality evaluation are considered.

Keywords: Technological mineralogy · Methods of analysis · Testing · Rare metals · Minerals · Ores · Specific features · Quality assurance · Quality assessment

1 Introduction

Variety of rare metal raw materials is determined, firstly, by a significant amount of industrial minerals, among which there are more than 20 main and about 30 minor and accessory minerals of different genesis; secondly, by the diversity of their genesis: magmatic, pegmatite, greisen, scarn, metamorphic, hydrothermal, sedimentary, hypergeneous. The rare metal-bearing minerals are the ones that contain Tantalum, Niobium, Bismuth, Tellurium, Zirconium, Hafnium, Yttrium, Scandium, Lanthanides, Lithium, Beryllium, Cesium, Rubidium, Strontium, Barium. A lot of rare metal ores, mainly tantalum-niobium, rare-earth phosphate, carbonate and silicate, and frequently zirconium, are radioactive. The vast majority of rare-metal ores are complex, and industrial minerals can be both main, and secondary, subordinate ones. General features of the rare metal-bearing ores are as follows:

- complex texture-structural pattern (a significant number of fine, metacolloidal spots formed by minerals and aggregates of micro-and nanometer size);
- polymineral composition associated with the simultaneous presence of minerals developed in different paragenetic associations;
- variations in chemical composition of the ore-forming minerals caused by chemical elements isomorphic substitutions in their crystal structure;

S. Glagolev (Ed.): ICAM 2019, SPEES, pp. 119–122, 2019.
https://doi.org/10.1007/978-3-030-22974-0_27

– mineral grains phase heterogeneity of various origin, namely decay of solid solutions, syngeneic inclusions, zonal growth, multiple stages of generation, partial
recrystallization, secondary solid-phase transformations, etc.;
– ore minerals with radioactive elements can undergo transformations resulting in
metamictic forms origin (disrupted crystal structure due to radiation damage) or
partially metamictic (damaged crystal structure due to radiation).

2 Methods and Approaches

The variety and complexity of rare metal ores mineralogical features identify the
necessity to use a set of modern physical research methods of analysis to get reliable
data on their composition and structure. This complex of mineralogical methods is
individually selected depending on ores features and research tasks.

Mineralogical study of rare metal ores is necessary at all stages of deposits exploration
and development. We particularly note the importance of ore mineralogical and analytical
study of at the early stages of geological exploration, that allow to carry out technological
assessment of raw materials with minimal investment, and at deposit exploration,
involving geological and technological mapping for the detailed study of the mineralization zoning, minerals and mineral associations distribution, variations in the ore-
bearing phases properties and characteristics, identification of the technological types and
species whiting the geological margins of the deposit. The research practice proves great
contribution of mineralogical study in deposits investigation and quality assessment.

When mineralogical investigation is the result of a set of implemented methods,
including not only usually traditional ones (optical microscopy, radiography), but also
precise analysis (analytical electron microscopy, microprobe). Mineral and technological mapping challenge today the Zashikhinsky, Tomtor, Chuktukon and other
deposits of rare earth elements.

3 Results and Discussion

The characteristic feature of technological mineralogy is integration/conjunction of
research methods and modern technical means/units. It is especially important when
studding the rare-metal mineralization, because it is not always possible to uniquely
identify industrially valuable minerals, to establish the mineral form of useful components, to identify and study characteristics of minerals appearing in the fine aggregates.

Pyrochlore-Mmonazite-Crandallite ores of the Tomtor deposit differ in specific
composition, are rich in content and reserves of REE, Niobium, Yttrium, Scandi- um,
Phosphorus and are a non-standard type of rare-metal raw materials. Ores features are
the variable granular composition, often high dispersion, polymineral composition,
various forms of occurrence the rare earths and niobium-bearing minerals, the character
of their localization in close association with Alumiium- Silicate minerals. The ores are
formed by polymineral aggregates with variable content of Crandallite and Kaolinite of
fine-grained structure. The aggregate cement contains also Siderite, Ilmenorutil,

Anatase, Pyrochlore, iron hydroxides, and other minerals. Most often the aggregates dispose earthy-type structure.

Pyrochlore is the main Niobium mineral, that form both individual octahedral crystals, fragments of rounded and angular forms, as well as aggregates of tiny grains in size ranging from 1 μm to 0.5 mm, in varying degrees transformed by hypergene processes. The rock is distributed in the form of microinclusions, forming a "rash" in the cementing mass of aluminophosphates and silicates. According to microprobe data (X-ray microspectral microanalysis), the main feature of the hypergenic alteration of Pyrochlore is the replacement of Ca and Na by Sr, Ba and Pb; the altered Pyrochlore varieties significantly dominate in the ores. Hypergeneous transformation of Pyrochlore was accompanied by textural transformations, typically clearly manifested in the disintegration of its large crystals into small blocks (Fig. 1). Often the cracks between the separate Pyrochlore individuals are filled by Crandallite group minerals and Apatite, rarely by sulfides.

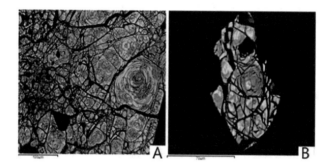

Fig. 1. Disintegrated crystals of Pyrochlore. TEM, image in back scattered electrons

Minerals of Crandallite group mostly form polymineral aggregates being dominant phase in these aggregates. There is a weak individualization of isometric and round shape grains, closely associated with fine Pyrochlore, Monazite, iron hydroxides, their grain size in the ores is often much less than 5 micrometers. According to microprobe analysis, the Crandallite group minerals have a mixed composition. According to X-ray powder diffraction analysis (XPD), the interplanar distances $d(hkl)$ reliably identificate Gorceixite (2.96, 3.55, 5.72 Å) and the intermediate Goyazite-Florencite series (2, 94, 3, 53, of 5.71 Å).

Mineralogical features of a Pyrochlore-Monazite-Crandallite ores (variable granular composition, often with high dispersion; polymineral composition due to the simultaneous presence of different paragenesis assemblages, different form of Niobium and rare earths presence, vide range of isomorphic substitutions in the structure of minerals, the proximity of their physical properties) determine the impossibility of these ores processing by methods of deep enrichment. Therefore, the prospects of such ores processing should be associated with hydrometallurgy.

Typical features of the Chuktucon Deposit rare earth ores were established on electron microscopic study. The main ore minerals are Pyrochlore, Monazite and the

Crandallite group minerals. All of them are superfinely dispersed and are in close assemblage/association with Iron and Manganese oxides and hydroxides. The latter form complex types of accretions with Pyrochlore (corrosive), Monazite (envelope of Goethite around Monazite grains), Crandallite group minerals (thin jointing), which negatively affects their disclosure and does not allow to identify and study these minerals by traditional methods of optic microscopy. Microprobe study indicated varieties of secondary Pyrochlore (bearing Cerium, Barium, Strontium and mixed type), and variable chemical composition of the Crandallite group minerals. Almost constant presence of Iron and Manganese mechanical impurities was established in all ore minerals. An independent mineral form of Cerium – Cerianite has been identificate by X-ray powder diffraction data.

4 Conclusions

The main challenging tasks for the rare metal ores investigation and quality evaluation should be considered:

- predictive mineral and technological assessment of raw materials of natural and man-made origin;
- geological and technological mapping using an Arsenal of methods of technological Mineralogy;
- forecasting of technological properties of ores at various stages of processing and development of methods of their directed change;
- increase of complexity of development of deposits and deep processing of ores;
- identification and involvement in the industrial use of non-traditional types of rare metal raw materials;
- assessment of environmental consequences of industrial development of deposits.

Therefore, the main task of technological mineralogy in the study of the rare metal ores is today their comprehensive study for quality assessment at all stages of geological research and development of mineral deposits.

Topochemical Transformations in Sodium-Bismuth-Silicate System at 100–900 °C

A. Pavlenko[✉] and R. Yastrebinskiy

Belgorod State Technological University named after V.G. Shukhov,
Belgorod, Russia
yrndo@mail.ru

Abstract. The authors have developed a method for producing highly dispersed sillenite bismuth silicate in the system Na_2O-Bi_2O_3-SiO_2 (NBS) from water solutions of organosilicon monomers (sodium methylsiliconate) and bismuth nitrate. The paper studies the phase composition and microstructure of the synthesized NBS material at different temperatures. The morphology of crystals in the NBS material and the peculiarities of its thermal-oxidative breakdown are investigated. X-ray diffraction spectra obtained using a cuK$_\alpha$-source are used to evaluate the crystal lattice spacing and to analyze the broadening of the maximum-intensity diffraction line for this crystal with due consideration of crystal indices h, k, l by the approximation method to determine the dimensions of the coherent scattering region and microdistortions of the crystal lattice $\Delta a/a$. The authors established that the silicate shell on $Bi_{12}SiO_{20}$ particles is close to the silicates with continuous chain radicals $[SiO_3]_\infty^{2-}$, and a part of them are bridges between the bismuth silicate particles.

Keywords: Sillenite · Phase composition · Microstructure ·
Crystal morphology · Thermal treatment · Crystal lattice microdeformation

1 Introduction

The development of highly dispersed metal-organosiloxane fillers with modified surface allows solving a multitude of important problems in the field of radiation materials science (Pasechnik 2006). The promising approach is to use water-soluble chemically active organosiloxanes as the basis for production of metal oligomers. A new technological approach to the solution of the stated complex problem is required.

As of today, the chemistry of organosiloxane compounds of bismuth attract particular attention. This is conditioned by multiple valuable properties of organosilicon compounds (high thermal stability, hydrophobicity, dielectric characteristics and resistance to a range of aggressive media). Besides, bismuth atoms have large capture cross-section of gamma-radiation, which is almost the same as for lead atoms in a wide energy spectrum. The presence of vacant 3d-orbitals in silicon atoms conditions high reactivity of bond $\equiv$Si-OH in silicate minerals.

© The Author(s) 2019
S. Glagolev (Ed.): ICAM 2019, SPEES, pp. 123–126, 2019.
https://doi.org/10.1007/978-3-030-22974-0_28

2 Materials and Methods

The authors have developed a method for producing highly dispersed sillenite bismuth silicate in system Na_2O-Bi_2O_3-SiO_2 from water solutions of organosilicon monomers (sodium methylsiliconate) and bismuth nitrate (Yastrebinskii et al. 2018).

The amounts of the components were calculated with the aim of producing stable bismuth silicate $Bi_{12}SiO_{20}$ ($6Bi_2O_3 \cdot SiO_2$).

At 100 °C we have obtained highly dispersed (0.2–0.3 μm) hydrophobic NBS material that is insoluble in water (NBS wetting angle is 122°). The density is 3780 kg/m³.

According to mass-spectroscopy, NBS material had the following composition (expressed as oxides), wt%: Na_2O - 23.83; Bi_2O_3 - 59.70; SiO_2 - 16.47.

Differential thermal analysis (DTA) and thermogravimetric analysis (TGA) of specimens were performed on a STA-449 F1 Jupiter derivatograph (Germany). X-ray diffraction (XRD) analysis of phases and structure was performed on a ARL™ X'TRA Powder Diffractometer (Switzerland) with Cu_{kx} source (λ_{kx} = 1.542 Å) using a nickel filter. The infrared spectra were obtained on a Specord-75IR spectrometer (Germany). The material microstructure was studied by raster electron microscopy (REM) in the modes of reflected (back-scattered) and secondary electrons.

3 Results and Discussion

Phase Composition and Microstructure of Mineral Phases in NBS Material Synthesized at 100 °C. XRD phase analysis using the Powder Diffraction File (PDF) and literature data (Gorshkov and Timashev 1981; Gorelik et al. 2002) allowed detecting the formation of three amorphous-crystalline mineral phases:

1. Metastable bismuth silicate Bi_2SiO_5 (d = 3.0379 Å/I = 100%; 3.7169 Å; 2.7223 Å) with tetragonal crystal system (a = 3.802; c = 15.134 Å), with the amorphous ring of about 3 Å.
2. Bismuth oxide α-Bi_2O_3 (d = 3.2596 Å/I = 100%; 3.2596 Å; 1.9625 Å) with monoclinic crystal system (a = 5.8499; b = 8.1698; c = 7.5123 Å) with the amorphous ring of about 3 Å.
3. Bismuth organosilicate $H_3C(Si_xBi_yO_z)Na$ with the amorphous ring of 10–12 10–12 Å and clear X-ray reflection at d = 11.4513 and 5.7090 Å. However, the precise determination of this composition using PDF failed.

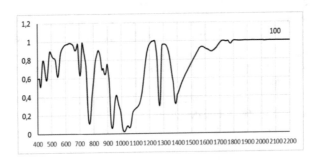

Fig. 1. IR spectrum of synthesized NBS material

The results of IR-spectroscopy, the silicate phases in NBS powder synthesized at 100 °C have linear structure. The splitting of the absorption bands in the range of 1000–1100 cm^{-1} that is typical for the siloxane bond indicates the presence of several types of siloxane phases (Fig. 1).

Morphology of Crystals in Synthesized NBS Material. According to REM, NBS material synthesized at 100 °C contained particle agglomerations of irregular shape with the size of 0.8–2.5 µm (Fig. 2).

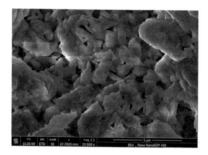

Fig. 2. Electron microphotographs (REM) of NBS material synthesized at 100 °C

Defectiveness of Crystals in NBS Material Subjected to Thermal Treatment. X-ray diffraction spectra obtained using a CuK$_\alpha$-source were used to evaluate the crystal lattice spacing and to analyze the broadening of the maximum-intensity diffraction line for this crystal with due consideration of crystal indices h, k, l by the approximation method to determine the dimensions of the coherent scattering region and microdistortions of the crystal lattice $\Delta a/a$.

At 100–300 °C XRD analysis detected amorphous-crystalline bismuth organosiliconate $H_3C(Si_xBi_yO_z)Na$ with the amorphous ring of 10–12 Å and clear X-ray reflection at d = 11.4810 Å and 5.7020 Å.

At the temperature of 200 °C, bismuth silicate Bi_2SiO_5 in the mixture of minerals in terms of X-ray parameters approaches to the benchmark silicate of this composition (card no. 36-288 PDF: d = 3.0400 Å (I = 100%, hkl = 103).

In the temperature interval of 300–400 °C, metastable bismuth silicate $Bi_{12}Si_{0,87}O_{20}$ with cubic crystal lattice in the synthesized dry mix transforms into stable bismuth silicate $Bi_{12}SiO_{20}$ also with cubic lattice.

In the temperature interval of 300–500 °C, the density of dislocations in the structure of bismuth silicate $Bi_{12}SiO_{20}$ crystals was lowering, while at the temperature higher than 650 °C, it was conversely rising up. The increase of the temperature from 300 to 500 °C improves the parameters of elementary crystal lattice of bismuth silicate $Bi_{12}SiO_{20}$ by 0.0357 Å and 0.0268 Å, as compared to the benchmark crystal. The volume of elementary crystal cell in this temperature interval grows by 2% and amounts to 1040.5870 Å^3.

4   Conclusion

In the study, a method for producing highly dispersed sillenite bismuth silicate in the system Na_2O-Bi_2O_3-SiO_2 (NBS) from water solutions of organosilicon monomers (sodium methylsiliconate) and bismuth nitrate was developed. The paper studied the phase composition and microstructure of the synthesized NBS material at different temperatures. The crystalline structure of the substance and the presence of silicate amorphous phase in it were discovered. The paper revealed the morphology of crystals in the synthesized NBS material and the peculiarities of its thermal-oxidative break-down; the silicate shell on the particles of $Bi_{12}SiO_{20}$ was close to continuous chain radicals $[SiO_3]_\infty^{2-}$ and a part of them played the role of bridges between bismuth silicate particles. The determination of physicochemical properties of modified $Bi_{12}SiO_{20}$ sillenite crystals was of appreciable significance.

Acknowledgements. The work is realized in the framework of the Program of flagship university development on the base of Belgorod State Technological University named after V.G. Shukhov, using equipment of High Technology Center at BSTU named after V.G. Shukhov, the project within strategic development program No. a-51/17.

References

Gorelik SA, Skakov YuA, Rastorguev LN (2002) X-radiography and electron-optical analysis (in Russian). MISIS, Moscow, 360 p

Gorshkov VS, Timashev VV (1981) Methods of physicochemical analysis of silicates (in Russian). Higher School, Moscow, 335 p

Pasechnik OF (2006) Study of properties and structure of polyamide films after interaction of low-orbit space factors. Obninsk, p 113. (in Russian)

Yastrebinskii RN, Bondarenko GG, Pavlenko VI (2018) Synthesis of stable bismuth silicate with sillenite structure in the Na2O–Bi2O3–SiO2 system. Inorg Mater Appl Res 9(2):221–226

Ag-Bearing Mineralization of Nevenrekan Deposit (Magadan Region, Russia)

E. Podolian[1,2(✉)], I. Shelukhina[1,2], and I. Kotova[2]

[1] "RMRL" Ltd. (Raw Materials Researching Laboratory),
Saint Petersburg, Russia
podolyan@lims-lab.com
[2] Department of Mineral Deposits, Saint Petersburg State University,
Saint Petersburg, Russia

Abstract. Nevenrekan deposit, located in Magadan region, Russia, is a perspective Au-Ag deposit with average silver content 445 ppm and gold 7.4 ppm. The study of the ores reveals the main Ag-bearing minerals: Au-Ag alloys, tetrahedrite-tennantite series and new phase – $Ag_8SnSe_2S_4$. New phase contains the main part of the silver of the deposit – 97.6%.

Keywords: Silver · Ag-sulphosalts · Ag-bearing minerals · Epithermal · Argyrodite

1 Introduction

The Nevenrekan Au-Ag deposit is located in the Severo-Even district of the Magadan region, Russia. The site is located within the central part of the Okhotsk-Chukotka volcanogenic belt. The main structural element is cretaceous-paleogenic Nevenrekan intrusive dome, which is crossed by quartz-adularia and quartz-carbonate veins with hydrothermal origin. These veins contain ores with Au-Ag mineralization. This work is motivated by mineralogical criteria for understanding ore-forming processes which are essential for future efficient exploration.

2 Methods and Approaches

Methods of research include: (1) chemical analyses and its interpretation (include X-Ray diffraction, atomic absorption analysis, inductively coupled plasma mass-spectrometry (ICP-MS) analysis for micro components and X-Ray fluorescence analysis for macrocomponents); (2) optical researches including petrographic research of thin sections and mineragraphic research of polish sections which are provided by electron microprobe analyses of chemical composition of ore minerals; (3) technological experiments (gravity и floatation methods of ores separation).

© The Author(s) 2019
S. Glagolev (Ed.): ICAM 2019, SPEES, pp. 127–128, 2019.
https://doi.org/10.1007/978-3-030-22974-0_29

3   Results and Discussion

The main types of rocks at the deposit are clastolaval rhyodacites and ignimbrite rhyodacites, which are host rocks, and quartz-adularia veins, which are gangue rocks. Ore mineralization of Nevenrekan deposit consists of 4 ore associations which are characterized by different temperature of formation: 1. rutile-kassiterite; 2. pyrite-arsenopyrite; 3. sphalerite-chalcopyrite-galena-Au-Ag alloys-stannite series-$Ag_8SnSe_2S_4$-tennantite-tetrahedrite series with Ag; 4. hydrohematite. The most important paragenesis is the third one because it includes the main mineral concentrators of silver and gold. Technological and following chemical and mineralogical researches of beneficiaries demonstrate that the main concentrator of silver is sulphosalts, especially new phase $Ag_8SnSe_2S_4$, which contains 97.6% of all silver. According to crystallographic researches, phase $Ag_8SnSe_2S_4$ is orthorhombic and should belong to argyrodite group (Zhai et al. 2018). Nevenrekan deposit belongs to low-sulfidation type in classification of epithermal deposits (White and Hedenquist 1995).

4   Conclusions

The main Ag-bearing minerals at the Nevenrekan deposit are Au-Ag alloys, tennantite-tetrahedrite series and $Ag_8SnSe_2S_4$ sulphosalt of argyrodite group, which contain 97.6% of silver. The mineral composition of ores and the mineral balance of silver determine the prospects of flotation flow sheet for ore beneficiation.

Acknowledgements. We thank Polymetal Engineering Company for the ore samples and RMRL Ltd. for sponsorship of the participation in 14th International Congress for Applied Mineralogy.

References

Zhai D, Bindi L, Voudouris P, Liu J, Tombros S, Li K (2018) Discovery of Se-rich canfieldite, Ag8Sn(S,Se)6, from the Shuangjianzishan Ag-Pb-Zn deposit, NE China: a multimethodic chemical and structural study. Miner Mag 1–21
White NC, Hedenquist JW (1995) Epithermal gold deposits styles, characteristics and exploration. SEG Newsl 23:9–13

Th/U Relations as an Indicator of the Genesis of Metamorphic Zircons (On the Example of the North of the Urals)

Y. Pystina[✉] and A. Pystin

Institute of Geology Komi SC UB RAS, Syktyvkar, Russia
pystina@geo.komisc.ru

Abstract. Having studied polymetamorphic complexes of the Urals, including its northern part, for many years we have collected material on the basis of which we attempted to make generalizations concerning both the morphology of zircons and their geochemical features, allowing the mineral to be used in the reconstruction of specific metamorphic events and the interpretation of geochronological data (Pystina et al. 2017; Pystina and Pystin 2002). In recent years we also obtained new results on the morphology and geochemistry of zircons from granitoids in the northern part of the Subpolar Urals (Pystina and Pystin 2002). Together, this made it possible to compare different morphological types of magmatic and metamorphic zircons.

Keywords: Polymetamorphic complexes · Geochronological · Morphology · Geochemistry of zircons

1   Introduction

Precambrian formations, especially Pre-Riphean, which underwent metamorphism, as a rule, experienced it repeatedly, i.e. were polymetamorphic. Accordingly, the zircons of the newly formed, or transformed from the previously existing ones, in the course of these events should have acquired some new properties, expressed in changes in the morphology of the crystals, the internal structure, the geochemical composition. That is what we see in zircons from various polymetamorphic complexes of the Urals, including those located in its northern part: Nyartin, in the Subpolar Urals, and Harbey in the Polar Urals, where up to five morphological types of this mineral are distinguished (Pystina and Pystin 2002).

2   Results and Discussion

Detritic zircons (type 1) determine the metamorphic affiliation to one or another source formation. Zircons of the "soccer ball" type or, as is customary in the Urals, after Krasnobayev (1986), call them "granulitic" (type 2), and also "migmatite" (type 4) fix several age levels of occurrences of high-temperature rock transformations. Zircon of irregular shape like "cauliflower" (type 3) is typical for rocks metamorphosed under

© The Author(s) 2019
S. Glagolev (Ed.): ICAM 2019, SPEES, pp. 129–132, 2019.
https://doi.org/10.1007/978-3-030-22974-0_30

conditions that do not exceed the low to medium stages of the amphibolite facies. In more high-temperature conditions, it is found in the rocks of the basic series. The reason for the emergence of such intricate forms of zircon may be the absence or deficiency of silicate melt. Therefore, when P-T reaches the conditions of metamorphism sufficient for the development of migmatization processes, such zircon can continue to crystallize only in metamorphic mafic composition, for which, as is known, the migmatization temperature is higher. Opaque zircons (type 5) are associated with the manifestation of medium temperature diaphtoresis. The internal structure of all morphotypes of zircons is characterized by the presence of nuclei of irregular or rounded shape, in the "migmatitic" type, oscillatory zonality is usually noted, the "granulitic" type is the most homogeneous. Among the isolated morphotypes of zircons in polymetamorphic complexes, the "granulite" and "migmatite" types prevail. If, by morphological features and internal structure, the zircons of polymetamorphic complexes are surely divided into morphotypes, which can be associated with certain metamorphic events or processes, then the geochemical composition of the scattered elements does not make a clear separation. On the one hand, this is due to the extremely low content of the elements themselves, most often zircon is enriched only in Hf, Y, U, P. C, on the other hand, the nature of the distribution of these elements in the crystal, their quantitative variations do not give grounds to separate some zircon morphotypes from others. Although in some cases it is possible. E.g., in "migmatite" zircons (type 4) from the gneiss of the Harbay Complex, the distribution of Hf decreases from the center of the crystal to the edge, while in other morphotypes of zircons from the same rocks it increases. At the same time, in the gneisses of the Nyartin Complex in all the selected morphotypes, including the "migmatite" type, the content of Hf increases from the center of the crystal to the edge. The same picture, according to our data, is observed in zircons from the rocks of the metamorphic complexes of the Southern Urals: the Alexandrov and Ilmenogorsky (Selyankin block). The content and distribution of the scattered elements according to the data available today with morphological types of zircons are not clearly correlated. In this regard, the Th/U ratio deserves interest, which for magmatic zircons are, as a rule, > 0.5 (Skublov et al. 2009), and for metamorphic, significantly lower – 0.1–0.3 (Pystin and Pystina 2015a), according to Rubatto < 0.07 (Rubatto 2002), although according to other data it can be > 0.5, for example, 0.73 in zircons from the eclogite of the Maksyutov Complex (Pystin and Pystina 2015b). But, despite some rebounds in Th/U values, the average indices for metamorphic zircons, as well as for magmatic ones, are fairly consistent. In the metamorphic zircons of the gneisses of the Nyartin Complex, obtained from 9 crystals, they vary from 0.02 to 0.39, but among these values, two are essentially out of the general picture – 0.75 and 0.68. Such values of the Th/U ratio correspond to zircons of magmatic origin. In our case, these are zircons of prismatic habitus, which we have isolated into the "migmatite" type. The isotope age, obtained by the U-Pb SHRIMP-II method, is 503 ± 8 Ma and 498 ± 8 Ma, respectively (Pystin and Pystina 2008). Their formation was associated with the metamorphism of the amphibolite facies and the associated granitization. Therefore, the conditions, under which this morphotype was formed, were similar to the conditions of granite formation, hence we assume high Th/U ratios.

An even clearer picture is obtained for zircons of the "granulitic" type from the Alexandrov Complex in the Southern Urals, where the Th/U ratio varies from 0.23 to

0.31, which is quite consistent with the metamorphic zircons. This is confirmed by the isotopic age of all the crystals, which is close and approximately 2.1 Ga (Pystin and Pystina 2015b, Pystina and Pystin 2002).

Our studies show that accessory zircons from rocks of different granitoid complexes in the northern part of the Subpolar Urals, which have different geological positions and isotopic age, differ in the set of morphotypes, their quantitative ratios and geochemical features (Pystina et al. 2017). Recently we studied morphological features of zircons from the granites of Nikolayshor (PR1), Kozhim (PF2–3), Badyayu (RF3–V) and Yarota (RF3–V) Massifs. Accessory zircons are very diverse in form, character of zonality, presence of inclusions, color, degree of metamictism and other features. All the main morphological types of zircons according to I.V. Nosyrev (Nosyrev et al. 1989) were observed: zirconium, hyacinth, spear-shaped, torpedo-shaped and citrolite. All of the above morphological types can relate to the generation of zircons of either synpetrogenic or superimposed genetic types.

In the granites of the Nikolashor Massif hyacinth, spear-shaped, torpedo-shaped, and detrital zircons were found. Granitoids of the Kozhimsky, Badyayusky and Yarotsky Massifs are characterized by the presence of three morphotypes of zircons, but if in the granitoids of the Badyayu and Yarota Massifs they are similar (zircon, hyacinth and torpedo-shaped morphotypes), while in the rocks of the Kozhim Massif, they are pattern, torpedo-shaped, torrent, and the beginning of torpedo-shaped morphotypes. Common to the granitoids of all the massifs is one morphotype – torpedo. The spear-shaped zircon is found only in the rocks of the Nikolashor Massif. There are also detrital zircons, which are absent in the granitoids of other massifs. Granites of the Kozhim Massif are distinguished from other granitoids by the presence of zircon of the citrolite morphotype. The presence of this type of zircons is a sign of the metasomatic (or metamorphic) processing of rocks. In contrast to the granites of the Badyayu and Yarota Massifs, they lack hyacinth type zircons.

The Th/U ratio in zircons from the granitoids of the northern part of the Subpolar Urals – Nikolashor, Kozhim, Khatalambo-Lapchin and Lapchavozh Massifs, is on average 0.73; 0.61; 0.51; 0.79, respectively. These values are maintained and observed in all zircons of the studied granitoids. In some cases, in zircons from the granitoids of the Khatalambo-Lapchin Massif, the Th/U values are knocked out of the overall picture, amounting to 0.22 and 0.15, which is not at all characteristic of magmatic zircons. If we take into account that the age of these zircon crystals obtained by the U-Pb SHRIMP-II method is 703.9 ± 8 Ma and 795 ± 41 Ma, and the rest of the zircons is 550–580 Ma, we can assume that the formation of ancient zircons The early stages of granite formation and the increased Th/U ratio are explained by the subsequent metamorphism of early generation granites.

3 Conclusions

Thus, we have to state the validity of the fact that "the only obvious systematic difference between the magmatic and metamorphic zircon is the Th/U ratio ..." [Hoskin and Schaltegger 2003, p. 48]. It allows not only to distinguish between

magmatic zircons from metamorphic, but taking into account the morphological features of individual crystals and isotopic age dating, it is more reliable to restore the
history of the formation of specific metamorphic and igneous complexes.

Acknowledgements. This work was supported by the Basic Research Program of the Russian
Academy of Sciences No. 18-5-5-19.

References

Hoskin PWO, Schaltegger U (2003) The composition of zircon and igneous and metamorphic
petrogenesis. Rev Miner Geochem 53:27–62

Krasnobayev A (1986) Zircon as an indicator of geological processes. Nauk, Moscow, 152 p

Nosyrev IV, Robul VM, Esipchuk KE, Orsa VI (1989) Generation analysis of accessory zircon.
Nauka Press, Moscow 203 p

Pystin A, Pystina J (2015a) The early precambrian history of rock metamorphism in the Urals
segment of crust. Int Geol Rev 57(11–12):1650–1659

Pystin AM, Pystina YI (2008) Metamorphism and granite formation in the Proterozoic-
Earlypaleozoic history of the formation of the Polar Ural segment of the earth's crust.
Lithosphere (6):25–38

Pystin AM, Pystina YI (2015b) The Archean-Paleoproterozoic history of metamorphism of rocks
of the Ural segment of the earth's crust. Works of the Karelian Research Center of the Russian
Academy of Sciences, no 7, Geology of Dokembriya, pp 3–18

Pystina YI, Denisova YV, Pystin AM (2017) Typomorphic signs of zircons as a criterion for the
dissection and correlation of granitoids (by the example of the northern part of the Subpolar
Urals). Bulletin of the Institute of Geology of Komi Scientific Center of the Ural Branch of
the Russian Academy of Sciences, no 12, pp 3–15

Pystina Y, Pystin A (2002) Zircon chronicle of the Ural Precambrian. UrD RAS Press,
Ekaterinburg, 167 p

Rubatto D (2002) Zircon trace element geochemistry: partitioning with garnet and the link
between U-Pb ages and metamorphism. Chem Geol 184:123–138

Skublov SG, Lobach-Zhuchenko SB, Guseva NS, et al (2009) Distribution of rare-earth and rare
elements in zircons from miaskite lamproites of the Panozersky complex of central Karelia.
Geochemistry (9):958–971

Phenomenon of Microphase Heterogenization by Means of Endocrypt-Scattered Impurity of Rare and Noble Metals as a Result of Radiation by Accelerated Electrons of Bauxites

I. Razmyslov[1(✉)], O. Kotova[1], V. Silaev[1], and L. A. Gomze[2]

[1] Institute of Geology Komi SC UB RAS, Syktyvkar, Russia
razmyslov-i@mail.ru
[2] University of Miskolc, Miskolc, Hungary

Abstract. During radiation-thermal transformation of Timan ferrous bauxites we discovered a previously unknown phenomenon of microphaseheterogenization, which can contribute to the extraction of many valuable impurities during processing of relatively poor quality bauxite raw.

Keywords: Bauxite · Radiation-thermal transformation ·
Microphase heterogenization · Profitability of bauxite raw processing

1 Introduction

The problem of processing of aluminum ores is related to the fact that bauxite-forming minerals are characterized by similar values of density, dispersion and fine mutual intergrowths of individuals, non-recoverability of many valuable microelements. Therefore, the development and improvement of methods for the enrichment and processing of bauxite remains highly relevant (Borra et al. 2015; Borra et al. 2016; Davros et al. 2016). In terms of their iron index, the studied Timan bauxites are subdivided into three mineral types: hematite-boehmite, hematite-berthierine-boehmite, and berthierine-boehmite (Vakhrushev 2011, 2012; Vakhrushev et al. 2012). The results of our studies showed that in these bauxites in the state of endocrypt scattering there are a lot of valuable elements-impurities, which extraction by modern technologies is either time consuming or not yet possible. Our experiments on heating of bauxites in combination with their irradiation with high-energy electrons lead to a change in the phase composition of bauxites and, as a result, to the improvement of their technological properties (Rostovtsev 2010; Kotova et al. 2016).

2 Methods and Approaches

We carried out experiments with thermal (heating to 500 and 600 °C with quadruple exposure by 60 min) and radiation-thermal (heating to 500–600 °C with double exposure by 20 min with irradiation by an electron beam with energy of 2.4 MeV

S. Glagolev (Ed.): ICAM 2019, SPEES, pp. 133–135, 2019.
https://doi.org/10.1007/978-3-030-22974-0_31

using ILU-6 industrial unit at the Institute of Nuclear Physics of the Siberian Branch of the Russian Academy of Sciences) modification of Timan iron bauxites.

3   Results and Discussion

Heating up to 500–600 °C with exposure to 60 min without irradiation led to almost complete dehydration of Al and Fe oxyhydroxides with the formation of γAl_2O_3 (spinelide with a defective structure) and hematite with relatively small alteration of structure and magnetic properties of the rocks. The gross chemical composition of the annealed samples remained almost unaltered, and the content of trace elements increased markedly in% to the original: Cu by 70–250; Zn by 20–25; Cd by 20–380; Zr by 2–20; Nb by 15–20; Sc by 25–40; Y by 35–70; Ce by 25–60; Nd by 1–10. Obviously, the latter is conditioned bya significant thermal dehydration of the studied bauxites.

The radiation-thermal treatment of ferrous bauxites led not only to dehydration of the original minerals, but also to chemical disproportionation of the original substance, its radical magnetic restructuring resulted from the presence of strong magnetic phases (maghemite, magnetite) and the formation of valuable trace elements of many new minerals with individuals varying in shape from isometric to needle-like and in size from submicronic to 0.5 mm due to endocrypt scattered impurity. The newly formed phases include native metals (Au, Pb, Al, Cu, Zn); sulfides (pyrite, galena); oxides of Sn, Ta, Nb, Zr, lanthanides; silicates (zircon, kaolin); rare sulphates, etc. Thus, the heating, combined with irradiation with high-energy electrons, resulted not only in transformation of primary minerals in the ferrous bauxites, but also in microphase heterogenization with the formation of new minerals (Fig. 1). It can be assumed that this kind of transformation can contribute to extraction of many valuable impurities and increase the profitability of processing of relatively low-quality bauxite raw.

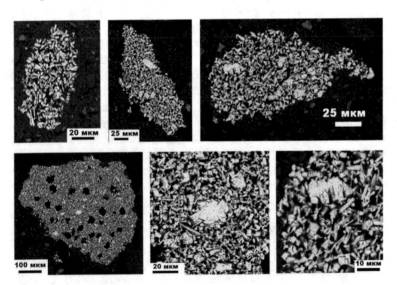

Fig. 1. Micro pocket segregations of Ce-Gd oxides in radiation-thermally modified Timan bauxites

Acknowledgements. This research was supported by UB RAS project 15-18-5-44 and project AAAA-A19-119031390057-5 "The main directions of integrated assessment and effective use of geo-resources in the Timan-North Ural-Barents Sea region".

References

Borra CR, Mermans J, Blanpain B, Pountikes YB, Gerven T (2016) Selective recovery of rare earths from bauxite residue by cjmbination of sulfation, roasting and leaching. Min Eng 92:151–159

Borra CR, Pontikes Y, Binnemans K, Gerven T (2015) Leaching of rare earts from bauxite residue (red mud). Min Eng 76:20–27

Davros P, Balomenos E, Panias D, Paspaliaris I (2016) Selective leaching 0f rare earth elements from bauxite residue (red mud). Hydrometallurgy 164:125–135

Kotova OB, Razmyslov IN, Rostovtsev VI, Silaev VI (2016) Radiation-thermal modification of ferruginous bauxites during processing. Enrich Process 4:16–22 (in Russian)

Rostovtsev VI (2010) Theoretical foundations and practice of using electrochemical and radiation (accelerated electrons) effects in the processes of ore preparation and enrichment of mineral raw materials. Vestnik of the Chita State University, vol 8, pp 91–99. (in Russian)

Vakhrushev AV (2012) Crystal chemistry of bauxite minerals from the Vezhayu-Vorykvinskoe deposit. Structure, substance, history of the lithosphere of the Timan-Northern Ural segment. Geoprint, Syktyvkar, pp 32–34. (in Russian)

Vakhrushev AV, Lyutoev VP, Silaev VI (2012) Crystal-chemical features of ferrous minerals in bauxite at the Vezhayu-Vorykvinskoe deposit (Middle Timan). IG Komi Science Center UB RAS, pp 14–18. (in Russian)

Gold Extraction to Ferrosilicium, Production of Foam Silicate from Processing Tails of the Olimpiada Mining and Processing Complex Gold Processing Plant (Russia, Krasnoyarsk Territory)

A. Sazonov[1]([✉]), V. Pavlov[2], S. Silyanov[1], and E. Zvyagina[1]

[1] Siberian Federal University, Krasnoyarsk, Russia
Sazonov_am@mail.ru
[2] Special Design and Technology Bureau "Science", Krasnoyarsk, Russia

Abstract. The paper describes the studies of processing tails of the Olimpiada Mining and Processing Complex with the methods of chemical, mineralogical, electronic microscope, deep reductive melting with division of the melt into a silicate and metal parts. It is demonstrated that 85% of the processing tails consist of the oxides: SiO_2, CaO, Al_2O_3, Fe_2O_3, H_2O and CO_2. The distribution of gold and silver is provided by the size classes of the initial blend, after melting of which in reduction conditions re-distribution of gold to the metal phase of the melt occurs. The silicate part of the melt when released into water in the "thermal shock" mode forms a light porous X-ray amorphous material "foam silicate", which also serves as a resource for stable chemical composition for production of a wide range of import substituting ceramic materials.

Keywords: Pyrometallurgy · Processing tails · Ferrosilicium · Gold extraction

1 Introduction

The development of the methods of flotation, gravity and metallurgic processing of gold bearing ores and man-made products today is aimed at increase of the recovery of the valuable product (Pavlov 2005; Pavlov et al. 2015; Meimanova and Nogayeva 2014; Bogdanovich et al. 2013; Tselyuk and Tselyuk 2013; Algebraistova et al. 2017; Amdur et al. 2015). There are almost no technologies for comprehensive use of processing tails to produce an additional marketable product both for task of comprehensive use of processing tails for production of foamed resources for ceramic items production with associated gold recovery to the ferrosilicium matrix is vital.

2 Methods

The paper proposes the pyrometallurgic approach to the solution of the problem of comprehensive wasteless processing of man-made gold-bearing resources with the method of deep reductive melting with division of the melt into the deferrized silicate

S. Glagolev (Ed.): ICAM 2019, SPEES, pp. 136–139, 2019.
https://doi.org/10.1007/978-3-030-22974-0_32

part with its further chilling in the thermal shock mode (Pavlov et al. 2015) and the metal part with associated re-distribution of gold in the ferrosilicium matrix. The specimens of the silicate part of the melt were prepared for study using the powder technology; the metal specimens were covered with epoxy resin with subsequent polishing. The chemical phase composition was studied with the use of an X-ray fluorescent S2 RANGER analyzer and the scanning electronic microscope (SEM) Hitachi S-3400N with EMF Bruker.

3 Samples

The input material for pyrometallurgic extraction of gold is represented by the tails of the gold processing plant of the Olimpiada Mining and Processing Complex, which is the leader in Russia in terms of the ore processing volumes and marketable gold production. The gold-antimony and gold-arsenic ores of the Olimpiada deposit have a complex mineral composition and are extremely refractory for gold recovery from them. Most of gold is in the form of thin dissemination in sulfides (Kirik et al. 2017; Novozhilov et al. 2014). No more than 50% of gold is extracted by direct cyanation. At the gold processing plant the flotation concentrates of the ores are exposed to biooxidation using the BIONORD® technology and subsequent leaching to release gold capsuled in sulfides.

The processing tails have loose consistence, mainly sand-aleuritic-clayey size, with the content of (-0.071) mm fraction of about 25%. The chemical composition is dominated by SiO_2, CaO, Al_2O_3, H_2O and CO_2, the share of which is about 90 wt.%. The concentrations of S are 0.69%, C—2.35%, Sb—0.11%, As—0.19%, and Ag — $< .1\%$. The gold grade in individual samples as per the fire assay data varies from 0.2 to 0.97 g/t, with the average metal grade in the bulk sample received for study of 0.6 g/t. The main minerals of tails are quartz, calcite and stratified silicates making 98–99%. Sulfides, oxides, hydroxides and sulfates of iron, arsenic, antimony and tungsten are the impurities. In addition to the native highcarat gold, aurostibite ($AuSb_2$) and jonassonite ($AuBi_5S_4$) are present in single grains. Copper, antimony and mercury impurities are noted in some gold particles. Native gold particles are less than 0.045 mm in size (90%). Most of gold is noted in the form of micron inclusions in sulfides, quartz, carbonate and micas. Successive chemical leaching in the tail material using the method (Antropova et al. 1980) identified about 20% of mobile gold forms (water-soluble, sorbed, ferriforms) and sulfide and telluride forms $\sim 15\%$.

4 Results

In the process of the experiment the flotation tails sample with the weight of 400 g was mixed with lime ($Ca(OH)_2$) and brown coal. The blend was exposed to reductive melting at the temperature of 1500–1550 °C, in the process of which melt division into a metal and silicate parts occurred. The silicate part of the melt was poured into water with production of foamed amorphous material (foam silicate). The recovery of foam silicate was 150 g. The metal phase recovery was 26 g (4.4%). Phase composition of the metal aggregate (wt.%): ferrosilicum (FeSi) 82.8; xifengite (Fe_5S_3) 8.97; wustite ($Fe_{0.974}O$) 0.79; wollastonite ($CaSiO_3$) 4.24; calcic clinoferrosilite ($Fe_{1.5}Ca_{0.5}(SiO_3)_2$) and elemental iron (Fe) 0.49. Chemical composition of the produced foam silicate (wt. %) – SiO_2—43.7; TiO_2—0.7; Al_2O_3—7.79; Fe_2O_3—0.19; MgO—2.22; CaO—42.5; K_2O—1.5; SO_3—0.77; Cl—0.27, and metal phases: Si—23.5; Al—1.6; Fe—66.2; Mn—2.22; Mg 0.47; Ca—1.97; S—0.42; As—0.68; Cl—0.14; P—1.0; Co—0.35; V—0.28; Cu—0.21; Au—0.2; and Cr—0.18. The optical research of the metal alloy showed non-uniform aggregate composition. Six individualized metal phases of more complex composition have been identified in the iron and silicon alloy matrix. Gold is a part of the alloy consisting of (wt.%): Au—0.25–5.11; Sb—0.4–0.7; Sn—0.57–3.30; As—up to 9.47; Cu—4.51–32.07; Fe—59.8–33.8; Mn—7.2–1.33; Ga—0.24–9.38; and Si—14.53–12.96.

5   Conclusions

Therefore, silicate and metallic semi-finished products have been produced as the result of deep reductive melting of processing tails: (1) ferrosilicium, which is a gold collector; (2) foam silicate material as an additional product of the main production, can be used for production of ceramic materials for different purposes. The use of the method of pyrometallurgic processing of processing tails allows mitigating their adverse effect on the environment.

References

Algebraistova NK, Makshanin AV, Burdakova EA, Markova AS (2017) Processing of precious - metal raw materials in centrifugal devices. Non-ferrous Met 1:18–22

Amdur AM, Vatolin NA, Fyodorov SA, Matushkina AM (2015) Movement of disperse drops of gold in porous bodies and oxide melts during heating. Rep Acad Sci 465(3):307–309

Antropova LV, Shuraleva AZ, Farfel LF, Aizenberg FM, Priyemov GA (1980) Forms of gold occurrence in the rock. Explor Met Eng 136:5–21

Bogdanovich AV, Vasilyev AM, Shneyerson YaM, Pleshkov MA (2013) Gold extraction from stale processing tails of pyritic copper-zinc ores. Ore Process 5:38–44

Kirik SD, Sazonov AM, Silyanov SA, Bayukov OA (2017) Study of disordering in the structure of natural arsenopyrite with the X-ray diffraction analysis of polycrystals and nuclear gamma resonance. J Siberian Federal Univ Eng Technol 10(5):578–592

Meimanova ZhS, Nogayeva KA (2014) Study of flotation processability of stale tails from the Solton-Sary processing plant. Nauka i novye Tekhnologii 2:15–16

Novozhilov YI, Gavrilov AM, Yablokova SV, Arefyeva VI (2014) Unique commercial gold antimony deposit Olimpiada in upper proterozoic terrigenous deposits. Ores Met 3:51–64

Pavlov VF (2005) Physical bases for the technology of production of new materials with set properties on the basis of creation of the system for comprehensive use of man-made and barren resources. Siberian Branch of the Russian Academy of Sciences, Novosibirsk

Pavlov MV, Pavlov IV, Pavlov VF, Shabanova OV, Shabanov AV (2015) Features of processes of pyrometallurgic processing of polymetallic ores. Chem Benefit Sustain Dev 3:263–266

Tselyuk OI, Tselyuk DI (2013) Prospects of use of gold heap leaching for involvement into commercial development of stale tails of the Eastern Siberia gold processing plants. In: Proceedings of the Siberian Department of the Section of Earth Sciences of the Russian Academy of Natural Sciences, vol 1, no. 42 pp 103–110

Predictive Assessment of Quality of Mineral Aggregates Disintegration

S. Shevchenko, R. Brodskaya$^{(\boxtimes)}$, I. Bilskaya, Yu. Kobzeva,
and V. Lyahnitskaya

Karpinsky Russian Geological Research Institute, St. Petersburg, Russia
Rimma_Brodskaya@vsegei.ru

Abstract. The paper is supposed to discuss one of the most important technological properties of mineral aggregates: their ability to disintegrate, to be
destroyed for the subsequent redistribution and extraction of a useful component.
The strength of a mineral aggregate is determined by the strength of accretion of
composing individuals and depends on many factors, including the energy
(kinetics) of formation conditions and the subsequent transformation conditions.
The energy of the boundary of accretion of mineral grains in aggregate cannot be
directly measured since currently, there are no reliable and proved methods for
measuring the surface energy of solids. It is only possible to calculate quantity,
proportional to the surface energy of the fusion of mineral grains – this is the
calculation of the atomic density of the surface, parallel to the boundary of
accretion of a pair of mineral grains. The methodology and calculation of this
characteristic have been developed and published. The use and obtaining of
results are possible at the stage of preliminary mineralogical study by methods of
geometrical analysis of the structure of the mineral aggregate. The effectiveness of
the proposed method is determined by the knowledge of the quality and features
of destruction of the aggregates using different physical methods of disintegration: traditional mechanical, ultrasonic, electro-pulse, electro-hydraulic.

Keywords: Technological properties · Mineral aggregate ·
Aggregate disintegration · Repartition · Strength · Accretion boundaries ·
Energy of boundaries

1 Introduction

The mineral aggregate is considered as a structured system consisting of mineral
individuals and the accretion boundaries between them. The formation of the aggregate
results from cooperative thermodynamic processes of synthesis of substances, corresponding to the composition of minerals, as well as the crystallization of mineral
individuals and the formation of crystal-grain boundaries, their aggregation. The
aggregation or accretion of mineral individuals with each other is carried out along the
forming boundaries of accretion. The process of aggregation of grains in the mineral
aggregate may somewhat lag behind the synthesis and crystallization of individuals,
but their parallel development is also possible. The accretion boundaries do not always

© The Author(s) 2019
S. Glagolev (Ed.): ICAM 2019, SPEES, pp. 140–142, 2019.
https://doi.org/10.1007/978-3-030-22974-0_33

correspond to the boundaries of the crystallization of individuals, since they are formed in a different energy situation.

2 Methods and Approaches

The accretion boundaries of mineral individuals represent a certain transition zone between crystal lattices of accreted grains. The orientation of the accretion boundaries with respect to the symmetry of crystalline lattices of mineral grains depends on the local energy potentials during the formation and transformation of the mineral individuals and the aggregate. Local thermodynamic potentials of the forming mineral system directly affect the safety of an individual through the energy of its boundary, the choice of the orientation of the boundary relative to the symmetry of its crystal lattice and the orientation of the accretion boundary with the nearest stable grain. Thus, a compromise boundary is established for the accretion of each pair of mineral individuals, which is stable in a specific range of thermodynamic forces and flows.

During evaluation of sequence and quality of opening of the accretions, it is necessary to take into account the distribution of mineral individuals in the aggregate which possess a very perfect cleavage.

It is possible to assess the degree and quality of opening of the accretions at the stage of preliminary mineralogical analysis when studying the structure of the mineral aggregate in thin sections using geometric analysis methods.

3 Results and Discussion

The boundaries of mineral individuals, as well as the accretion boundaries, possess a certain stability reserve, ensuring their functioning in a certain range of thermodynamic and mechanical parameters. The stability margin of the boundaries is the wider, the higher the kinetics of the cooperative processes that form them, their genesis. The accretion boundaries of mineral grains are the more stable in the field of destructive forces, the more surface energy they accumulate.

The stability of the accretion boundary of mineral grains depends, in addition to the above said, on the degree of its balance by the value of energy saturation of each of the mineral grains in the accretion region.

The open boundaries of the mineral grains of accretions after relaxation of the crystal lattice, as well as remaining unopened accretions, have different elemental composition, amount of energy, and different floatability.

4 Conclusions

The accretion boundaries determine the grain system as a mineral aggregate and determine its stability, as both the strength of the accretions of mineral grains, and the strength of the mineral aggregate as a whole.

The strength of mineral aggregates is understood as resistance to destructive forces on the system of mineral grains forming the aggregate due to the presence of their "accretion" surfaces.

The accretion boundaries of the mineral grains can be balanced or unbalanced by value of energy of accretion surfaces of individuals.

The accretion of grains on surfaces, which atomic density or crystallographic indices correspond to the edges of habit forms, will collapse faster and earlier by the same destructive force, than the accretion of energy-intensive boundaries with a low atomic density and high crystallographic indices.

References

Brodskaya RL, Makagonov EP (1990) Determination of the spatial orientation of crystalline individuals in microstructural analysis and prospects for its use in stereometric petrography. Notes All-Union Mineral Soc 119(4):84–93

Brodskaya RL, Bil'skaya IV, Lyakhnitskaya VD, Markovsky BA, Sidorov EG (2007) Boundaries of accretions between mineral individuals in an aggregate. Geol Ore Deposits 49(8):669–680

Development of Methods for Anti-filtration Formations Destruction Inside a Heap Leach Pile

H. Tcharo[✉], M. Koulibaly, and F. K. N. Tchibozo

Department of Mineral Developing and Oil&Gas Engineering,
Engineering Academy, RUDN University, Moscow, Russia
honoretcharo@yahoo.com

Abstract. This article discusses the new technical solutions that increase the restoration of the quality of pregnant solutions flowing out from the heap leach pile.

Keywords: Anti-filtration formations · Heap pile · Pipe · Sprinklers

1 Introduction

The presence of low permeable soils is one of the most difficult problems of metal extraction during heap leaching (Ozhogina et al. 2017; Vorobyov et al. 2017).

The development of new models will reduce the degree of their negative impact on the intensity of heap leach process.

2 Methods and Approaches

A wide range of methods were used for the current research: the analysis of the earlier conducted researches, the mathematical and physical simulation, determination of dependences, the calculations and the control, the experimental studies and measurements in accordance with the conventional standard.

For the experience, we used a section of a rectangular shaped heap leach pile 5×5 m size. We have installed an air injection pipe near a homogeneous low-permeable layer. Assuming that the bottom and sides are sealed to air leakage, we used the SVAirFlow software to simulate the air/oxygen flow through a uniform anti-filtration layer. In addition, the temperature inside the mass is constant and equal to 20 °C. The air supply pressure through the pipeline is 121 kPa, and the upper part has an atmospheric pressure of 101 kPa.

3 Results and Discussion

The results of the model are shown in Fig. 1. We've noticed that the air pressure in the whole low-permeable layer varied between 109 and 120 kPa. This means that the pores are slightly wider open.

© The Author(s) 2019
S. Glagolev (Ed.): ICAM 2019, SPEES, pp. 143–145, 2019.
https://doi.org/10.1007/978-3-030-22974-0_34

The second technical solution, which also allows to increase the intensification of heap leach process is the displacement of spraying devices in the direction of weakly affected by technological solutions places.

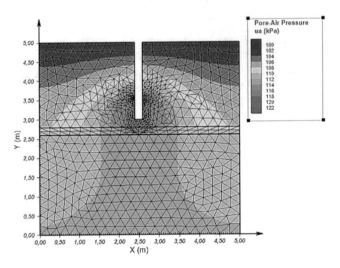

Fig. 1. Model showing the impact of injecting air/oxygen

4 Conclusions

According to our study results, we can conclude that supplying the air under much higher pressure than the atmospheric one in the places where anti-filtration layers are formed, along with the displacement of the sprinklers, will enhance more seepage of pregnant solutions through a heap leach pile.

References

Ozhogina EG, Shadrunova IV, Chekushina TV (2017) Mineralogical rationale for solving environmental problems of mining regions. Gornyi zhurnal 11:105–110

Vorobyov AE, Lyashenko VI, Tcharo H, Vorobyev KA (2017) Efficiency increase for gold-bearing ore deposits with respect to the influence of sulfide surface nanobarrier on metal adsorption. Sci Tech J "Metall Min Ind" 7:28–36

Microtomographic Study of Gabbro-Diabase Structural Transformations Under Compressive Loads

L. Vaisberg[1]([⊠]) and E. Kameneva[2]

[1] «Mekhanobr-Tekhnika» REC, St. Petersburg, Russia
gornyi@mtspb.com
[2] Petrozavodsk State University, Petrozavodsk, Russia

Abstract. Transformations of the porous space structure for gabbro-diabase under compressive loads is studied. Quantitative assessment of respective structural parameters for the pore space is ensured through the application of X-ray computer microtomography, enabling visualization of the internal three-dimensional structure of each sample and a detailed quantitative analysis of the pore space structure for both separate sections and the entire sample volume. Differences in the number, sizes, shapes, connectivity and spatial distribution of pores are established. It is shown that, when the sample is destroyed, the structure of the pore space in its fragments is transformed as follows: intracrystalline pores are partially closing, with the simultaneous emergence of new pores of large capillary sizes, concentrated in the cracks. In terms of their structure, these cracks represent a system of interconnected pores containing micron-size mineral particles.

Keywords: Computer microtomography · Gabbro-diabase · Pore space

1 Introduction

In the current theory for the mechanism of disintegration of rocks, destruction is a process that develops in time. The formation and development of microscopic failures begin upon application of a loading force, either dynamic or static, and resume for the entire period the rock remains under load until fracture (Zhurkov 1980; Krivtsov 2007).

The existing works on rock destruction generally identify the most probable structural elements, along which the destruction processes tend to develop. These are pores, tiny fractures, intergrowth boundaries of mineral phases and intergrain boundaries. In this regard, porosity, understood as the sum of all cavities enclosed in the rock, including pores, pore channels, and tiny fractures, becomes a useful feature, linking the strength of a rock with the defects in its structure. The physical or total porosity of a rock, determined by calculation using known values of mineral and bulk densities, enables only indirectly assessing the transformations occurring in the rock microstructure during destruction. Unbiased data may only be obtained if the structure of the pore

© The Author(s) 2019
S. Glagolev (Ed.): ICAM 2019, SPEES, pp. 146–151, 2019.
https://doi.org/10.1007/978-3-030-22974-0_35

space is taken into account, including its dimensions, pore shapes, connectivity and spatial orientation (Romm 1985).

The purpose of this work is to study the transformations occurring in the microstructure of a rock during destruction. X-ray computer microtomography was selected as the method for the quantitative assessment of respective pore space structural parameters of rocks at the microscopic level.

2 Methods and Approaches

X-ray tomography (X-ray micro-CT) is a non-destructive method for studying the internal structure of solid materials, based on the dependence of the linear coefficient of X-ray radiation attenuation on the chemical composition and density of the substance analyzed. Computer processing of shadow projections obtained by x-ray scanning of samples allows visualizing the internal three-dimensional structure of each sample and performing a detailed analysis of its morphometric and density characteristics both at separate sections and throughout the entire sample volume, obtaining quantitative values of respective parameters. The non-destructive nature of the method is an important advantage of X-ray microtomography, as it enables subsequent application of the same samples for other types of analysis, in particular, for establishing their strength characteristics.

A gabbro-diabase sample of cylindrical shape (L = d) without visible defects was prepared for the studies. The gabbro-diabase sample was characterized by a massive texture and a uniform medium-grained structure. The size of mineral grains was up to 2–3 mm.

The experiment included X-ray tomography of the gabbro-diabase sample with identification of the following parameters of the pore space: dimensions, shape, volume, specific surface, pore connectivity, spatial orientation, distribution, and pore density, both in separate tomographic Sections (2D system) and in the entire volume of the sample (3D system). The sample was then subjected to axial loading using a manual hydraulic press until fracture. The loading force was 192 MPa, the lateral pressure was 1 atm, at t = 20 °C, the loading time was 121 s. The resulting fragment was then subjected to x-ray tomography.

The tomographic studies were carried out using SkyScan-1172 (Belgium) with resolutions of 0.5 to 27 μm. In the experiments, the samples were carefully oriented on the table with respect to the optical axis of the instrument; the tube was supplied with 100 mA and 100 kV; the X-ray power was 90 W. The pixel size at the maximum magnification (nominal resolution) was 3.9 μm, which generally allows identifying pores of 4 μm or larger. The table with the sample was rotated by 360° in 0.25 increments. The subsequent reconstruction work was performed using Nrecon, CTan, and CTvol SkyScan software.

3   Results and Discussion

The original sample contains crystals of 0.2–2.0 mm, represented by plagioclase (41.5%), actinolite (47.0%), quartz (3.6%), sphene (5.7%) and biotite (2.2%).

The porosity of the initial sample is 0.7%. The pores in the sample volume are unevenly distributed. Plagioclase crystals have the highest porosity that is 4.1 to 4.9 times higher than the value for the entire sample (2.9–3.4% vs 0.7%). In quartz, actinolite and biotite crystals, only individual pores are observed.

The concentration of pores in the original sample is 84.55 mm^{-3}.

The largest pore size is 32–34 μm. In quantitative terms, pores of up to 10 μm prevail. The largest pore size is 32–34 μm with their median value of 5.6 μm. In quantitative terms, pores of up to 4–5 μm (80.7%) prevail in this group.

In the fragment after fracture, the total pore concentration increases to 163.9 mm^{-3}, super-capillary pores of 179 to 180 μm are observed, and the porosity increases to 1.8% (Table 1).

Table 1. Results of microtomography (3D-system)

Parameters	Values	
	Original sample	Fragment after sample fracture
Pore fraction in sample volume, %	0.7	1.8
Concentration of pores, mm^{-3}	84.55	163.9
Largest pore size, μm	32–34	178–180
Pore connectivity, %	2.81	28.88

Pore sphericity decreases from small pores to large pores. Pores of 4–5 μm are close to spherical in shape. Pores of 110–180 μm have elongated shapes, with their sphericity not exceeding 0.15 to 0.3.

It can be seen on the tomographic sections (Fig. 1) that the fragments formed upon sample fracture contain incomplete cracks (not leading to the formation of fracture surfaces) with the length of L = 7–8 mm and width of h = 40–150 μm (for main cracks) and of h = 1–10 μm (for feather cracks).

The study of the structure of the cracks shows that these are linear-plane sections consisting of interconnected cavities containing micron-size mineral particles (Fig. 2).

The availability of these cracks satisfactorily explains the increase in pore connectivity. In the original sample, the pore connectivity is 2.81%; in the fractured fragment, it is an order of magnitude higher (27.88%). The increase in connectivity indicates the association of small pores into larger pores.

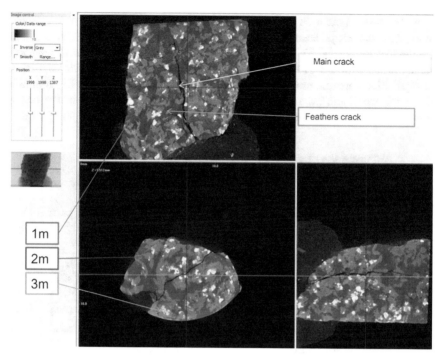

Fig. 1. Cracks in Gabbro-Diabase sample fragments (2D System) 1 m—*Plagioclase*; 2 m—*Actinolite*, 3—*Sphene*

Fig. 2. Structure of cracks

At the same time, a comparative tomographic study of the most porous mineral component, the plagioclase crystal, shows a decrease in pore concentration under compressive loads (Fig. 3). The plagioclase crystal porosity values are 3.23% for the original sample and 1.11% for the fractured fragment, which indicates that the intracrystalline pores are closing. Not only the finest pores, but also the larger ones are joined: the maximum pore size in the plagioclase crystal is 32 µm in the original sample and 16 µm in the fragment.

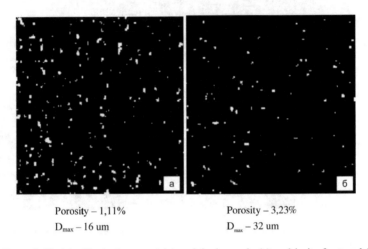

Porosity – 1,11%   Porosity – 3,23%
D_{max} – 16 um D_{max} – 32 um

Fig. 3. Pores (white) in Plagioclase crystal in original sample (a) and in its fractured fragment (b) in 2D system

The increase in the number of pores in the volume of the fragment formed after fracture, with the closing of intracrystalline pores, suggests that the newly formed pores are mainly concentrated in the cracks. In terms of their structures, these cracks represent a system of interconnected pores containing micron-size mineral particles.

4   Conclusions

When a gabbro-diabase sample is fractured under the action of a compressive load, the structure of the pore space is transformed in the fragments formed as follows: intracrystalline pores are partially closing, with the simultaneous emergence of new pores of large capillary sizes, concentrated in the cracks.

Acknowledgements. The study was supported by the grant of the Russian Science Foundation (project No. 17-79-30056).

References

Krivtsov AM (2007) Deformation and fracture of solids with microstructure, Fizmatlit, Moscow

Romm YeS (1985) Structural models of pore space of rocks, Nedra, Leningrad

Zhurkov SN (1980) On the question of the physical nature of the strength. J Solid State Phys 22 (11):13–15

Process Mineralogy as a Basis of Molybdoscheelite Ore Preparation

L. Vaisberg[1(✉)], O. Kononov[2], and I. Ustinov[1]

[1] REC Mekhanobr-Tekhnica, St. Petersburg, Russia
gornyi@mtspb.com
[2] Moscow State University, Moscow, Russia

Abstract. The Tyrnyauz ore field (Big Tyrnyauz) is one of the largest and most geologically complex deposits of tungsten and molybdenum. The main valuable mineral of the deposit is molybdoscheelite, a representative of the sheelite—powellite isomorphic series, as well as molybdenite and respective accessory minerals. The main problem in the processing of ores of the Tyrnyauz ore field consists in the high variability of mineral associations of its rocks, including the availability of ores and nonmetallic minerals with similar physical and chemical properties, as well as in the availability of ores of various natural types. Selective mining and processing of various geological and industrial types of ores with a wide use of vibrational technologies for the selective disintegration of raw materials is a promising approach to the development of the Tyrnyauz concentrator.

Keywords: Molybdoscheelite · Tungsten · Molybdenum · Disintegration · Separation · Flotation

1 Introduction

The Tyrnyauz ore field is located in the North Caucasus and is one of the largest and most geologically complex deposits of tungsten, molybdenum and associated metals. The ore bodies of the Tyrnyauz deposit occur on the platform of Eljurta granites and diabases. In the central part, these have undergone intensive metamorphism and converted into amphibole-biotite hornfelses, overlain by numerous contact-metasomatic and hydrothermal formations. The main valuable minerals of the deposit are molybdoscheelite $Ca(W,Mo)O_4$, a representative of the isomorphous series of scheelite $CaWO_4$—powellite $CaMoO_4$, and molybdenite.

The deposit had been intensively developed starting from the middle of the XX century. Flotation had always been the main concentration process used at the plant. When the global market for tungsten changed at the end of the XX century, the Tyrnyauz concentrator was shutdown. By that time, the Main Skarn with tungsten and molybdenum mineralization and a significant part of the amphibole-biotite hornfelses with molybdenum mineralization had been mined out. At present, an upgrade of the Tyrnyauz concentrator is required. After the enterprise is repeatedly put into operation, its ore base will be represented by high-calcite skarn ore bodies (up to 10–15%) and

S. Glagolev (Ed.): ICAM 2019, SPEES, pp. 152–156, 2019.
https://doi.org/10.1007/978-3-030-22974-0_36

skarned marbles with the calcite content of 40–70%. It is known that calcite and molybdoscheelite have similar flotation properties (Barskyi et al. 1979).

Due to the high calcite to molybdoscheelite ratios, their selective flotation with the use of traditional reagent regimes is impractical. Moreover, at lower mining horizons, scheelite is almost completely replaced by molybdoscheelite, its mineral variety with greater brittleness and lower hardness. These circumstances call for a new mineralogical and technological assessment of the Tyrnyauz deposit. Another independent promising type of ores are talcose hornfelses with high molybdenite contents and associated gold mineralization. However, the availability of a pair of naturally hydrophobic minerals (molybdenite and talc) also hinders their flotation concentration (Tarasevitch et al. 2014).

2   Methods and Approaches

Several hundred small and several dozens of large samples of promising ore types were selected and analyzed. Additionally, during the period of active operation of the deposit, the material composition and ore concentration indicators had been continuously analyzed for the respective production processes.

The grain-size distribution of molybdoscheelite (MSh) of different compositions and powellite is due to the regular decrease in hardness and increase in brittleness with the increase in the molybdenum content in the scheelite—powellite series. The higher hydrophobicity of powellite and high-molybdenum MSh (with powellite inclusions), which increases for particles of fine classes due greater specific surface values, predetermines their accumulation in the sulfide-molybdenum concentrate. In contrast, the higher affinity between scheelite and fatty acids, as compared to MSh and powellite, gives certain advantages in the process of rougher scheelite flotation and in respective final treatment.

The dependence of flotation properties of MSh in these operations on the composition, chemical and phase inhomogeneity of the particles enables the use of MSh composition and dispersion values to evaluate the efficiency of flotation processes.

Chemical sampling data on the contents of Mook and WO3, their ratio and related dispersion and the results of MSh local composition analysis serve as the criteria for assessing the chemical inhomogeneity of MSh. The color of luminescence, spectroscopic and kinetic parameters of excitation and radiation represent rapid test indicators for MSh. The color of luminescence is effective when assessing the content of pure scheelite with high hardness and flotability (blue luminescence) and of all varieties of MSh with powellite characterized by low hardness and poor flotability (yellow luminescence). When predicting the processing properties of MSh, dispersion assessment data is of greater significance. Therefore, the criteria used for identifying the processing varieties of ores with account of the composition of MSh should include variations in the composition of pure scheelite, molybdoscheelite and powellite. Similar distribution of molybdenite oxidation values in the ore biotite hornfelses (with almost zero MSh content) occurring in parallel to the Slepaya deposit may be used to assess the content and distribution of powellite.

Skarns of the Slepaya deposit do not form independent bodies. These either form part of the metasomatically altered skarned hornfelses, layered and massive marbles, or are represented by relics in the areas of development of later metasomatic mineral associations. In both cases, the skarns are represented by thin lens-shaped and clustered units (with the thickness of several centimeters to several tens of centimeters) having a fine to medium-grained structure with the characteristic light greenish-brown color.

Marbles in the skarn bodies are white, light gray and dark gray massive rocks with a spotted, brecciated or banded texture caused by the alternating interlayers of dark and light marbles of different granularity, alternating interlayers of light marbles and thin banded light green pyroxene-plagioclase hornfelses and new units in the form of wollastonite-fluorite, pyroxene-fluorite and quartz-pyroxene-calcite-garnet veins, veinlets and lenses. The marbles are composed of calcite. The sizes of polysynthetic calcite grains containing fine scattered inclusions of graphite and pyrrhotite range from 0.08 to 1.6 mm.

In the skarned marbles, the distribution of structural, textural and color varieties of marble is non-uniform. Within the contours of the ore body, fine, medium and non-uniformly crystalline and, respectively, dark and light sulfur varieties predominate (in quantitative terms). The lighter and white varieties ("clarified" marbles) form among them a system of veinlets, lens-shaped interlayers, clusters and blocks and are part of the parallel-banded varieties of marbles with calcite bands with a thickness of 1 to 5 cm. Their formation is associated with recrystallization along the fractures and increased permeability zones, as well as in tectonic deformation processes. In addition, lighter zones are always observed at the interfaces between marbles and skarns and around quartz-silicate veins and veinlets.

Luminescent properties of the ore minerals and host rocks of the Tyrnyauz deposit were studied since it is possible to use these luminescent properties as indicators for rapid diagnostics and as separation features in ore separation.

Preliminary concentration of the Tyrnyauz ore, which may be implemented using X-ray luminescent separation (RL), is required due to the processing of comparatively low-grade ores with large amounts of diluting materials and high mineralization contrasts. The contrasts in the luminescent properties of minerals and, therefore, of rocks and ores, may also be used for the rapid diagnostics of various natural and industrial types of ores in the process of their transportation and sorting, for optimizing their grinding conditions, as well as in their subsequent processing. The RL spectra were registered in the continuous and staged scanning modes. In the continuous scanning mode, with the constantly enabled source of excitation, the entire spectrum was recorded. This is the traditional and most informative form of data recording. However, under the influence of X-rays, the luminescence intensity of a substance usually increases. After a certain period of time that is specific for respective centers of luminescence, the saturation level is reached, when the luminescence intensity practically ceases to change or changes so slowly that it produces no effect on the spectrum image at any moment of the scanning time.

Therefore, reproducible and comparable results in the continuous scanning mode may only be obtained when the samples are irradiated to the saturation state, i.e. for relatively long periods (tens of minutes or more, depending on the mineral). In this connection, a method was developed for expressly obtaining the required information on the spectra by means of high-speed analysis in pulsed mode with staged scanning of the spectrum. Spectral and kinetic RL characteristics were studied for the following minerals: scheelite and molybdenite, calcite, wollastonite, quartz and plagioclase. The resulting RL spectra were interpreted by comparison of the data obtained with the reference data. For all minerals, respective types of luminescence centers were determined and the issues of variability in luminescent properties were considered. For the spectra obtained, the intensities, positions of the maximum and half-widths of the intrinsic luminescence band associated with the [WO4] and [MoO4] groups were established, as well as the wavelength and the intensity of the luminescence lines for impurity ions of rare-earth elements isomorphically replacing calcium ions in the scheelite structure.

3 Results and Discussion

Visual observation of the luminescence for the molybdoscheelite samples studied showed that many of them are characterized by a non-uniform color glow (blue and yellow), which indicates joint availability of several varieties of scheelite. In this case, the experimentally studied RL spectrum is composed of individual spectra of these varieties and represents an averaged characteristic for the scheelites of the sample. The scheelite and molybdoscheelite samples studied are characterized by varying degrees of chemical composition inhomogeneity, which is manifested in different colors of their fluorescent luminescence. The half-width of the band of intrinsic luminescence may be used as the inhomogeneity measure.

The observed diversity of the spectra may be explained by the presence of different types of MSh in different sample groups with varying content ratios and by the presence of their spectra. Based on the spectroscopic data obtained, at least four varieties with different luminescent properties may be distinguished for the scheelite samples studied. Signs of vertical zoning are observed in the distribution of MSh with different types of luminescence spectra. At the uppermost horizons, the luminescence intensity is usually higher.

The luminescence and afterglow spectra for all calcite samples studied consist of a single broad band with the luminescence maximum of 630 nm (orange-red). It was also found that luminescence intensity is variable and depends on a number of parameters, such as the duration of x-ray irradiation, the concentration of manganese, and the color of the samples. An important element of the promising processing technology for the Tyrnyauz ores is the use of selective vibration in disintegration and classification of raw materials.

4 Conclusions

These studies in the field of process mineralogy of molybdoscheelite ores of the Tyrnyauz deposit form the basis for the design of a new combined process flow using various selective vibration methods for disintegration and preliminary lump separation.

Acknowledgements. The study was supported by the grant of the Russian Science Foundation (project No. 17-79-30056).

References

Barskyi LA, Kononov OV, Ratmirova LI (1979) Selective flotation of calcium minerals. Nedra, Moscow
Tarasevich Y, Aksenenko EV (2014) The hydrophobicity of the basal surface of talc. Colloid J 76(4):526–532

Crystallomorphology of Cassiterite and Its Practical Importance

I. Vdovina[(✉)]

Nizhny Novgorod Institute of Education Development,
Nizhny Novgorod, Russia
viann@inbox.ru

Abstract. The methods of applied mineralogy are used to assess the scale of the mineralization of tin occurrences and deposits. One of such methods is the crystallomorphological method for estimating the level of tin ore bodies developed by N.Z. Evzikova. According to the crystallomorphological features of cassiterite it allows one to estimate the possible depth of distribution of mineralization and its magnitude. Using the example of three studied tin-ore districts of the Far East the possibilities of using crystallomorphology of cassiterite to predict and evaluate the prospect of mineralization of both the ore region as a whole and individual ore occurrence at the exploration stage are shown.

Keywords: Applied mineralogy · Crystallomorphology ·
Crystallomorphological method · Cassiterite · Scale of mineralization ·
An assessment of prospects

1 Introduction

The task of screening ore occurrences has always been and remains one of the important tasks of the geological survey. One of the achievements of applied mineralogy in the history of its development is participation in metallogenic researches and the development of the theory of mineral deposits prediction. Among the tasks solved by the methods of applied mineralogy (in particular the search for crystallomorphology) there are tasks aimed at determining the scale of mineralization (its vertical extent and the magnitude of the erosion-denudation slice). The methods of the search crystal morphology are based on two important postulates: the ontogenic development of the mineral form and the doctrine of typomorphism. One of such methods is crystallomorphological method for estimating of tin ore occurrences developed by N. Z. Evzikova on the example of tin deposits.

2 Methods and Approaches

Crystallomorphological method for evaluating of tin ore occurrences (Evzikova 1984) is based on the difference in the natural appearance of cassiterite crystals in different parts of the ore-bearing system as they move away from the source. The deeper parts of

© The Author(s) 2019
S. Glagolev (Ed.): ICAM 2019, SPEES, pp. 157–161, 2019.
https://doi.org/10.1007/978-3-030-22974-0_37

the ore bodies correspond to short-prismatic, isometric forms of cassiterite (I and II types), the higher—elongated forms of crystals (IV and V). The scale of the mineralization and the degree of its preservation reflect the crystallomorphological criteria: scoring represented by the statistical predominance of types IV and V and values from 50 to 200; degree of elongation of crystals (object's class) reflecting the completeness of the crystallomorphological evolution; the homogeneity of the morphology of cassiterite in different height sections of the ore object. The scoring factor is calculated by the formula B = [2 (V) + (IV)] − [2 (I) + (II)], where I, II, IV, V is the percentage of crystal morphological types in the sample.

The testing of this method at various times took place at tin ore deposits of the Far East where for the first time this method was used to evaluate the prospects for tin ore mineralization in exploration. The author has researched the crystallomorphological features of cassiterite from the ore manifestations of three Far Eastern tin ore regions (the Komsomolsk, Badzhal, Yam-Alin regions) in order to obtain a crystallomorphological estimate of the prospect of their mineralization.

3   Results and Discussion

Ore districts are located within the Khingan-Okhotsk volcano-plutonic belt and the same-name tin-bearing region. The Khingan-Okhotsk volcano-plutonic belt represents a vast magmatic area of the northeast strike superimposed on a heterogeneous foundation formed in the atmosphere of the transform continental margin. The Badzhal region is located in the junction zone of ancient Archean-Proterozoic cratons and Mesozoic accretionary folded belts and is characterized by the presence of a Cretaceous volcano-tectonic formation in the central part of the granitoid batholith.

Komsomolsk and Yam-Alin regions are located in accretionary folded systems spatially coincides with the eponymous volcanic zone (Rodionov 2005).

The Komsomolsk Ore Region. Several hundreds of ore mineralization zones, mainly cassiterite-silicate formation, have been established within the Komsomolsk ore region. By the beginning of 90s of last century crystallomorphological researches covered about 40 manifestations of tin mineralization in which a fairly representative number of samples were analyzed. 25 ore manifestations of this number were evaluated at deep horizons in the process of prospecting and evaluation (14) or exploration which made it possible to determine the reliability of forecasts obtained earlier by the crystallomorphological method.

When sorting out 25 ore occurrences by crystallomorphological method 11 mineralization zones received a positive forecast. Subsequent testing at the exploration stage confirmed this forecast for 6 ore occurrences. In five positively evaluated by the subsequent preliminary exploration the forecast was not confirmed.

Among the ore occurrences that received a negative assessment on the depth 10 objects were studied. The convergence of the results in these cases turned out to be higher. A negative forecast was confirmed by 8 occurrences. And only in two zones contrary to expectations industrial ores have been established.

The Badzhal Ore Region. The crystallomorphological researching of cassiterite from the ore occurrences of the Badzhal ore district began with the first explorations in the middle of 70s of the last century (Vdovina 2005).

In the process of research at the exploratory stage the cassiterite ore occurrences of the "Badzhal axis" structure were studied. Crystallomorphological criteria indicated a high level of slice and insignificant prospects for ore occurrences. No detailed work has been done.

The Pravourmiyskoe deposit was the most studied in details. The morphology of cassiterite crystals of this deposit was researched at all stages of prospecting and exploration (Vdovina 1987, 2008).

At the first stage of work the crystallomorphological mapping was carried out on surface mine workings. Evaluation of the mineralization's prospect was as follows:

- the level of the denudation slice is high enough;
- the gradient of the vertical crystallomorphological variability is large; the deposit is characterized by the proximity of the upper and lower boundaries of mineralization;
- the estimated length of the mineralization to the depth is 200–220 m;
- the morphology of the crystals suggests a large length of mineralization to the depth and the occurence of industrial ores in the middle part of the deposit.

Material was obtained from the deep horizons of the deposit in the detailed exploration. The results of the crystallomorphological analysis of cassiterite of the Pravourmiyskoe deposit were confirmed by the results of exploration. The length of the mineralization to a depth was no more than 250 m. The level of the denudation slice is really high. In addition the results of cassiterite of the Pravourmiyskoye deposit's crystallomorphological features researching confirmed the regularity of the spatial and temporal distribution of the morphological types of cassiterite. The same sequence of the cassiterite crystals changing from type I to type V was established at the stockwork type of the cassiterite-quartz formation. The deposit is currently being exploited.

About a dozen have been studied *within the Verkhnebadzhalsky ore cluster.* Most detailed and from the surface and from the depth Blizhnee deposit has been explored (Vdovina 2011).

The Blizhnee Deposit. Cassiterite is of a short-prismatic almost isometric habit, IV and less than V type, the score is 92–130. According to the crystallomorphological features the prospect of mineralization to a depth is small. Exploration data did not confirm the depth of the deposit.

The Talidzhak mine is the most perspective one. Cassiterite is predominantly IV and V types, of medium degree of elongation, the scaling varies within 100–200. The gradient of crystal morphological variability is weak. Ore occurrences are characterized as poorly eroded. The vertical extent of mineralization is assumed to be 500 m. The detailed exploration has not been carried out.

The Yam-Alin Ore Region. Ore occurrences were researched only at the exploration stage. There is the Sorukan deposit. Cassiterite is situated in the zones of quartz-chlorite composition in the form of scattered and nested impregnation and quartz-cassiterite lenses and veinlets. Pyramid-prismatic crystals of type IV of crystallomorphological type prevail in their composition, crystals of type V are found in a small amount. There are a lot of prismatic fragments of crystals of very different lengths including conical. All type IV crystals are characterized by a moderate and strong degree of elongation, a zonal structure; growth zones fix the development of cassiterite in types II–IV. The deposit's class development is 4. The deposit is poorly eroded and perspective on depth.

The ore occurrences of the Shiroky, Exan and Bastion sections from the different ones are of interest. Cassiterite is very small, almost colorless, slightly yellowish, often observed in star splices and aggregates. It is presented by IV and V crystallomorphological types with a predominance of type V. The crystals are characterized by short heads (dull pyramid), moderate and strong degree of elongation. The crystals are transparent, zonal; the observed growth phantoms show their development in type IV and III. Often there are simple articulated twins. The presence of cassiterite of different physical and crystallomorphological properties gives the right to assume the development of tin mineralization in several stages. In general the district is prospective, poorly eroded.

According to the results of the crystallomorphological method the most promising sections include the majority of the studied sections.

4   Conclusions

The author believes that the use of crystallomorphological analysis to determine the denudation slice, the extent of mineralization to the depth and generally for the industrial evaluation of tin mineralization is appropriate at the exploration stage. But the credibility and reliability of the assessment will increase if it is used in combination with others, in particular, at the first stage with structural and morphostructural (Vdovina 2008), and at later ones with more "direct" mineralogical, geochemical and other researches.

References

Evzikova NZ (1984) Prospecting cristal morphology, Moscow

Rodionov SM (2005) Metallogeny of tin of East of Russia, Moscow

Vdovina IA (1987) Crystallomorphology of cassiterite as one of the criteria for industrial evaluation of the tin ore deposit in notes all-union. Mineral Soc Part XVI, I:60–65

Vdovina IA (2005) Crystallomorphological features of cassiterite occurrences in the Badzhal ore region. Nat Geogr Res, Komsomolsk-na-Amure, pp 4–13

Vdovina IA (2008) Morphostructural-crystallomorphological evaluation of the prospect of mineralization (on the example of the Badzhal tin ore region) in Fedorov session 2008. Thesises of international scientific conferences, Saint-Petersburg, pp 246–249

Vdovina IA (2011) About the crystallomorphology of cassiterite. In: X international conference «New ideas in the Earth sciences». Reports in 3 volume. Russian State Geological Exploration University, Moscow, vol 1, pp 107–108

Modal Analysis of Rocks and Ores in Thin Sections

Yu. Voytekhovsky$^{(\boxtimes)}$

Saint-Petersburg Mining University, Saint-Petersburg, Russia
Voytekhovskiy_YuL@pers.spmi.ru

Abstract. The article is devoted to the history and justification of the modal analysis of rocks and ores with a microscope. It is shown that the Delesse-Rosiwal-Glagolev ratios do not follow from the Cavalieri principle. They do not allow one to find the exact volume of the minerals in rocks or ores, but give only their average estimates. It is also shown that the volume fractions of convex mineral grains in rocks and ores, taken equal to the fractions of their flat sections, are always underestimated if compared with the matrix. Due to the wide variety and complexity of forms of mineral grains, the methods of stereological reconstruction lead to integral equations with a difficult to define form factor. Most likely, tomography methods should come to replace the modal analysis of rocks and ores in thin sections.

Keywords: Rocks and ores · Modal analysis · Stereological reconstruction

1 Introduction

Modal (quantitative mineralogical) analysis of rocks and ores in thin sections is one of the first fundamental quantitative methods of mineralogy (including technological mineralogy), petrography (in classifications of rocks and ores, and petrological reconstructions), and lithology (i.e., petrography of sedimentary rocks). That is why its rigorous justification is of fundamental importance owing to the Delesse-Rosiwal-Glagolev ratios, as well as to stereological reconstruction. It makes sense to summarize the history of these methods in Russia and abroad, and to formulate the conclusion about their prospects.

2 The Cavalieri Principle and the Delesse-Rosiwal-Glagolev Ratios

First of all, we point out that the relations suggested by Delesse (1848) $dV_i = dS_i$, Rosiwal (1898) $dS_i = dL_i$, and Glagolev (1932) $dL_i = dN_i$, decreasing the dimension of space (namely, equating the volume fractions of minerals to areal, areal to linear, linear to point-like), have no relation to the Cavalieri principle: $S_{1i} = S_{2i} \rightarrow V_1 = V_2$ (if the areas of all arbitrarily close parallel sections of two bodies are pairwise equal, then their volumes are also equal, Fig. 1).

S. Glagolev (Ed.): ICAM 2019, SPEES, pp. 162–166, 2019.
https://doi.org/10.1007/978-3-030-22974-0_38

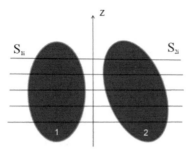

Fig. 1. To the justification of the Cavalieri principle

In recent notation, this principle, historically preceding integral calculus, has a clear meaning: $V_1 = \int S(z)\, dz = V_2$, where $S(z)$ is a continuous function of the areal fraction of a certain mineral along the z axis being normal to the sections. But, the modal analysis of rocks and ores in thin sections, which accumulates the statistics of areal, linear or point fractions of minerals from section to section, has nothing to do with the integration procedure. It only leads to an assessment of their average values. In this case, it can be argued that the volume of any mineral in a rock or ore is within the certain interval:

$$S(z)_{min}\Delta z = \int S(z)_{min} dz < V < \int S(z)_{max} dz = S(z)_{max}\Delta z$$

where Δz is the thickness of rock sample under study.

Despite of this contradiction, which was also considered in the works (Krumbein 1935; Chayes 1956), the method became firmly established in practice because of its apparent simplicity and was step by step automated (Shand 1916; Wentworth 1923; Hunt 1924; Dollar 1937; Hurlbut 1939) up to the use of modern computers for image analysis of thin sections. The list of parameters characterizing the cross-section of minerals, and the speed of processing have grown many times. But, in terms of the reconstruction of the true metric characteristics of mineral grains from those of their flat or even linear sections, the ideology remains the same.

Companies that produce computer structure analyzers offer software packages without discussing the fundamental problems. The analysis of 2D images does not use the available chapters of mathematics. For example, the distances between the mineral grains in thin section are replaced by the Euclidean distances between the points taken within the grains, whereas there is a more complicated, but easily programmable Hausdorff's metric which allows one to do this procedure correctly. In turn, it gives us the possibility of calculating space covariograms between the mineral grains of different species and their various clusters in the rock.

3 Stereological Reconstruction

A new line of research, i.e. stereological reconstruction, arose from the obvious observation that an arbitrary cross section of a spherical shape is always less than its characteristic cross section (Fig. 2, above). And it follows from this that the volume fraction of the convex mineral phase in the rock and ore, equal to the fraction of its flat sections, is always underestimated if compared with the host matrix. The corresponding general problem – finding the distribution of true particle sizes from the size distribution of their random sections – belongs to the inverse problems typical for geophysics. It is analytically solved only for spherical and ellipsoidal particles due to relatively simple description of these forms (Wicksel 1925, 1926). But, even in this case, the practical use of the theory requires the selection of the best solution and an estimate of the errors (Fig. 2, below). For more complex forms of mineral grains this can't be done without mathematical modeling on powerful computers.

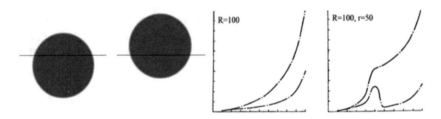

Fig. 2. Left: the size of the cross section of a convex grain is always less than the characteristic one, in the modal analysis its volume fraction is underestimated. Right: to recognize the true size of spherical particles by the size of circular sections; R = 100 – particles with a radius of 100 arbitrary units; R = 100, r = 50 – two sets of particles of the same type, indistinguishable in sections (for example, two generations of one mineral); horizontal scale – section radii from 0 to 100 bits per 10 classes; vertical scale – frequencies by classes (lower curves) and accumulated frequencies (upper curves).

The history of this area in Russia can be found in the following incomplete list of works: Zhuravsky A.M. Mineralogical analysis of thin section in terms of probabilities. Moscow-Leningrad: Gosgeolizdat, 1932. 20 p.; Glagolev A.A. Quantitative mineralogical analysis of rocks with the microscope. Leningrad: Gosgeolizdat, 1932. 25 p.; Glagolev A.A. On the geometric methods of quantitative mineralogical analysis of rocks. Moscow-Leningrad: Gosgeolizdat, 1933. 47 p.; Glagolev A.A. Geometric methods for quantitative analysis of aggregates with a microscope. Moscow-Leningrad: Gosgeolizdat, 1941. 263 p.; Chayes F. An elementary statistical appraisal. New York: John Wiley & Sons, Inc., 1956; Shvanov V.N., Markov A.B. Granulometric analysis of sandstones in thin sections//Geology and exploration. 1960. N 12. P. 49–55; Ivanov N. V. A new direction in testing ore deposits. Moscow: Gosgeolizdat, 1963. 179 p.; Chernyavsky K.S. Stereology in metallurgy. Moscow: Metallurgy, 1977. 375 p.; Ivanov O.P., Ermakov S.F., Kuznetsova V.N. Improving the accuracy of determining the weight particle size distribution of ore minerals from measurements in thin sections//Proc. CNII of Tin. Novosibirsk: Science, 1979. p. 10–14; Gulbin Yu.L. On stereological reconstructions of grain sizes in aggregates//Proc. Rus. Miner. Soc. 2004. N 4. P. 71–91.

4 Conclusions

Thus, due to the extraordinary diversity and complexity of the forms of mineral grains in rocks and ores, the methods of stereological reconstruction lead to integral equations with an analytically difficult-to-define form factor. The practical application of the theory is drowning in the selection of the best solution to the inverse problem and complex estimates of measurement errors. It seems that the modal analysis of rocks and ores in thin sections should be replaced by tomography methods. Standardizing modal analysis of rocks and ores in thin sections by creating their artificial counterparts with previously known volume fractions of mineral grains and a wide range of petrographic structures can serve as an inter-laboratory comparison of the accuracy of the method. But it does not solve the problems in essence.

References

Delesse M (1848) Procede mecanique pour determiner la composition des roches. In: annales des mines. De memoires sur l'exploitation des mines. 4me serie. T. XIII. Carilian-Goeury et Dalmont, Paris, pp 379–388

Dollar ATJ (1937) An integrating micrometer for the geometrical analysis of rocks. Mineral Mag 24:577–594

Hunt WF (1924) An improved Wentworth recording micrometer. Am Mineral 9:190–193

Hurlbut CS Jr (1939) An electric counter for thin-section analysis. Am J Sci 237:253–261

Krumbein WC (1935) Thin-section mechanical analysis of indurated sediments. J Geol 43:482–496

Rosiwal A (1898) Über geometrische Gesteinanalysen. Ein einfacher Weg zur ziffermässigen Feststellung des Quantitätsverhältnisses der Mineralbestandtheile gemengter Gesteine. In: Verh. der k.-k. Geol. Reichsanstalt. Verlag der k.-k. Geol. Reichsanstalt, Wien, pp 143–175

Shand SJ (1916) A recording micrometer for geometrical rock analysis. J Geol 24:394–404

Wentworth CK (1923) An improved recording micrometer for rock analysis. J Geol 31:228–232

Wicksel SD (1925) The corpuscle problem: a mathematical study of a biometric problem. Biometrica 17:84–99

Wicksel SD (1926) The corpuscle problem: case of ellipsoidal corpuscles. Biometrica 18:151–172

Quality Assurance Support (QA/QC System) of Mineralogical Analysis

O. Yakushina[✉], E. Gorbatova, E. Ozhogina, and A. Rogozhin

FSBE "All-Russian Institute on Mineral Raw Materials" (VIMS),
Moscow, Russia
vims@df.ru

Abstract. Mineralogical studies are an integral part of the exploration and development of solid mineral deposits, the effectiveness of which directly depends on the quality of the measurements. Moreover, five decades QA/QC system of mineralogical analysis (UKARM) for Russian geological survey is developed. Its specific features and tasks are discussed. The system provides to obtain complete, reliable, metrologically evaluated and legally valid information about the material composition and structure of rocks and ores. QA/QC of mineralogical analysis ensures the coordination of testing laboratories, starting with the resources of testing laboratories, stuffing at last. This system covers the entire process of mineralogical research, starting at the resources of testing laboratories as stuff and equipment, through the research procedure as the selection of testing object and its preparation for analysis, the accuracy of analysis rank, the testing method and technique, to metrology data and the quality assessment of the results obtained.

Keywords: Mineralogical analysis · Quality management · Assurance · Control system · Metrology · Reference materials

1 Introduction

Mineralogical analysis is an integral part of mineral deposits exploration and development; its effectiveness directly depends on the quality of testing (ISO/IEC 17025:2005; JORC Code 2012). The quality of analysis directly affects efficiency and reliability of the whole investigation. The main requirement for mineralogical research is to obtain reliable, complete, metrologically evaluated and legally valid information about the studied matter composition and structure, namely of rocks, ores and technologically processed products. The quality management system of mineralogical works ensures the coordination of testing laboratories, starting with the resources of testing laboratories, and ending with the processes occurring in them. Laboratory mineralogical studies with varying degrees of depth of mineral substance analysis are carried out at all stages of geological exploration from geological prospecting to operational exploration and development of mineral deposits. The main requirement for laboratory mineralogical studies is to provide all spheres of activity of the Ministry of Natural Resources and Ecology of the Russian Federation with reliable,

S. Glagolev (Ed.): ICAM 2019, SPEES, pp. 167–171, 2019.
https://doi.org/10.1007/978-3-030-22974-0_39

reliable, standardized, metrologically evaluated and legally valid information obtained as a result of the use of a wide range of laboratory methods and equipment (Ozhogina et al. 2017a, b).

2 Methods

The variety and complexity of objects of natural and man-made origin, the widespread use of quantitative methods of mineralogical analysis, the presence of a large number of mineralogical laboratories with different instrumentation and methodological base, personnel composition, determines the need to improve and develop the Quality Management System QA/QC for Mineralogical works UKARM, established at VIMS in the 1980th (Ginsburg et al. 1985) and renovated today. Developed by VIMS the UKARM System coordinates laboratory studies, preparation for testing operations and also monitors the quality of research. Scientific-methodical support of the UKARM functioning carries out the Scientific Council on methods of mineralogical methods of analysis (NSOMMI). The main components of UKARM mineralogical QA/QC system include as follows:

- make the General concept of development of mineralogical service and deliver priority directions of its improvement;
- preparation of proposals for the program and coordination plan of scientific research aimed at solving mineralogical laboratory research methods of testing;
- develop the industry system of standard samples, coordinate their production, registration and use;
- the NSOMMI Scientific Council review, update and elaborate new guidance documents on methods of all kinds of laboratory mineralogical works, the system and means of these documents verification and approval;
- ensuring the functioning of a unified quality management system of laboratory mineralogical work, organization and carrying the interlaboratory comparative tests (ICT), internal and external laboratory analysis control;
- methodological assistance support for the basic and regional laboratories and research centers in certification of mineralogical methods of analysis;
- implementation of complex and unique mineralogical and analytical studies;
- development of operational information system for mineralogical research.

UKARM QA/QC system is based on the requirements of the requirements of ISO/IEC 17025 standard in relation to mineralogical research, starting with the resources as stuff and equipment, ending with metrology and data quality assessment (Fig. 1). The testing laboratory stuff must have the necessary competence to perform at a high level of mineralogical works. UKARM QA/QC includes workshops, advanced training courses for laboratory stuff. For example, every year, since 2011, VIMS arrange annual Seminar "Mineralogical school – Current problems and modern methods". Seminar members discuss the basic state-of-art concepts of mineralogy, prospects of development, mineralogical support of geological exploration, methods of mineralogical analysis, environmental problems, specific features of man-made raw materials study, mineralogy for enrichment of ores, the nature of minerals technological properties and their behavior in geological and technological circulation, mineralogical works metrological and methodological support, etc. Specialists of various affiliation took part in the Seminar.

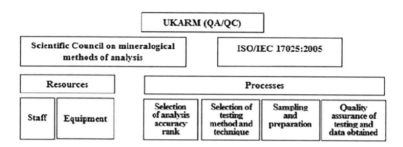

Fig. 1. UKARM QA/QC system structure

Testing laboratories are equipped with the necessary units: measuring instruments, software, standards, reference data, reagents and consumables. Also the methodical base is developed or adapted to technical base of analysis current state, corresponding metrological assessment and its legal approve. The certification and registration of standard samples of phase composition and properties, and artificial mixtures is also a necessary element of QA/QC. The choice of mineral substance testing method and mode is determined by substance peculiarities in composition and structure, the hardware capabilities, the availability of relevant methodological documents and the staff competence. In 2018 the Register on mineralogical studies of natural or techno-geneous mineral raw matter includes 3 industry standards, 45 instructions, 31 practical guidelines and 174 recommendations.

Today, the methods of quantitative phase analysis of rocks, ores and man-made substance are especially in demand. UKARM QA/QC includes the classification of quantitative phase analysis laboratory methods, depending on their reproducibility, are divided into six categories: I – particularly accurate quantitative analysis, II – full phase analysis with increased accuracy, III – ordinary quantitative analysis, IV – ordinary analysis with reduced accuracy requirements (express), V – semi-quantitative analysis, VI – qualitative phase analysis. Any testing may/should be characterized by a level of accuracy (Ozhogina et al. 2017a, b).

3   Results and Discussion

The quality of measurements is provided by measuring units' verification and calibration tests; the availability of measurement guidance and their strict observance; internal and external laboratory control. Organization of the control and dispatching service for mineralogical analysis and interlaboratory comparative tests, carrying out the control of phase composition and properties by standard samples, certified mixtures; participation in analysis to certify the standard samples of phase composition and properties, certification of laboratories with mandatory experimental evaluation.

We state the leading role and the one of the most effective forms of external control are *Interlaboratory Comparative testing* (ICT). The last one allow to evaluate the reliability of the results obtained in each individual laboratory, and to obtain information about the real accuracy of measurement techniques in general. In 2016–2018 VIMS conducted a series of ICT on X-ray powder diffraction (XPD) and petrography analysis. The samples for control were artificial mixture of minerals (2), and synthetic corundum powder (1), igneous (1) and sedimentary (1) rocks thing-sections. Each ICT involved from 8 to 12 participants, totally 46 laboratories. The obtained results are considered satisfactory, they can be used by laboratories during accreditation procedures and confirmation of technical competence. The ICT comparative testing provide an opportunity to assess the quality of measurements in different laboratories, to carry out corrective actions to ensure the uniformity of measurements and to show the technical competence of the laboratory.

4   Conclusions

Mineralogical study of any substance should be based on a QA/QC System. Testing laboratory, which performs mineralogical studies supporting geological exploration and technological works in order to meet the requirements of mineralogical work quality management should have the relevant stuff, modern test methods, proper equipment, capabilities and means of verification and calibration, be supplied by industry techniques procedures, guidelines, as well as strictly follow them, and comply with the requirements of a unified quality control system.

References

Ginsburg AI, Vikulova LP, Sidorenko GA (1985) On some typical mistakes at mineralogical investigation (studies) J. In: Zapiski RMO (Proceedings of RMS), vol 3, pp 324–333

JORC Code: 2012 Edition. http://jorc.org/

ISO/IEC 17025:2005: General requirements for the competence of testing and calibration laboratories

Ozhogina EG, Gorbatova EA (2017a) System of mineralogical quality management. J Actual Probl Min 1:3–7

Ozhogina EG, Lebedeva MI, Gorbatova EA (2017b) Interlaboratory comparison tests in the mineralogical works. J. Stand. Samples 2:37–47

Mineral Preparation in Geological Research

T. Yusupov$^{(\boxtimes)}$, A. Travin, S. Novikova, and D. Yudin

V.S. Sobolev Institute of Geology and Mineralogy SB RAS, Novosibirsk, Russia
yusupov@igm.nsc.ru

Abstract. Paper deals with discussion of minerals' preparation requiring high purity monoproducts, this has special importance for minerals –geochronometers. Widely used methods including gravitation, magnetic separation, floatation provide fractions with 90% of targeted mineral. Further monominerality increase requires special separation methods; one of them – "Strat" is perspective. It is based on separation in organic liquids under gradual density change. Combination of bromoform with d – 2.89 g/sm^3 and dimethyle formamide with d – 0.8 g/sm^3 is used; density gradations till 0.001 g/sm^3 are possible therefore isomorphic inclusions could be separated. Another direction is presented by trybotreatments under higher energies in planetary mills with a centrifugal factor to 40–50 g. Exotic surface substances presented mainly by kaolinite, muscovite, calcite, gothite are removed as trybotreatment result. Special planetary mills - classifiers are used for processing of big samples. This method together with minerals opening in disintegrator under destruction by the free pulse is recommended for wide application.

Keywords: Monominerality · Geo chronometers · Separation · Surface · Organic liquids · Trybotreatment

1 Introduction

Many mineralogical and geochemical researches are based on mono mineral products' studies; acquisition of these products is based on research intensive processes of minerals' revealing and extracting. This assumption is relevant to geochemical, geophysical, lithologic, petrochemical and other studies; geochronological definitions became wider during recent years.

The task to extract minerals – chronometers of mono mineral purity is very important and complicated. The problem becomes much more complicated when chronometers are extracted from geo objects which have precious and rare metal character. This happens due to their very low content, thin dispersion and occurrence in genetic association (intergrowth) with usually rock-forming minerals.

Range of rock-forming and ore minerals which are used for rock dating is widening. Analysis is concentrated on such minerals as plagioclases, olivine, ortho and clino pyroxene, phlogopite, tourmaline, sphalerite, volframite, tin spar, pyrite, pyrrhotite, pentlandite and others. Special methods should be developed for the extraction of many of these minerals (Isotope…, 2015; Methods…, 2018).

© The Author(s) 2019
S. Glagolev (Ed.): ICAM 2019, SPEES, pp. 172–175, 2019.
https://doi.org/10.1007/978-3-030-22974-0_40

2 Methods, Approaches, Results and Discussion

We develop new section in sample preparation – mineral preparation, which includes number of new research and methodological aspects.

1. Preliminary concentrating of minerals – chronometers from objects with very low their content. Only intermediate products which are used for mono mineral fractions extraction could be obtained by traditional types of separation – gravity, magnet and floating.

   Gravitation methods are helpfully used in the situations where differences between density of extracted mineral and monaural basis are not lower than 3 kg per sm^3. Losses of target mineral are significant under other combinations of densities. Magnet methods are more effective when differences in magnet sensitivity of separating components are sufficient for separation. This is true for the case of quartz from -0.40 to 0.10 and biotite from $+46.7$ to 86.7 109 m^3/g. Under lower differences in magnet sensitivity of minerals which are contained in samples at the level of accessory units extraction is extremely difficult.

   Great perspectives in the extraction of minerals from extremely poor subsurface rocks are related with floatation process which makes it possible to get minerals with less than 0.1% content in sample. Obtained products of preliminary concentration should either be further grinded in order to open intergrowths or be dressed with the help of special concentrating methods till mono mineral state. Combine schemes with mono mineral and similar products extractions at the initial stage are often used; then intergrowths minerals opening and repeated concentrating of targeted minerals take place (Berger 1962).

2. Opening of minerals from intergrowth stage done with the help of mechanical treatments may be accompanied by significant structural chemical changes. It is necessary to avoid high temperatures, local high pressures and if possible to use dry process. High energetic free pulse realized in desintegrators is effective method of minerals – chronometers' opening. Prospectivity of desintegrator's use for ore preparation is proved on the cases of different minerals: spodumene, apatite, sulfides and others. Higher preservation of crystal structure and lower over grinding are considered to be main advantages here (Yusupov et al. 2015). Positive aspects of mineral preparation were revealed under disintegrating of quartz – feldspar associations (Yusupov et al. 2018). Disintegrated sample preparation is recommended for wide use.

3. Obtained concentrates were dressed by methods of mono mineral fractions extraction with extraction of products with 90% of targeted mineral (Methods… 1985). Further increase of mono mineral character is reached by the help of special methods. Gradual separation in organic liquids on density – "Starts" method and trybo treatment – surface attrition under higher energies of mechanical treatment are wide used. Mixture of bromform with d – 2.89 g/sm^3 and dimethyle formamide with d – 0.8 g/sm^3 is used as separation media. Different density gradations till 0.001 g/sm^3 are possible here. Potential of the method is shown on the example of quartz – feldspar associations with – 0.3 +0.2 mm size (Table 1).

Table 1. Reparability of minerals of non electromagnetic quartz feldspar product

Fraction density g/sm^3	Output, %	Elements content, %					
		SiO$_2$	Fe$_2$O$_3$	Al$_2$O$_3$	K$_2$O	Na$_2$O	Li$_2$O
Initial		81.1	0.03	12.3	2.48	5.18	0.028
2.44–2.55	7.881	65.3	0.02	19.3	13.3	1.28	0.097
2.55–2.58	3.751						
2.58–2.61	5.109						
2.61–2.63	43.795	71.0	0.03	18.4	0.51	9.82	0.004
2.63–2.65	35.737	98.7	0.03	1.4	0.12	0.62	0.017

Potassic feldspar product with 7.88% output was extracted under 2.44–2.55 g/sm^3 density. Fraction of sodium feldspar is concentrated under higher density d – 2.61–2.63 g/sm^3 with output of 43.79%. Density of quartz fraction is more close to similar indicator for sodium spar with density interval being 2.63–2.65 g/sm^3 and output 35.73%.

These results confirm high effectiveness of density method; separation of 0.05 mm and lower size products looks possible (Yusupov et al. 2015). Method is successfully used for not only quartz and feldspar separation but also for muscovite, biotite, glauconite (Katz 1977).

Trybo attrition impacts are under investigated though they are important for mineral's homogeneity increase. They enable to remove inclusions of tramp substances presented mainly by kaolinite, muscovite, calcite, gothite as well as by remaining floating reagents.

It is important to take into account that after removal of surface layers of 0.1–10 mcm thickness surface is characterized by different structural imperfections. They could vary from practically unchanged state to totally crystal and chemically destroyed surface (Yusupov et al. 2018).

Trybo treatment looks as important way to increase mono mineral character. Hand attrition in jet and jasper stamps is widely used in institute's analytical practice. Method is applied for treatment of biotite, glauconite, amphibolites, tourmolin spinels, phosphates, sulphides and other minerals with monomimeral coefficients being about 100%.

Facilities of PMK type are effective for large samples with weight more than 1 kg trybo treatment. Material here is exposed by planetary rotating movement of ore mass. Such technological regime selectively destroys impurity substances and increases monomineral properties of extracting products. Improved version of this mechanism is being developed at CJSC "Itomak", Novosibirsk.

Trybotreatment not only removes impurity substances but also changes surface defectiveness character and mineral heterogeneity type. Heterogeneity management is under investigated methodologically however its role in ores processing is constantly growing. For example concentrate with 0.05% ferrous oxide was got from quarts with 1.5% of this component. This result could not be reached by other methods. It is important to provide transfer to homogeneous state when minerals have similar character of structural defects. Type and level of trybo treatment make it possible to solve these problems to certain extent.

3 Conclusions

Methodical bases of samples preparation in processes of mineral products extraction are reviewed. Taking into account orientation of these methods it is suggested to name this approach mineral preparation.

Possibilities of minerals separation by "Strat" method which is based on use of organic liquids with different density and surface trybo treatment are discussed. Mono mineral products of high quality are obtained as a result of methods application.

These methods as well as disintegrated minerals opening are recommended for wide application in analytical practice.

References

Berger GS (1962) Minerals' floatating. Gortechizdat, Moscow

Ginzburg AI (1985) Mineral research methods Nedra, Moscow

Isotope dating of geological processes: new results, approaches and perspectives (2015) VI Russian conference in isotope geochronology. IPGG RAS, Saint Petersburg

Katz MY (1977) Minerals' heterogeneity analysis. Nauka, Moscow

Methods and geo chronological results of isotope geometric minerals systems and ores studies (2018) VII Russian conference in isotope geochronology, Moscow

Yusupov TS, Baksheeva II, Rostovtsev VI (2015) Analysis of different-type mechanical effects on selectivity of mineral dissociation. J MiningSci 51:1248–1253

Yusupov TS, Travin AV, Yudin DS, Novikova SA, Shumskaya LG, Kirillova EA (2018) Improvement of methods of minerals – geo chronometers extraction from ores for geological processes isotope dating. In: Shchiptsov VV (ed.) Fundamental and applied aspects of technologica lmineralogy, Petrozavodsk. Russian conference in isotope geochronology, pp 86–92

Industrial Minerals, Precious Stones, Ores and Mining

Impact Diamonds: Types, Properties and Uses

V. Afanasiev[1]([✉]), N. Pokhilenko[1], A. Eliseev[1], S. Gromilov[2],
S. Ugapieva[3], and V. Senyut[4]

[1] VS Sobolev Institute of Geology and Mineralogy, Siberian Branch,
Russian Academy of Sciences, Novosibirsk, Russia
avp-diamond@mail.ru
[2] Nikolaev Institute of Inorganic Chemistry, Siberian Branch,
Russian Academy of Sciences, Novosibirsk, Russia
[3] Diamond and Precious Metal Geology Institute, Siberian Branch,
Russian Academy of Sciences, Yakutsk, (Sakha) Yakutia, Russia
[4] Joint Institute of Mechanical Engineering of the NAS of Belarus,
Minsk, Belarus

Abstract. Popigai is the world largest crater produced by an impact event. Abundant graphite in the target rocks underwent martensitic transformation into a mixture of high-pressure phases (an aggregate of nanometer cubic diamond and hexagonal lonsdaleite crystals), and some amount of graphite survived as a residual phase. They are of two types: (i) diamonds extracted from tagamites as chips of grains crushed during processing; (ii) yakutites in placers inside and around the crater, which formed at the impact epicenter and dispersed during the event. The impact diamonds possess exceptional abrasive strength, 1.8 to 2.4 times greater than in synthetic diamonds. The outstanding wear resistance, a large specific surface area and a thermal stability (200–250 °C greater than in synthetic diamonds) are favorable for main technological uses. With these properties, impact diamonds are valuable as material for composites and tools.

Keywords: Impact crater · Impact diamond · Yakutite · Abrasion strength

1 Introduction

The Popigai impact crater located in Yakutia (Russia) stores unlimited amounts of impact diamond with exceptional technological properties.

Studies of the Popigai impact crater have a dramatic history. The research was active from 1971, when V.L. Masaitis first proved the impact origin of the crater, to 1986, when the work stopped unexpectedly. Since neither geological surveys, nor engineering testing were undertaken; the testing was impossible because few diamond samples were available. In 2010 we got a collection of impact diamonds (about 3000 carat) from previous work at Popigai and resumed the tests. In this paper we characterized the diamonds as valuable raw for advanced technologies.

S. Glagolev (Ed.): ICAM 2019, SPEES, pp. 179–182, 2019.
https://doi.org/10.1007/978-3-030-22974-0_41

2   Popigai Impact Diamonds

A huge crater, about 100 km in diameter, formed 35.7 Ma when a large bolid hit the Earth at the Anabar shield in Siberia (Masaitis et al. 1998). The target Archean gneisses of the Khapchan Group containing abundant crystalline graphite were broken, remolten and partly dispersed outside the crater. Thereby the graphite underwent martensitic transformation into a mixture of high-pressure cubic diamond and hexagonal lonsdaleite; some amount of graphite survived as a residual phase either in aggregates with high-pressure phases or as separate particles. Upon conversion, the graphite reduced in volume by 1.6 and the resulting diamond formed as aggregate of $n * 10 - n * 100$ nm crystals (Walter et al. 1992). Impact diamonds possess exceptional abrasive strength, greater than synthetic diamonds by 1.8 to 2.4 depending on relative percentages of phases. The ultrahard impact diamonds can be successfully used for composites and tools.

3   Types of Impact Diamonds

The Popigai diamonds comprise yakutites in placers and diamonds hosted by tagamites, the primary impact rocks.

Yakutites formed at the impact epicenter and dispersed during the event. Currently they are found in placers inside and around the crater. The farthest dispersed findings of yakutites occur 550 km away from the crater or even more. The yakutite aggregates of nanometer grains are 0.7–0.8 mm to 1.0 mm in size and have a shapeless morphology or sometimes preserve hexagonal contours of primary graphite. They formed at the highest pressure and consist of cubic diamond and hexagonal lonsdaleite. The presence of graphite appears in Raman spectra but is undetectable by X-ray diffractometry, possibly, because of minor contents.

Tagamite-hosted diamonds are extracted by crushing from very hard host rocks and thus have angular shapes and particle sizes from a few microns to 1 mm. During subsequent flotation, large amounts of fine graphite are extracted along with diamond. These diamonds are likewise aggregates of nanometer diamond, lonsdaleite and graphite grains. Diamond chips look laminated like the precursor graphite. Impact diamonds from tagamite are of two species. A-diamonds are colorless or yellowish, almost fully consisting of the cubic phase; they are the most resistant to wear. Those of B species comprise cubic diamond, lonsdaleite and graphite and are slightly less resistant. Most of impact diamonds, including those we analyzed, are from the Skalnoye deposit in the southwestern flank of the Popigai crater, away from the epicenter. They formed by shock waves while the target rocks were melting and remained enclosed in tagamite, i.e. they originated at lower pressures than yakutites.

Thus, placer yakutites and tagamite-hosted diamonds share the same impact origin but formed at different pressures and evolved in different ways afterwards. The highest-pressure yakutites were dispersed and quenched, and preserved their primary structure, whereas diamonds in tagamites underwent prolonged annealing in tagamite melt. The annealing explains the presence of nitrogen impurity (N3V) which lacks in yakutites.

We explored the Udarnoye and Skalnoye sites within the Popigai crater. The Skalnoye deposit stores up to 100 carat diamond per ton of tagamite, while the total estimated resources exceed 162 billion carat (Masaitis et al. 1998). The resources in the whole crater are actually inexhaustible. The amount of yakutites in placers around the crater may reach 1.5 billion carat (Masaitis et al. 1998).

4   Possible Uses of Impact Diamonds

Possible uses of impact diamonds are based on their exceptional wear resistance. Only tagamite-hosted diamonds were studied. Micropowders were made from a collection of impact diamonds (mixed A and B species) at the Institute of Superhard Materials (Kyiv, Ukraine). The impact diamond micropowder of any grain size shows greater abrasive strength than synthetic diamonds (Table 1).

Table 1. Comparative abrasive strength of impact synthetic diamonds

Impact diamond micropowders		Synthetic diamond micropowders	
Grain size, µm	Abrasive strength, relative units	Grain size, µm	Abrasive strength, relative units
+60	5,05	+60	3,67
60/28	6,53	60/28	3,69
40/20	5,89	40/20	3,54
28/14	5,70	28/14	3,33
14/7	4,85	14/7	2,91
10/5	3,80	10/5	2,16
7/3	2,98	7/3	1,71
5/2	2,20	5/2	1,21

The extremely hard impact diamonds can be superior substitutes for synthetic diamonds in industry. Furthermore, a large specific surface area and relatively high adsorption and thermal stability 200–250 °C greater than in synthetic diamonds, make impact diamonds excellent raw for composites. Sintered composites of "impact diamond - Fe-Ti bond" powdered in a planetary mill to 5 to 50 µm granules show better polishing properties compared with similar powders based on synthetic diamond DSM 20/14 (Table 2).

Table 2. Comparative polishing properties of impact diamonds – Fe-Ti bond and similar powder of synthetic diamonds

Composite powders		Polished material	Weight loss (mg/min)	Stability (min)
Fe-Ti/impact diamond	5/50	Silica	35,4	>30
Fe-Ti/ACM	5/50	Silica	17,8	14

The impact diamond-based abrasive powders used for magnetic-abrasive polishing of silica plates have 1.5–2 times greater strength and 2 times greater stability (lifetime) than their counterparts made from synthetic diamonds (Table 2).

The technological properties of yakutites remain poorly investigated. They obviously have a very high wear resistance, possibly, higher than in tagamitic diamonds, while their grain sizes are suitable for using them in blade tools.

Meanwhile, apart from abrasive properties, there is a large scope of potential uses associated with structure, phase composition, etc. Currently, the only problem is the shortage of diamond specimens for testing. Sufficient amounts of such specimens can be obtained by building a pilot plant for extraction of diamonds.

5 Conclusions

The Popigai impact diamonds possess exceptional abrasion strength, large specific surface area and high thermal stability superior over the respective properties in synthetic diamonds. With these technological properties and actually unlimited resources, the impact diamonds are valuable industrial raw. The development of the Popigai diamond deposit can be successful due to progress.

Acknowledgements. The work was supported by grants 16-05-00873a and 17-17-01154a of the Russian Foundation for Basic Researchers and was carried out as a part of the Project No. 0330-2016-0006.

References

Masaitis VL, Mashchak MS, Raikhlin AI, Selivanovskaya TV, Shafranovskiy GI (eds) (1998) Diamond-bearing impactites from the popigai impact crater. VSEGEI, St. Petersburg

Walter AA, Eryomenko GK, Kvasnitsa VN, Polkanov Y (1992) Carbon minerals produced by impact metamorphism. Naukova Dumka, Kiev

Authentic Semi-precious and Precious Gemstones of Turkey: Special Emphasis on the Ones Preferred for Prayer Beads

E. Çiftçi[1][(⊠)], H. Selim[2], and H. Sendir[3]

[1] Department of Geological Engineering, Faculty of Mines, ITU, 33469 Maslak, Istanbul, Turkey
eciftci@itu.edu.tr
[2] Faculty of Engineering, Department of Jewellery Engineering, ITICU, 34840 Küçükyalı, Istanbul, Turkey
[3] Faculty of Engineering and Architecture, Department of Geological Engineering, EOGU, 26480 Odunpazarı, Eskişehir, Turkey

Abstract. There are 6 semi-precious gemstones in Turkey that are the most significant in terms of abundance and authenticity. These include smoky quartz, blue chalcedony, chrysoprase (aka Şenkaya emerald), diaspore (aka sultanite/zultanite), sepiolite (aka meerschaum/Eskişehirstone), and jet (aka Oltustone). The smoky quartz occurs in the south of Büyük Menderes Basin within metamorphic rocks of the massif. Chalcedony occurrences are in Çanakkale and Sarıcakaya (Eskişehir). Chrysoprase is acquired in Bursa, Ala-şehir (Manisa) as yellowish green in color, Biga (Çanakkale), Sivrihisar (Eski-şehir) and Şenkaya (Erzurum) as dark green in color. Diaspore occurs in Milas (Muğla), Söke (Aydın), Tire (İzmir), Bolkardağı Gerdekkilise area and Saimbeyli (Adana). Sepiolite occurrences are limited to Kıbrısçık-Köroğlu (Bolu), Eskişehir and Konya. Jet (aka Oltustone) occurs in Oltu area (Erzurum). These are the major varieties that has been used to produce both jewelry and ornamental objects among which the prayer beads are one of the most indispensable object of oriental cultures in particular. These are also considered to be the authentic gems of Turkey and exported as raw and processed in into many forms. As for the prayer beads, lightness, color, durability, hardness, and cost are the main criteria. Thus, the Oltustone has been the major source for decades. However, the Eski-şehir stone and blue chalcedony are becoming popular as well.

Keywords: Turkish gems · Oltustone · Eskişehirstone · Diaspore · Smoky quartz · Blue chalcedony

1 Introduction

Today, the jewelry sector is living its golden age. With increasing technical capacity and manual aptitude, gold, silver, platinum etc. metals can be embroidered with the precious and semi-precious gems and delivered to consumers. This sector is a fashion-driven dynamic sector and subjected to rapid changes and innovations that are con-trolled and dictated by the fashion trends and demands of the consumers. Companies

© The Author(s) 2019
S. Glagolev (Ed.): ICAM 2019, SPEES, pp. 183–188, 2019.
https://doi.org/10.1007/978-3-030-22974-0_42

are forced to keep up with those and both develop and manufacture the products accordingly. Use of the semi-precious and precious gems is likewise chosen according to the demand by the fashion and modern trends. Consumers religiously follow those two facts. Some of those are considered to be semi-precious gems of Turkey including smoky quartz, chalcedony, chrysoprase, diaspore, sepiolite, and jet (Selim 2015).

2   Methods and Approaches

Representative samples that acquired from the deposits were prepared for further analytical studies for comprehensive characterization. Polished sections and powdered samples were prepared for Transmitted and Reflected Light Microscopy integrated with Optical Cathodoluminescence, Electron Probe Microanalysis (EPMA) and Secondary Electron Microscopy-Energy Dispersive Spectroscopy (SEM-EDS), X-ray Diffraction (XRD), X-ray Fluorescence (XRF) and Inductively-Coupled Plasma Mass Spectroscopy analyses.

3   Results and Discussion

3.1   Smoky Quartz

Smoky quartz is one of the varieties of quartz with dark gray-black color (Fig. 1). It occurs rarely in the Alps, Colorado and California (North America). Its occurrence in Turkey is limited to the Menderes Massif, one of the metamorphic complexes of Turkey. It occurs in metamorphic rocks composed of mafic minerals that expose in southern portion of the massif in Great Menderes basin (İlhan 2012). Location is the city of Aydın and covers a few rustic villages within Çine, Ovacık and Koçarlı county limits. Host-rocks are variable degree metamorphic rocks including slate, phyllite, schist, and gneiss. Smoky quartz occurs locally within the fractures of these rocks. Its major use is as jewelry and ornamental objects. Type locality is Mersinbelen village in the area.

Fig. 1. Macroscopic views of natural smoky quartz and blue chalcedony (Selim 2015)

3.2 Chalcedony

Chalcedony occurs in a few locations including Bolu (Kıbrısçık), Ankara (Beypazarı), Afyon (Bayat), İzmir (Aliağa, Bergama, Seferihisar-Yukarıdoğanbey), Tokat (Zile) in lilac-purple colors, Yozgat (Çekerek) with amethyst as botryoidal masses, Çanakkale (center) and Eskişehir (Sarıcakaya) in blue color. Druzy chalcedony occurs in Sivas (Yıldızeli) (İlhan 2012). Chalcedony is a silica variety mineral. Its color may vary from deep blue to faint blue, purple in places, translucent and opaque (Fig. 1). Among these, blue chalcedony occurrence in Turkey is located in Mayıslar village (Sarıcakaya-Eskişehir) owned and exploited by Kalsedon Co. (Hatipoğlu et al. 2010a). Blueish color is either due to the water content or abnormally high barium content. This is also considered to be the type locality. Its major use is as jewelry and ornamental objects.

3.3 Chrysoprase

Chrysoprase or chrysoprasus is a gemstone variety of chalcedony that contains small quantities of nickel. The color varies from faint light green to apple green (Fig. 2). It is translucent and opaque. It worldwide occurrences are reported from USA (California, Oregon, and Arizona) and Eastern India. In Turkey, it occurs in Alaşehir (Manisa) in yellowish green, Dikmendağı area in Biga (Çanakkale) deep green, Sivrihisar (Eskişehir) in deep green, İkizce village (Bursa), Zümrüt and Turnalı villages of Şenkaya county (Erzurum). In Erzurum, due to it locality and most probably to the color, it is erroneously named as *"Şenkaya Emerald"*. It has naturally formed patterns and quasi figures that make small objects even more fascinating.

Fig. 2. Macroscopic views of natural chrysoprase and diaspore (Selim 2015)

3.4 Diaspore

Diaspore is one of the minerals forming the bauxite ore. It is the least hydrous Al-oxyhydroxide mineral in the bauxites. Diaspore becomes the main, even only mineral of bauxites when the bauxite ore gets metamorphosed. Such ores are better called as diasporite, since those are monomineralic. This is very common occurrence especially in the Menderes Massif (western Anatolia). Diaspore crystals sometimes grow into really gigantic sizes. Then becomes gem material. Diaspore naturally occur in yellowish green and light green colors (Fig. 2). In addition to the metabauxites, diaspore occurrences are also reported in laterites and Al-rich clays, in recrystallized limestones, with corundum, in some alkali pegmatites as late stage hydrothermal mineral, and in magmatic rocks as a result of recrystallization of Al-rich xenoliths (Hatipoğlu et al. 2010b; Kumbasar and Aykol 1993). However, they hardly grow into gem size and quality. In Turkey, type locality is Milas county (Muğla) (Lake Bafa vicinity and area in between Milas-Yatağan counties). There are reports on its existence in Söke (Aydın), Tire (İzmir), in central Anatolia (Gerdekkilise-Bolkardağ) and Saimbeyli county (Adana), however, these are not at mineable scale. Gem quality diaspores are mostly used in jewelry. Two names were adopted, also trademarks, for such diaspores, *"Sultanite or Zultanite"* in honor of Ottoman Sultans and *"Csarite"*. It has very unique phenomenal color changing property and rarity thus becoming more popular. It is considered as "diamond of the future" by many.

3.5 Sepiolite

Sepiolite, also known as meerschaum, is a soft white clay mineral - a complex magnesium silicate, a typical chemical formula ($Mg_4Si_6O_{15}(OH)_2 \cdot 6H_2O$). It can be present in fibrous, fine-particulate, and solid forms. Sepiolite is opaque and off-white, grey or cream color, breaking with a conchoidal or fine earthy fracture, and occasionally fibrous in texture (Fig. 3). There many reports on its occurrence worldwide. It is generally associated with carbonate/evaporate sequences, sedimentary in origin. However, it also occurs in Kıbrısçık (Bolu) as a result of hydrothermal alteration of glassy tuffs of Middle Miocene Deveören volcanics in Köroğlu (Galatya) volcanic belt (Irkeç and Ünlü 1993). Although there are occurrences worldwide, most of the sepiolite of commerce is obtained chiefly from Sepetçi, Margı, Sarısu, Kayı, Gökçeoğlu, İmişehir, and Türkmentokat villages of the city of Eskişehir in Turkey. Nemli (Kütahya) is also important for its alike occurrences. It is mostly used to make pipes and small objects, paring beads. Calcite, gypsum and dolomite often company sepiolite.

Fig. 3. Macroscopic views of worked and raw Eskişehirstone and raw Oltustone (Selim 2015)

3.6 Jet/Oltu Stone

Jet (literal name) or Oltustone (locality name), is an opaque black coalified fossilized drift wood of trees of the family Araucariaceae which is 180 million years old (Fig. 3).

It is formed from the high-pressure decomposition of wood. Jet is chemically related to brown coal, or lignite, but it is more solid and tough, looks more like obsidian than a coal (Toprak 2013). Jet has been used both as a talisman and a jewel for over four thousand years. In Turkey, it occurs and mined in Oltu county (Erzurum) where more than 300 locations around Tutlu (Lispek), Yeşilbağlar (Norpet), Güllüce, and Güzelsu villages. Jet is also found in other locations around the world. However, only the British and Spanish deposits in addition to the Turkish occurrences have been worked commercially. Jet is called *"Oltustone"* in Turkey due to its type locality (Ciftci et al. 2004). It is mostly used in jewelry and to make praying beads. Framboidal pyrites and clay pockets commonly occur in jet.

4 Conclusions

Among the many, 6 semi-precious stones are the most common and renowned: smoky quartz, Eskişehirstone, Oltustone, sultanite, blue chalcedony, and Şenkaya emerald. Major use of these include jewelry (sultanite, Oltustone, smoky quartz, blue chalcedony, and Şenkaya emerald), ornamental objects (Eskişehirstone, Oltustone, smoky quartz, blue chalcedony, and Şenkaya emrald), prayer beads (Oltustone, blue chalcedony, and Eskişehirstone). Sultanite (diamond of the future) is a rising star in the precious stones due to its only occurrence in Milas area (Muğla) and unique properties.

References

Ciftci E, Yalcin MG, Yalcınalp B, Kolaylı H (2004) Mineralogical and physical characterization of the Oltustone, a Gemstone Occurring around Oltu (Erzurum-Eastern Turkey). In: 8th ICAM, São Paulo-Brazil, pp 537–538

Hatipoğlu M, Babalık H, Chamberlain SC (2010a) Gemstone deposits in turkey. Rocks Miner 85 (2):124–133

Hatipoğlu M, Türk N, Chamberlain SC, Akgün AM (2010b) Gem-quality transparent diaspore (Zultanite) in bauxite deposits of the Ilbir Mountains, Menderes Massif. SW Turk Miner Deposita 45(2):201–205

İlhan NN (2012) Türkiye'nin Mücevher Taşları Haritası. I. Türkiye Mücevher Taşları Sempozyumu, Bildiriler Kitabı, İstanbul, pp 33–38

İrkeç T, Ünlü T (1993) Volkanik kuşaklarda hidrotermal Sepiyolit oluşumuna bir örnek: Kıbrısçık (Bolu) Sepiyoliti. MTA Dergisi, Ankara, vol 115, pp 99–118

Kumbasar I, Aykol A (1993) Mineraloji. İstanbul Teknik Üniversitesi Kütüphanesi, Sayı: 1519, İstanbul

Selim HH (2015) Türkiye'nin değerli ve yarı değerli mücevher taşları. İstanbul Ticaret Odası (İTO) Yayınları No: 2014/4, 102s, İstanbul

Toprak S (2013) Petrographical properties of a semi-precious coaly stone, Oltu stone, from eastern Turkey. Int J Coal Geol 120:95–101

Biooxidation of Copper Sulfide Minerals

Yu. Elkina[1,2(⊠)], E. Melnikova[2], V. Melamud[2], and A. Bulaev[1,2]

[1] Faculty of Biology, Lomonosov Moscow State University, Moscow, Russia
yollkina@mail.ru
[2] Research Center of Biotechnology RAS, Moscow, Russia

Abstract. The effects of temperature and the presence of NaCl on bioleaching of chalcopyrite, enargite, and tennantite were studied. Rate of copper extraction from all minerals depended on temperature and was the highest at 45–50 °C. NaCl addition increased rate of copper extraction from chalcopyrite but led to the decrease in copper extraction from enargite and tennantite.

Keywords: Bioleaching · Chalcopyrite · Enargite · Tennantite · Acidophilic microorganisms

1 Introduction

Copper and zinc are mainly extracted from sulfide ores using pyrometallurgical techniques. Pyrometallurgical processing of arsenic containing ores is a problem due to the emission of toxic gases (Filippou et al. 2007). Biohydrometallurgy is widely used to process gold bearing concentrates, and may also be used to extract non-ferrous metals arsenic-containing concentrates (Neale et al. 2017). The goal of the present work was to study copper bioleaching from arsenic-containing minerals and chalcopyrite at different temperatures and in NaCl presence.

2 Methods and Approaches

Chalcopyrite ($CuFeS_2$), enargite (Cu_3AsS_4), and tennantite ($Cu_{12}As_4S_{13}$) as well as mixed culture of acidophilic microorganisms oxidizing ferrous iron and sulfur compounds were subjects of the study. The experiments were carried out in flasks with 100 ml of nutrient medium supplemented and 2 g of milled minerals (P_{100} 75 μM) on a rotary shaker at temperatures from 40 °C to 60 °C for 30 days. In one variant of the experiment, nutrient medium was supplemented with100 mm NaCl.

3 Results and Discussion

The results of the experiments (rates of copper extraction) are shown in Table 1.

© The Author(s) 2019
S. Glagolev (Ed.): ICAM 2019, SPEES, pp. 189–191, 2019.
https://doi.org/10.1007/978-3-030-22974-0_43

Table 1. Rate of copper extraction from the minerals for 30 days (%)

Variant of the experiment	Chalcopyrite (CuFeS$_2$)	Enargite (Cu$_3$AsS$_4$)	Tennantite (Cu$_{12}$As$_4$S$_{13}$)
40 °C	14.33 ± 0.08	8.18 ± 0.65	15.33 ± 0.24
45 °C	17.84 ± 1.69	12.89 ± 1.79	26.16 ± 1.33
50 °C	25.29 ± 4.15	14.04 ± 0.95	18.43 ± 0.84
50 °C, 100 NaCl	33.25 ± 0.12	5.91 ± 1.09	13.04 ± 0.03
55 °C	26.75 ± 1.57	14.39 ± 0.01	14.84 ± 0.01
60 °C	17.32 ± 1.01	5.86 ± 2.18	12.83 ± 0.17

We showed that the rate of copper extraction from all minerals depended on temperature and was low at 40 °C. Extraction rate at 60 °C was also low as this temperature might inhibit microbial activity. Addition of NaCl increased rate of copper extraction from chalcopyrite that was a well known phenomenon. At the same time, NaCl addition led to the decrease in copper extraction rate.

4 Conclusions

The results obtained suggest that the efficiency of copper sulfide minerals depended on temperature, while NaCl addition did not allowed increasing the rate of copper bioleaching from arsenic-containing minerals in contrast to chalcopyrite. This fact should be taken into consideration when planning laboratory scale trials on bioleaching of copper sulfide concentrates.

Acknowledgements. The work was supported by the President Grant of the Russian Federation, grant No. MK-6639.2018.8.

References

Filippou D, St-Germain P, Grammatikopoulos T (2007) Recovery of metal values from copper – arsenic minerals and other related resources. Miner Process Extr Metall Rev 28(4):247–298
Neale J, Seppälä J, Laukka A, van Aswegen P, Barnett S, Gericke M (2017) The MONDO minerals nickel sulfide bioleach project: from test work to early plant operation. Solid State Phenom 262:28–32

Genetic Problem of Quartz in Titanium Minerals in Paleoplacers of Middle Timan

I. Golubeva[1], I. Burtsev[1], A. Ponaryadov[1], and A. Shmakova[1,2(✉)]

[1] Institute of Geology Komi SC UB RAS, Syktyvkar, Russia
alex.sch92@yandex.ru
[2] Karpinsky Russian Geological Research Institute (VSEGEI),
Saint-Petersburg, Russia

Abstract. Titanium ore in the Devonian paleoplacers of Middle Timan is predominantly represented by leucoxene, less frequently by modified ilmenite (pseudorutile). Other titanium minerals are found in small amounts (or have a sharply subordinate significance). All titanium minerals have numerous inclusions of quartz, which create an intractable problem in the enrichment of titanium ore. Metamorphogenic porphyroblastic explains the presence of quartz in titanium minerals. Precambrian seric-chlorite clay weathering crusts are a supplier of titanium minerals Timan. Leucoxene and ilmenite form in paraschist under conditions of facies of green shale of regional metamorphism, poikiloblasts. In the poikiloblasts, the poikilite and helicitic structures are well defined, due to numerous poikilite inclusions of quartz. The poikiloblasts, poikilite and heliic structures are well represented, due to the numerous poikilite incorporating quartz.

Keywords: Titanium · Paleoplacer · Shale · Poikiloblastez · Leucoxene · Ilmenite

1 Introduction

The main resources of titanium in Russia are concentrated in identified (Pizhemskoe and Yaregskoye deposits) and are designed (Vodnenskoe and other manifestations) of the Devonian titanium paleoplacers Timan. Ores are represented by leucoxene, to a lesser extent altered under exogenous conditions by ilmenite (pseudo-ethyl). Ore is difficult to enrich due to the large number of quartz inclusions in titanium minerals. A large number of quartz inclusions in titanium minerals is an exceptional feature of Timan paleoplacers. The high content of quartz in titanium minerals is explained by its primary metamorphogenic genesis. Slates of the Precambrian folded basement of the Timan are the root source of titanium minerals. Titanium paleoplacers are formed due to the redeposition of the weathering crust on shale (Kochetkov 1967; Kalyuzhny 1972; Makhlaev 2006; Ponaryadov 2017).

2 Methods and Approaches

Photos of minerals were taken on a JSM-6400 scanning microscope with a Link ISIS-300 energy-dispersive spectrometer and a polarized microscope.

© The Author(s) 2019
S. Glagolev (Ed.): ICAM 2019, SPEES, pp. 192–194, 2019.
https://doi.org/10.1007/978-3-030-22974-0_44

3 Results and Discussion

Titanium minerals - ilmenite and leucoxene (rutile, anatase and quartz aggregate) crystallize the parachale under the conditions of regional metamorphism of the green slate facies. This process is widely developed in the Riphean schists of Timan. The content of titanium minerals varies from 1.5–3.0%, occasionally rising to 5%. Titanium minerals crystallize in the form of porphyroblasts saturated with numerous quikilic poikilite inclusions. Poikilite inclusions of quartz are fragments of aleurite sizes. The quikite and inclusions of quartz and sericite captured during porphyroblasty determine the helicocyte structure (Fig. 1a, b). Titanium minerals are easily separated from shale and are separated during physical and chemical weathering. They accumulate due to gravitational separation during transportation and redeposition of weathering products. In paleoplacer metamorphic structures of titanite in titanium minerals are well preserved (Fig. 1d, f). In titanium minerals, fragments of quartz veins recorded with the growth of titanium minerals are diagnosed in paleoplacer. Sometimes fragments of sericite-chlorite schists with ilmenite grains are found (Fig. 1e). The quartz inclusions in leucoxene upon lithification of the ore-bearing sandstone can be regenerated with an increase in the volume of inclusions. The amount of SiO_2 in titanium minerals in bedrock - shale is 6.6–11.47%, and in paleoplacer - 12.2–28.19% (Ignatiev 1997). The high content of silicon dioxide in the titanium minerals of paleoplacer Timan is a specific feature of ore in this area and is explained by the metamorphogenic porphyroblastic genesis. For comparison, the SiO_2 content in titanium minerals of another well-known Tuganov paleoplacer in Western Siberia is given: silica is 1.82% for

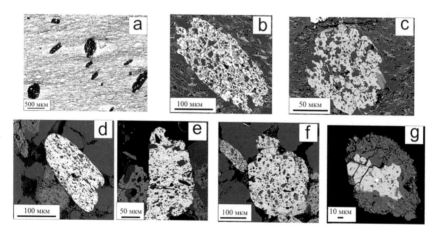

Fig. 1. Titanium minerals in shale and paleoplacers: a - leucoxene porphyroblasts in Riphean sericite-chlorite shale; b - helicitic structure in cross section of leucoxene porphyroblast in schist; c - poikilite inclusions in the longitudinal section of porphyroblast ilmenite in schist; d - helicitic structure in cross section leucoxene plate from paleoplacers; e - helicitic structure in leucoxene from paleoplacers; f - a longitudinal section of ilmenite with poikilitite inclusions of quartz from paleoplacers; g - a rounded fragment of sericite-chlorite shale with the inclusion of ilmenite from paleoplacers

ilmenite and 8.5.4% for leucoxene 6.5. The primary metamorphogenic nature of titanium minerals in Timan is indirectly indicated by the same increased content of MnO in porphyroblasts of leucoxene and ilmenite in native shale (2.5–3.63%) and in paleoplacer (to 2.33%).

4 Conclusions

The presence of a large amount of quartz in the titanium minerals of the paleoplacer Timan is explained by the porphyroblastic growth of shales under conditions of regional metamorphism of ilmenite and leucoxene. The increase in the percentage of silica in titanium minerals is also due to the processes of lithogenesis.

Acknowledgements. This research was supported by project AAAA-A17-117121270037-4 "Scientific basis for effective development and use of the mineral resource base, development and implementation of innovative technologies, geological and economic zoning of the Timan-North Ural region".

References

Ignatiev VD, Burtsev IN (1997) Timan's leucoxen: mineralogy and technology issues. Science, St. Petersburg 215 p

Kalyuzhny VA (1972) Geology of new placer formations. Science, Moscow 263 p

Kochetkov OS (1967) Accessory minerals in the ancient strata of Timan and Kanin. Science, Leningrad 200 p

Makhlaev LV, Golubeva II (2006) Ilmenite-containing metapelites as the most important source of the formation of giant and supergiant titanium placers. Titanium-zirconium deposits of Russia and prospects for their development. Moscow, IGEM RAS, pp 39–42

Ponaryadov AV (2017) Mineralogical and technological features of ilmenite-leucoxene ores of Pizhemskoe deposit, Middle Timan. Vestnik of the Institute of Geology, Komi SC UB RAS, no 1, pp 29–36. https://doi.org/10.19110/2221-1381-2017-1-29-36. (in Russian)

Gold and Platinum Group Minerals (PGM) from the Placers of Northwest Kuznetsk Alatau (NWKA) (South Siberia, Russia)

V. Gusev[1,2(✉)], S. Zhmodik[1,2], G. Nesterenko[2], and D. Belyanin[1,2]

[1] Department of Geology and Geophysics, Novosibirsk State University, Novosibirsk, Russia
vityansky@igm.nsc.ru
[2] Institute of Geology and Mineralogy SB RAS (IGM SB RAS), Novosibirsk, Russia

Abstract. Native gold and PGM from NW Kuznetsk Alatau (NWKA) (South Siberia, Russia) have been investigated. Applying the complex of advanced analytical, geo-chemical, and statistical techniques permits determination of motherlode types of noble mineralization (NM). For gold, it is mineralization of three types: 1 – the gold-sulfide-quartz type associated with dykes of the basic composition and fault-line zones (2), as well as the gold-skarn type (3). For PGM it is mineralization of the Ural-Alaskan (1) and the ophiolitic (2) types, as well as multicomponent alloys associated with layered intrusions (3). The presence of rims and inclusions is indicative of postmagmatic transformations of the minerals. In the meantime, Au and PGM from NW Kuznetsk Alatau placers retain genetic traits of motherlodes.

Keywords: Alluvial placers · Gold · Platinum Group Elements (PGE) · Altai–Sayan folded area · Kuznetsk Alatau

1 Introduction

Au placers with PGM (0.1–0.2% and more by weight of Au) are widespread on the territory of the Kelbes placer region NWKA. The placers were abandoned during last centuries, but motherlode deposits have not been discovered here. Au-quartz, Au-quartz-sulfide types, and Au-magnetite (skarn) have been discovered in NE and the central Kuznetsk Alatau region, as well as in Salair range, and PGE have been revealed in chromites and dykes of the basic composition. The contribution presents new data on the morphology, micro-inclusions, change types, and the composition of native Au and PGM from placers and their comparison with NM from motherlodes of adjacent regions in the South Siberia.

S. Glagolev (Ed.): ICAM 2019, SPEES, pp. 195–197, 2019.
https://doi.org/10.1007/978-3-030-22974-0_45

2 Methods and Approaches

Minerals were selected from heavy mineral concentrates under binocular microscope. An Axio Scope A1 (Carl Zeiss) microscope was used for determination of sizes and morphology of NM particles, and then polished preparations were made from these particles for microscopic investigations by methods of ore and electron microscopy. The composition and interrelations between minerals, micro-inclusions, and newly formed phases were studied by SEM (MIRA 3 LMU, Tescan Ltd.), and EMP methods (Camebax Micro) in the share use Center for Multielement Isotope Studies (Novosibirsk).

3 Results and Discussion

The examined gold has different morphologies (sizes, degrees of grain rounding, and deformation), it is characterized by the wide range of variations in the fineness (from 720‰ to 1000‰), by different degrees of chemical change, contains inclusions of quartz, magnetite, and clay minerals, and it is coated with material consisting of Fe hydroxides.

PGM are represented by ferroplatinum and rutheniridosmine associations, as well as by small amount of sperrylite and Pd minerals. More than 65% of grains belong to ferrous platinum and isoferroplatinum; about 30% are minerals of the Ru, Ir and Os system and about 3% are Pd minerals. The degree of grain rounding of ferroplatinum is higher than that of rutheniridosmines and sperrylite, but it is generally low which indicates a short range of their transfer. As impurities, Cu, Rh and Pd are commonly found, Au is found in rare cases.

4 Conclusions

The identified features of native Au and PGM indicate the presence of several types of motherlodes, and possibly, intermediate reservoirs, as well as various distances of placers from sources of supply. The data obtained are compatible with the assumption that input of gold from ledge ores of Au-quartz, Au-quartz-sulfide, and Au-magnetite types widespread in the eastern part of the Kuztetsk Alatau and on the Salair range, as well as from sediments of the Simonovskaya suite. The sources of PGM were rocks of the Ural-Alaskan and ophiolite complexes, fragments of which (in particular the Kaygadatsky massiv) have been established in NWKA.

Acknowledgements. This work was supported financially by the RFBR (Grant 19-05-00464) and government assignment (project No. VIII.72.2.3 (0330-2014-0016)).

Noble Metal Mineralization of the PGM Zone "C" of the East-Pana Layered Intrusion (Kola Peninsula)

O. Kazanov[1], G. Logovskaya[2(✉)], and S. Korneev[2]

[1] Moscow Branch, FSUE "All-Russian Scientific-Research Institute of Mineral Resources named after N.M. Fedorovsky", Moscow, Russia
[2] Institute of Earth Science, Saint-Petersburg State University, Saint-Petersburg, Russia
galkanuu@gmail.com

Abstract. The first study of the noble metal mineralization of the ore zone "C" of the East-Pana massif allowed to divide it into two types: early magmatic low-sulfide and late post-magmatic proper PGE associations.

Keywords: Fedorovo-Pana massif · East-Pana massif · Stratiform mineralization · Platinum metal mineralization · Early magmatic and late post-magmatic platinum mineral associations

1 Introduction

Since the late 1980s in Kola Peninsula the platinum content, associated with ultramafic-mafic magmatic complexes, has been studied. The Fedorovo-Pana early proterozoic layered intrusion of the peridotite-pyroxenite-gabbronorite formation has been recognized as the most promising object for discovering the industrial reserves of complex low sulfide PGE ores.

2 Methods and Approaches

The following research methods were used in the work: mineralogical description of sections of boreholes, study of samples of rocks, slides and polished sections, microprobe analysis of transparent polished thin sections from ore intervals, statistical treatment of chemical analysis data with program Statistica v.6.1 and then analysis and interpretation of results. These methods allowed to identify similarities and differences in the ore intervals and their mineralization, and to confirm the assumption of a various genesis of PGE mineralization in the ore zone "C" (Subbotin et al. 2012).

3 Results and Discussion

Based on the results of the study of ore mineralization within the ore zone "C", two associations of noble metal minerals were identified. The first, early, magmatic association is represented by platinum and palladium sulfides (Fig. 1), native gold and

© The Author(s) 2019
S. Glagolev (Ed.): ICAM 2019, SPEES, pp. 198–200, 2019.
https://doi.org/10.1007/978-3-030-22974-0_46

silver, moncheite, kotulskite and ferroplatinum, forming inclusions in the magmatic major sulphide and rock-forming silicate minerals. There is a significant positive correlation dependence of Pt, Pd, Au with Cu, Ni and S. It is indicated in the Sungiyok area. Later, postmagmatic association is represented by sperrylite (Fig. 1), mertieite-I, mertieite-II, temagamite, telargpalite, stibiopalladinite and kotulskite, associated with secondary silicate and later sulphide (millerite) minerals. They develop along cracks and veins in the rock-forming and early sulfide minerals. For the second type, correlations with Cu are not characteristic, with S there is a significant negative correlation. This association is observed in the Chuarvy area. Both types of mineralization are characterized by abnormally low values of Pd/Pt = 0.1–1.7 (Voytekhovich et al. 2008).

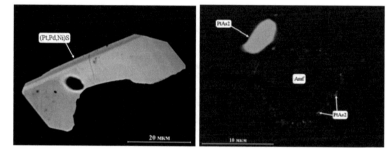

Fig. 1. Subhedral zonal aggregate of braggite ((Pt, Pd, Ni)S - early magmatic association of PGE mineralization (left); Spot edge around amphibole and micron streak of sperrylite (PtAs$_2$) - postmagmatic, hydrothermal metasomatic association of PGE mineralization (on the right).

4   Conclusions

Thus, based on the identified features of the noble metal mineralization of the ore horizon "C", it can be divided into two types: low-sulfide platinum and properly platinum. The first type is characterized by a close relationship with sulfide minerals: noble metal minerals form intergrowths with the main sulphides, are included in rock-forming silicate minerals. This type is typical for the ore horizons of the Sungiyok area. Minerals of noble metals are localized in cracks and streaks and are associated with minerals formed during the post-magmatic stage of intrusive evolution. This type is observed in Sunghiyok area and Chuarvy area.

Acknowledgements. Authors are grateful to V.V. Shilovskich for the research of minerals which carried out on the "Hitachi S-3400 N" scanning electron microscope at "Geomodel" Resource Centre of Saint-Petersburg State University.

References

Subbotin VV, Korchagin AU, Savchenko EE (2012) Platinometal mineralization of the Fedorovo-Pana ore cluster: types of mineralization, mineral composition, genesis peculiarities. Vestnik Kola Sci Cent Russ Acad Sci 1:55–65

Voytekhovich VS, Kazanov OV, Kalinin AA (2008) Report on the results of prospecting and assessment work on platinum metal mineralization in the eastern part of the PanaTundras-massif in 2006–2008. LLC Kola Mining and Geological Company, Apatity

Shungites and Their Industrial Potential

V. Kovalevski and V. Shchiptsov[(✉)]

Institute of Geology, Karelian Research Centre, Petrozavodsk, Russia
shchipts@krc.karelia.ru

Abstract. Shungite rocks are widespread in Zaonezhye, Republic of Karelia, where they constitute dozens of carbonaceous rock deposits of the Paleoproterozoic Onega structure with predicted carbon resources of more 4 billion tons. The lower age boundary is of 2.1 Ga. Shungite rocks belong to carbonaceous rock class. These rocks metamorphosed in greenshcist facies of muscovite-chlorite-biotite subfacies are unique natural, noncrystalline, non-graphitized, fullerene-like carbon. They have various structural-mineralogical levels: (a) supramolecular, (b) molecular, (c) electron-energetic, (d) structural-physical and (e) geologic-genetic (parametric). Shungite rocks contain shungite carbon (shungite matter) and a variety minerals, microminerals and nanominerals. The applications of shungite rocks are determined with regard for their natural types. Authors had shown their intergrated application in ore-thermal processes.

Keywords: Shungite rocks · Carbon · Metamorphism · Paleoproterozoic · Application

Shungite rocks, named so after a Karelian town, Shunga, have attracted attention for decades and have no counterparts in the Earth's geological evolution with respect to the mode of occurrence and tremendous reserves. They are unique from both scientific and practical points of view. Shungite rocks are part of Paleoproterozoic carbonaceous formation in Karelia. They are common in the Trans-Onega area, Russian Karelia, where 25×10^{10} tonnes of autochthonous organic matter have been formed over an area of about 9000 km^2. These complexes constitute dozens of carbonaceous rock deposits in the Paleoproterozoic Onega Structure with forecast carbon reserves of over 4 billion tones and are mainly confined to the rocks of the Ludicovian system with the lower age boundary of 2.1 Ga. The bulk of free carbon (Corg.) is in the Trans-Onega suite of the Ludicovian superhorizon. Phenomenal carbon accumulation in this superhorizon is responsible for the mineralogenic specialization of the rocks of the Trans-Onega suite. Their Corg. concentration varies from less than 1% to 70 wt% in rocks and up to 98 wt% in anthraxolite aggregates. Carbon concentration in the rocks of the Kondopoga suite of the Kalevian superhorizon is not more than a few percent. Thus, the black shale formation of the Onega Structure is formed of the rocks of the Trans-Onega, Suisari and Kondopoga suites. The world's largest Zazhogino Ore Field, covering an area of 3240 km^2, with two active quarries, Zazhogino and Maksovo, has been identified and relevant evidence was presented (Table 1).

Metamorphism and metasomatism have contributed markedly to the genetically distinctive shungite rocks. They belong to a metamorphogenetic class of shungite formation. Their industrial properties are controlled by metamorphism and a special

© The Author(s) 2019
S. Glagolev (Ed.): ICAM 2019, SPEES, pp. 201–204, 2019.
https://doi.org/10.1007/978-3-030-22974-0_47

Table 1. Comparative description of the parameters of major deposits in Zazhogino Ore Field

Ore body shape	Size		Maximum thickness	Average free carbon content, %
	Length, m	Width, m		
Shunga deposit				
Sheet-like	1400	300	5.2	41
Maksovo deposit				
Sheet-like-cone shaped	700	500	120	40
Zazhogino deposit				
Sheet-like-cone shaped	400	300	60	27
Zalebyazhskoe deposit				
Sheet-like	2000	700	38	35

contribution of K-Na alkaline metasomatism which has affected the formation of the natural types and varieties of shungite and, correspondingly, industrial types and varieties and applications of this unique raw material. On the P-T scheme, showing principal correlations between metamorphic facies and subfacies after S.A. Bushmin and V.A. Glebovitsky, shungite rocks were derived under greenschist-facies, muscovite-chlorite-biotite-subfacies conditions of metamorphism at a temperature of 325–450 °C and a pressure of 2–5 kbar (Bushmin and Glebovitsky 2016).

The mineral form of shungite (shungite matter) is non-graphitizable fullerene-like carbon, which differs from graphite at a supramolecular, atomic and zonal (electron) structure level. The main supramolar character of shungite is an ability to form spherical structures (empty globules). At an atomic level, in addition to hexagonal rings alone typical of graphite, pentagonal and heptagonal rings, characteristic of fullerene-like structures, are also observed. At a zonal structure level, the energy of collective excitations of valent (external) and framework (internal) π- and σ-electrons decreases relative to graphite, which is typical of fullerenes as well (Kovalevsky et al. 2016). Shungite from some deposits displays diamagnetic properties characteristic of fullerenes. The structure of shungite rocks is similar to that of vitreous-crystalline materials, where highly dispersed crystals are distributed in non-crystalline matrix.

Shungite rocks belong to a class of carbonaceous rocks varying in carbon content and mineral diversity and those are natural carbon-mineral composite materials. They contain shungite (shungite matter) and a variety of micro- and nanominerals. Silicate minerals are highly disperse and are evenly distributed in carbon matrix. Major rock-forming minerals are quartz, mica, albite and pyrite. High secondary and accessory mineral concentrations and a certain spectrum of layered and cluster impurities are observed. Natural types of shungite rocks display several textural and structural varieties and the non-uniform phase composition of carbon and geochemical characteristics.

Shungite applications are determined by natural types and varieties of shungite rocks (Kovalevsky et al. 2016). The structure and properties of shungite rocks are responsible for their application in oxidation-reduction processes: in blast furnace production of foundry (high-silicon) cast iron: in ferroalloy production; in yellow phosphorus production; in carbide and silicon nitride production; as a reinforcing component of groove masses; as a filler of non-stick paints.

The sorptive catalytical and reduction properties of shungite rocks are used: for treatment of high-quality drinking water in flow systems and wells; for removal of many contaminants from urban domestic and industrial sewage; for swimming-pool water treatment; for water treatment at heat power plants; for electrically conductive paint production; for electrically conductive concrete and brick production; for electrically conductive and plastering and masonry solutions; for electrically conductive asphalt production. These materials were used for developing and designing heaters: rooms screening electromagnetic radiation; a method for removal of ice from roads (warm pavements and roads).

Finely ground shungite can be mixed with any binders of organic and inorganic origin and can thus be used as: black pigment for oil and water paints; a filler for polymers materials (polyethylene, polypropylene, fluoroplastic, etc.); substitute for white soot and technical carbon in rubber production.

One essential feature of shungite rocks is that they can be modified with regards for their desirable application. Enrichment of shungite rocks and their division into micro- and nanosized components make a possible to activate shungite carbon and expand the potential applications of shungite rocks in science-intensive technologies, e.g. nanotechnologies.

There is only one geologo-industrial classification of shungite rocks (after Yu. Kalinin) based on industrial types distinguished with respect to their mineralogical composition and corresponding applications. However, the latest results of the large-scale practical application of shungite rocks shows that such a division into industrial types is clearly insufficient. Therefore, division into subtypes or varieties is required.

At the present stage of the study of Karelia's shungite rocks the goal of research and appraisal is not only to estimate the reserves of potential deposits and to assess the petrographic and structural-chemical characteristics of shungite rocks but to develop criteria and recommendations for the industrial application of rocks from potential deposits in innovative and science-intensive fields (Kalinin and Kovalevski 2013).

For this purpose it is proposed:

1. To re-assess earlier geological, physico-chemical and technological data on known deposits to make up a list of known shungite rocks suitable for particular applications.
2. To reveal promising shungite rock applications (e.g. metallurgy, tyre production, water disposal, the production of composite materials, etc.) and certification requirements to be met by raw material for each application.
3. To specify the critical properties of shungite rocks for each application (e.g. chemical composition, mineralogical composition, the structural parameters of carbon and rocks) and express methods for their analysis.
4. To develop certification requirements for shungite rocks with regard for their applications.

5. To analyze shungite rock samples and cores from each of the bodies and rock exposures and classify all shungite rocks into types and subtypes referenced to a major application, based on the criteria developed.
6. To correlate the geological parameters of the deposits and outcrops with the geochemical and structural-petrological characteristics of shungite rocks to forecast the prospecting of shungite rocks with preset properties for required applications.

A classification of the geologo-industrial types of shungite rocks will be worked out and the most promising shungite rock prospects for each application will be specified, based on the results of prospecting and prospecting-and-appraisal for shungite rocks. The accomplishment of the work planned will result in the efficient investment of the money spent in the cost of future deposits and the updating of innovative approaches to the use of Russia's unique carbonaceous raw material.

Shungite has unveiled many of his mysteries, has become known all over the world, has attracted the interest of experts by its great potential and yet has remained largely unknown and open to new discoveries.

Acknowledgements. The work is performed in the framework of the PFNI GAN research of IG KarRC RAS.

References

Bushmin SA, Glebovitsky VA (2016) Scheme of mineral facies of metamorphic rocks and its application to the Fennoscandian Shield with representative sites of orogenic gold mineralizaion. Trans KRC RAS 3–27

Kalinin Yu, Kovalevski V (2013) Shungite rocks: scientific search horizons. Nauka v Rossii 6:66–72

Kovalevsky V, Shchiptsov V, Sadovnichy R (2016) Unique natural carbon deposits of shungite rocks of Zazhogino ore field, Republic of Karelia, Russia. In: SGEM Conference Proceedings of International Multidisciplinary Scientific GeoConference Surveying Geology and Mining Ecology Management, pp 673–680

Gold-Silver Natural Alloy of Chromitites from the Kamenushinsky Massif (The Middle Urals)

A. Minibaev[✉]

Mining University, Saint-Petersburg, Russia
a.m.minibaev@yandex.ru

Abstract. A detailed study of the chromitites clinopyroxenite Kamenushinsky massif in the middle Urals has allowed to allocate from them the gold-silver alloy. The study of the chemical composition, morphology of gold-silver alloys, peculiarities of their placement allowed not only to determine their genetic relationship with platinum-bearing chromites, but also to make an assumption about the formation of chromitites as a result of a single geological process.

Keywords: Urals Platinum Belt · Kamenushinskiy massif · Chromite-platinum mineralization · Gold

1   Introduction

The history of the development of platinum placers of the Urals dates back nearly a two centuries and a distinctive feature is the permanent presence of gold and its natural alloys along with the extraction of the platinum groups minerals (hereafter - PGM) (Vysotsky 1913). It is known that the main source of the PGM is a process of a zonal massifs erosion of the Ural-Alyaskan type. Nevertheless, the question of a primary source of gold in placers is remained as a controversial issue. Thus, for example, in research works of N. Vysotsky (Vysotsky 1913) was given a support of the idea that the appearance of gold in platinum placers was due to the weathering of an acid rocks, but its absence in some zonal massifs of the Urals Platinum Belt (hereafter - UPB) keeps the matter in abeyance. In the 70s of the 20th century during setting deposit into exploitation, located within the Kachkanar zonal massif UPB, for the first time gold was determined and described from the bed rock (Fominykh et al. 1970). In spite of this fact, further question of the nature of the appearance of gold in the ores of the UPB massifs was not under active discussion.

Inside the Nysyamsky platinum placer field – one of the main leaders of the platinum mining of the Urals – also the presence of gold was reported (Vysotsky 1913). The source of the placer formations was Kamenushinsky dunite-clinopyroxenite massif which had a potential for of bed chromite-platinum mineralization (Ivanov 1997; Tolstykh et al. 2011, Minibaev et al. 2015). During the study platinum-bearing chromitites were divided into two petrographic types with characteristic features: vein-imbedded and massive (Minibaev 2018). In addition to study of PGM features from the

S. Glagolev (Ed.): ICAM 2019, SPEES, pp. 205–207, 2019.
https://doi.org/10.1007/978-3-030-22974-0_48

chromitites of both petrographic types, the scientific observation of the determined Au-Ag alloys is under a great interest

2 Methods and Approaches

A study of morphology and chemical composition of Au-Ag natural alloys and its bearing chromitites was carried out by the scanning electron microscope Carl Zeiss EVO (OPTEC, Moscow), equipped with attachment: EDS (energy dispersive X-ray spectrometer) and BSD (Backscattered Electron Detector).

3 Results and Discussion

Au-Ag natural alloys was determined as impurities of irregular crystallographic habitus in chromespinelides of vein-imbedded (Fig. 1a) and massive (Fig. 1b) chromitites.

The size of the aggregates is around 4–6 microns. The chemical composition of Au-Ag alloys (Table 1) corresponds to the compositions of similar objects founded in the ore chromospinelides of the Konder massif (Pushkarev et al. 2015).

The similar composition of Au-Ag alloys, its presence directly in chromespinelides, also the absence of traces of visible deformations of the mentioned indicates that its formation does not correspond to the processes of serpentinization or overlaid hydrothermal processes (as it was known before), it corresponds to the fact that inclusions were gained into chromespinelides during crystallization. Inclusions such as Cr and Fe in Au-Ag alloys are the evidence of inherited condition of mineralization and chemistry features of the ore-forming system.

Table 1. Chemical composition of Au-Ag alloys from the chromitites of the Kamenushinsky massif (mass. %)

№ point	Au	Ag	Cr	Fe	Sum
1	81,45	10,84	3,84	2,02	98,15
2	81,21	10,47	3,55	1,89	97,12

Notes: 1 - from vein-imbedded chromitites; 2 - from massive chromitites.

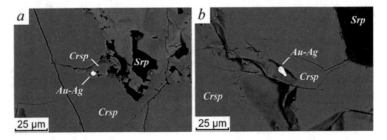

Fig. 1. Determination of grains Au-Ag natural alloys in chromitites: a - vein-imbedded; b - massiv

4   Conclusions

The obtained results are correlated with the conclusions about the syngenetic nature of the platinum-bearing vein-imbedded and massive chromitites, where the latter are characterized by a later origin relating to the final stage of evolution of the ore-forming system (Minibaev 2018). Also these results can be used not only to confirm the hypothesis about the common substance source of two petrographic types of chromitites, but also allow to make the conclusion that the formation of chromitites is generally resulted from a single geological process.

Acknowledgements. Work was done with the support of government program № 5.12856.2018/8.9.

References

Fominykh VG, Uskov ED, Volchenko YuA (1970) Gold in the ores of the Kachkanar massif in the Middle Urals. In: Questions of the Geology of the Gold Deposit, Tomsk, pp 31–36

Ivanov OK (1997) Concentric-zonal pyroxenite-dunite massifs of the Urals. Ed. Ural University, Ekaterinburg, 488 p

Minibaev AM (2018) About the origin of chromite-platinum mineralized zones of the clinopyroxenite-dunite Kamenushinsky massif of the middle Urals. In: Materials of the Eighth Russian Youth School with International Participation "New in the Knowledge of the Processes of Ore Formation". IGEM RAS, Moscow, pp 255–258

Minibaev AM, Stepanov SYu (2015) Perspectives of the identification of chromite-platinum mineralization in the rocks of the Kamenushinsky zonal clinopyroxenite-dunite massif (Middle Ural). In: Proceedings of the Fifth Russian Youth School with International Participation "New in the Knowledge of the Processes of Ore Formation". IGEM RAS, Moscow, pp 145–148

Pushkarev EV, Kamenetsky VS, Morozova AV, Hiller VV, Glavatskikh SP, Rodemann T (2015) Ontogeny of ore chromespinelide and composition of inclusions as an indicators of pneumatolyte-hydrothermal formation of platinum-bearing chromite of the Konder massif (Aldan shield). Geol Ore Deposits 57(5):394–423

Tolstykh ND, Telegin Y, Kozlov AP (2011) The native platinum of the Svetloborsky and Kamenushinsky massifs of the Ural platinum-bearing belt. Geol Geophys 52(6):775–793

Vysotsky NK (1913) Platinum deposits of Isovsky and Nizhne-Tagilsky regions of the Urals. In: Proceedings of the Geological Committee. New edit, no. 62, 692 p

Microbial Processes in Ore-Bearing Laterite at the Tomtor Nb-REE Deposit: Evidence from Carbon Isotope Composition in Carbonates

V. Ponomarchuk[✉], E. Lazareva, S. Zhmodik, N. Karmanov, and A. Piryaev

Institute of Geology and Mineralogy SB RAS, Novosibirsk, Russia
ponomar@igm.nsc.ru

Abstract. The unique Nb-REE deposit is located within the Tomtor complex of ultramafic alkaline and carbonatite rocks in the northern Sakha Republic (Yakutia) (Kravchenko and Pokrovsky 1995; Dobretsov and Pokhilenko 2010; Lazareva et al. 2015). Ores reside in three layers (Severny, Yuzhny, and Buranny sites) which fill depressions in subsided profiles of weathered carbonatites. Judging by stable isotope analysis, carbonates from laterite weathering profiles at the Tomtor Nb-REE-deposit formed by different mechanisms, including microbially mediated organic-clastic sulfate reduction and anaerobic oxidation of methane.

Keywords: Tomtor · Nb-REE-deposit · Laterite · Carbonates · C-isotopes · Microbial · Methane

1 Introduction

Rare earth elements have been broadly used in advanced technologies. They often occur in carbonatite deposits among which laterite profiles of weathered carbonatites are most attractive commercial targets. A unique Nb-REE deposit is located within the Tomtor complex of ultramafic alkaline and carbonatite rocks in the northern Sakha Republic (Yakutia) (Kravchenko and Pokrovsky 1995; Dobretsov and Pokhilenko 2010; Lazareva et al. 2015). Ores reside in three layers (Severny, Yuzhny, and Buranny sites) which fill depressions in subsided profiles of weathered carbonatites. Lately, evidence has been obtained that localization of elements in ore-bearing beds may have biotic controls (Lazareva et al. 2015). The possible role of microorganisms in ore formation has not been discussed yet in the literature on laterite profiles from carbonatite deposits, including the giants of Mountain Weld (Australia) or Araxa and Catalão (Brazil). In this respect, the Tomtor deposit is remarkable by a contribution of biogenic and bacterial processes to its formation, besides magmatism and high-temperature hydrothermalism that are common to all carbonatite deposits. Biogenic agents are usually identified from the carbon isotope composition of rocks controlled by their interaction with fluids. The discussed stable isotope compositions of carbonates from ore-bearing laterites at the Tomtor deposit provide evidence of possible microbial mediation.

© The Author(s) 2019
S. Glagolev (Ed.): ICAM 2019, SPEES, pp. 208–211, 2019.
https://doi.org/10.1007/978-3-030-22974-0_49

2 Methods and Approaches

The stable isotope compositions were analyzed in three samples (TM-592, TM-590, and BH 101) from laterite ore zones with $\sim 1\%$, 0.6%, and 1.6% REE, respectively. Extraction of monofractions for analysis is difficult because the laterite samples are fine grained while the carbonates are zoned. The problem can be solved by using selective acid extraction based on progressively slower reactions of carbonates with H_3PO_4 in the stoichiometric series: calcite $\rightarrow$ dolomite $\rightarrow$ ankerite $\rightarrow$ siderite $\rightarrow$ rhodochrosite. According to experimental evidence (Al-Aasm et al. 1990), the reaction duration sufficient for $\delta^{13}C$ determination is 1 h for calcite, 24 h for dolomite, and 5–7 days for siderite and rhodochrosite. The C and O isotope compositions of carbonates are determined by digestion in anhydrous H_3PO_4 followed by measurements on a FINNIGAN MAT-253 mass spectrometer with a GasBench II analyzer, in a stream of pure helium. The GasBench unit was also used for preconditioning of samples digested at 70 °C. The results are quoted in (‰) relative to the Vienna Peedee belemnite PDB standard for carbon and relative to SMOW for oxygen. The analytical errors were ±0.1‰ for carbon and 0.15‰ for oxygen. The composition and morphology of minerals were studied at the Analytical Center of IGM (Novosibirsk) on a Tescan MIRA 3 LMU scanning electron microscope with Oxford Instruments Nanoanalysis Aztec Energy/INCA Energy 450+ XMax 80 and INCA Wave 500 analyzers, applicable to scan nanometer particles.

3 Results and Discussion

Sample TM-590 is composed mainly of goetite, siderite (with a Mn impurity), calcite, and rhodochrosite. Goethite occurs either as dripstone, with concentric zonation, or as pseudomorphs after a disappeared mineral. The carbonates exist as zoned anhedral grains, with calcite being the latest phase. Fine apatite crystals, pyroxene, monazite, pyrochlore, and sphalerite are present in lesser amounts. The rock-forming phases in TM-592 are goethite, siderite, and calcite and the accessories are apatite and monazite. Apatite occurs as fine prismatic euhedral grains in calcite. Sample BH 101 consists of finely intergrown Fe-chlorite (chamosite) and apatite. The apatite-chamosite aggregate encloses clearly seen large rhodochrosite grains with siderite in their core. The two phases have a distinct boundary but have inherited orientations of the crystallographic axes. The accessory phases are TiO_2, submicrometer monazite platelets, sphalerite, and galena.

Synthesis of C and O isotope data from magmatic and postmagmatic (hydrothermal, metasomatic) carbonates in different carbonatite deposits world wide shows a large range of values from -10 to $+3‰$ (PDB) for $\delta^{13}C$ and from $+6$ to $+30‰$ (SMOW) for $\delta^{18}O$ (Deines 1989).

The $\delta^{13}C$ values obtained for the Tomtor samples (Table 1) are lower. The $\delta^{13}C$ patterns in carbonates are controlled by the isotope composition of bicarbonate. Low $\delta^{13}C$ in bicarbonate is due to oxidation of organic matter where $\delta^{13}C$ can reach $-32‰$ (Oleary 1988). In the course of microbially mediated sulfate reduction of geopolymers and biopolymers in aerobic conditions, the $\delta^{13}C$ values of bicarbonate that forms by reaction (1) and those of carbonate produced later by reaction (2) are inherited from the precursor component.

$$2CH_2O + SO_4^{2-} \rightarrow 2HCO^{3-} + H_2S \tag{1}$$

$$2HCO^{3-} + Me^{2+} = MeCO_3 + CO_2 + H_2O \tag{2}$$

There are two facts that implicitly support the above considerations: laterite contains organic remnants (Lazareva et al. 2015) that are electron donors in reaction (1); the samples contain sulfides, including framboidal pyrite, which formed at the account of H_2S released in reaction (1). However, this model can account only for the isotope composition of the calcite component in TM-590, while other $\delta^{13}C$ values ($-24.8‰$ in TM-590) are below $-32‰$, and may have formed with participation of methane. Biogenic methane has $-55‰$ to $-80‰$ $\delta^{13}C$, and the $\delta^{13}C$ values in bicarbonate resulting from microbially mediated methane oxidation (react/3) and in carbonate that formed later by reaction (2) will be low.

$$CH_4 + SO_4^{2-} \rightarrow HCO^{3-} + HS^- + H_2O \tag{3}$$

Table 1. $\delta^{13}C$ and $\delta^{18}O$ (‰) in different fractions of carbonates

Sample		1 h	24 h	7 days
TM-590	$\delta^{13}C$	−29.6	−39.2	−24.8
	$\delta^{18}O$	+12.0	+16.1	+14.6
TM-592	$\delta^{13}C$	−31.2	−44.1	−37.1
	$\delta^{18}O$	+7.6	+13.6	+15.1
BH 101	$\delta^{13}C$	−59.0	−54.6	−
	$\delta^{18}O$	+9.4	10.5	−

Note that a low value of $\delta^{13}C$ was previously reported (Pokrovsky 1990; Kravchenko and Pokrovsky 1995) at the Tomtor deposit, but the mineralogical and geochemical characterization of the sample is not given.

4   Conclusions

The reported isotope data show that organic clastic sulfate reduction and anaerobic methane oxidation with participation of microbial communities were among key mechanisms responsible for the formation of carbonates in laterite profiles of the Tomtor Nb-REE deposit.

Acknowledgements. The study was supported by grant 18-17-00120 from the Russian Science Foundation.

References

Al-Aasm IS, Taylor BE, South B (1990) Stable isotope analysis of multiple carbonate samples using selective acid extraction. Chem Geol Isot Geosci Sect 80(2):119–125

de Toledo MCM, de Oliveira SMB, Fontan F, Ferrari VC, de Parseval P (2004) Mineralogia, morfologia e cristaloquímica da monazita de Catalão I (GO, Brasil). Braz J Geol 34(1):135–146

Deines P (1989) Stable isotope variations in carbonatites. In: Bell K (ed) Carbonatites, Genesis and Evolution. Unwin Hyman, London, pp 301–359

Dobretsov NL, Pokhilenko NP (2010) Mineral resources and development in the Russian Arctic. Russ Geol Geophys 51(1):98–111

Kravchenko SM, Pokrovsky BG (1995) The Tomtor alkaline ultrabasic massif and related REE-Nb deposits, Northern Siberia. Econ Geol 90(3):676–689

Lazareva EV, Zhmodik SM, Dobretsov NL et al (2015) Main minerals of abnormally high-grade ores of the Tomtor deposit (Arctic Siberia). Rus Geol Geophys 56(6):844–873

Olary MO (1988) Carbon isotopes in photosynthesis. Bioscience 38(5):328–336

Pokrovsky BG, Belyakov AYu, Kravchenko SM et al (1990) Origin of carbonatites and ore-bearing rocks of the Tomtor massif, NW Yakutia, according to isotopic data. Geokhimiya (9):1320–1329. (in Russia)

Ribeiro CC, Brod JA, Junqueira-Brod TC, Gaspar JC, Petrinovic IA (2005) Mineralogical and field aspects of magma fragmentation deposits in a carbonate–phosphate magma chamber: evidence from the Catalão I complex, Brazil. J South Am Earth Sci 18(3–4):355–369

Peridot: Types of Deposits and Formation Conditions

S. Sokolov[(✉)]

FSBE "All-Russian Institute of Mineral Raw Materials" (VIMS),
Moscow, Russia
vims-sokol@mail.ru

Abstract. The generalized information about the peridot from deposits and ore occurrence of well-known formation type is presented in this paper. Peridots from specific deposits vary in the accompanying mineral associations. These gems, depending on the genesis, contain differing in phase composition inclusions of mineral-forming medium and have different crystallization temperatures.

Keywords: Peridot · Formations · Deposits · Origin · Temperature of formation

1 Introduction

The increased interest to peridot – jewelry varieties of olivine, was observed over the last years. This has resulted in numerous publications on the geology, mineralogy, and genesis of this gem. Peridot has been found in almost 30 countries and on all continents, including Antarctica (gem-quality mineral is present in the basalts from the Ross Island); furthermore it was diagnosed in several stony-iron meteorites. Peridot is present in different geological formations, which almost all (with the exception of exogenous placers) belong to derivatives of endogenous processes and often genetically related to ultrabasic and basic rocks.

2 Formation Types of the Deposits of Peridot

Diamond-bearing kimberlites, concentrating peridot in different quantities, are known in Yakutia (Udachnaya-Vostochnaya and Mir pipes), the Republic of South Africa (Kimberly and De Beers pipes), and in Tanzania (Mvadun pipe).

Ultrabasic-alkaline rocks and carbonatites – UAC complexes. Peridot installed at complexes of the Kola peninsula (Kovdor) and Polar Siberia (Kugda).

Basalts and basaltoids (normal and (sub) alkaline). Peridot phenocrysts in the nodules of peridotites and olivenites are situated in lava flows and volcanic craters at deposits and ore occurrence in Russia, USA, Hawaiian Island, the Czech Republic, Madagascar Island, Ethiopia, China, Pakistan, Australia, Antarctica.

Alpinotype hyperbasites. The most famous deposit of peridot – Zabargad Island is classed to this formation type. They are known also in Russia at the East Sayan and at the South Urals.

© The Author(s) 2019
S. Glagolev (Ed.): ICAM 2019, SPEES, pp. 212–213, 2019.
https://doi.org/10.1007/978-3-030-22974-0_50

Placer deposits. Eluvial-deluvial plasers are the most productive among the exogenous deposits of peridot. They are connected with different genetic types of primary deposits of this gem (Zabargad, San-Carlos, Kugda).

3   Origin and Formation Conditions of Peridot

Peridots from the deposits of different formation clearly differ in the associations of accompanying minerals.

A detailed study of the jewelry peridots from the Kovdor and Kugda deposits allowed us to determine the specific features of their micromineralogy, (Yarmishko and Sokolov 2005): crystalline inclusions, represented mainly by minerals accompanying to peridot; decomposition products (magnetite dendrites and magnetite-diopside plates; crystallized melt inclusions.

Fixation multiphase crystallized melt inclusions in peridots confirms the magmatic nature of the deposits some formation types. For example, olivine and calcite from kimberlites of the Udachnaya-Vostochnaya pipe contain the melt inclusions the total temperature range of homogenization of which is 1100–880 °C (Tomilenko et al. 2009). Peridots from basalts of the Hawaiian Islands contain the inclusion of glass (Gubelin and Koivula 1992). The melt inclusions in jewelry peridot from the Kovdor deposits were homogenized at temperatures of 970–930 °C. Crystallization of peridot at the Zabargad deposit occured in temperature range from 900 to 750 °C (Maaskant 1986).

References

Gŭbelin EJ, Koivula JI (1992) Photoatlas of inclusions in gemstones. ABC Edition, Zurich
Maaskant P (1986) Electron probe microanalyses of unopened fluid inclusions, semiquantitative approach. Neues Jahrb Miner 7:297–304
Tomilenko AA, Kovyazin SV, Dunlyansky YuV, Pokhilenko LN (2009) Primary melt and fluid inclusions in minerals from kimberlites of the Udachnaya-Vostochnaya pipe, Yakutia. ECROFI-XX. Abstracts, University of Granada, pp 255–256
Yarmishko SA, Sokolov SV (2005) Micromineralogy of peridots from rocks of alkaline-ultrabasic massifs. In: VII International Conference «New Ideas in Earth Sciences». Abstracts, vol 2, Moscow, p 76. (in Russian)

Mineralogical Analysis of Glacial Deposits and Titanium Paleoplacers of the East European Part of Russia

N. Vorobyov and A. Shmakova[✉]

Institute of Geology of Komi SC UB RAS, Syktyvkar, Russia
alex.sch92@yandex.ru

Abstract. Mineralogical analysis is one of the main methods to determine sources for paleoplacers and location of source glacial provinces. Our studies defined location of source glacier provinces and sources for titanium paleoplacers of the East European Part of Russia.

Keywords: Minerals · Reconstruction · Glacial deposits · Titanium paleoplacers

1 Introduction

Mineralogical analysis is the most significant method in reconstructing sources paleoplacers and glacial deposits (boulder loams), determining location of source for glacial provinces. The paper presents two objects - glacial deposits of polar and vychegda horizons, Middle Devonian and Middle Triassic titanium paleoplacers.

2 Results and Discussion

Quaternary deposits have been studied in two areas of Pechora lowland: in the northeast in the basin of the Padymeytyvys river and northwest in the basin of the Kui river. Titanium paleoplacers were studied at the Kydzarasyu river (Preural Foredeep) and Middle Timan (Pizhma paleoplacer).

Relations between mineral composition of Quaternary deposits and underlying bedrocks are very important. During active exaration activity of the glacier, the underlying rocks controlled composition of boulder loams. The formation of heteroaged horizons of boulder loams is associated with the Northeastern (Paykhoy-Ural-Novaya Zemlya) and Northwestern (Fennoscandinavia) terrigenous-mineralogical source provinces. The rocks of the eastern province are characterized by higher levels of epidote and ilmenite fraction, and the northwestern ones - amphiboles and garnets.

In the basin of the Padymeytyvys river two horizons of boulder loams were drilled. In the lower (Vychegodsky) horizon, epidote (19.9%) and siderite (21.2%) dominate in the heavy fraction. In the composition of the heavy fraction of the upper (polar) horizon, pyrite (22.1%) and epidote (20.7%) have maximum concentrations, siderite content (11.7%) and ilmenite (14.7%) are high. Pyrite and siderite indicates the relation between the glacier and Triassic and Permian underlying rocks, and ilmenite may be associated with Uralian rocks.

S. Glagolev (Ed.): ICAM 2019, SPEES, pp. 214–215, 2019.
https://doi.org/10.1007/978-3-030-22974-0_51

In the valley of the Kui river one horizon of boulder loam – polar – was drilled. The heavy fraction is represented by amphibole (13.6%) - garnet (18%) - epidote (20.8%) associations with the increased content of pyrite (7.7%). High concentrations of amphiboles (13.6%) and garnets (18%) may indicate the relation between the glacier and the rocks of Fennoscandinavia.

The number of amphiboles in boulder loams decreases from west to east and amounts to the first percent. High total contents of pyrite and siderite in the north-east of Pechora lowland are typical of local Mesozoic source province.

Heavy fraction of the paleoplacer at the Kydrasyu river is represented by ilmenite (45.19%), epidote (23.21%), magnetite (15.76%), amphibole (6.69%), chrome spinelide (5.65%), garnets (2.06%). Also zircon, leucoxen, rutile, hematite, kyanite, martite, pumpellyite, staurolite are present (0.04–0.4%). The high content of ilmenite indicates the relation between the paleoplacer and igneous rocks. Epidote, garnet, kyanite, staurolite, and pumpellyite found in the heavy fraction are of metamorphic origin. The morphology of zircons is characteristic of minerals formed in igneous rocks of medium composition. The peculiarity of these minerals is their good preservation. This may be due to the nearby source. The mineral composition indicates that the source of minerals was not far from its burial. Titanium minerals and satellite minerals were most likely of magmatic and metamorphic origin.

The mineral composition of titanium placers often contains stable minerals and there are no unstable ones, although both are present in the bedrocks. The main reason is that the paleoplacer formed due to erosion of the weathering crust. An example of a paleoplacer formed that way can be Pizhma paleoplacer (Middle Timan) formed due to erosion and redeposition of weathering crusts on Riphean shales. It's mineral composition in contrast to the paleoplacer at the Kydzarasyu river, is very poor. The main part is represented by leucoxenized ilmenite and leucoxene. Tourmaline, garnet and zircon are found in single units.

3 Conclusions

We determined location of the source glacier provinces and sources for the titanium paleoplacers of the East European part of Russia by the mineralogical analysis.

Oil and Gas Reservoirs, Including Gas Hydrates

A Bench Scale Investigation of Pump-Ejector System at Simultaneous Water and Gas Injection

S. Karabaev[(✉)], N. Olmaskhanov, N. Mirsamiev, and J. Mugisho

Department of Mineral Developing and Oil and Gas Engineering,
Engineering Academy, RUDN University, Moscow, Russia
simpleforfiza@mail.ru

Abstract. In this paper, a bench study of the pump-ejector system for simultaneous water and gas injection (SWAG) was conducted. For these purposes, a pump-ejector system stand was used. A differential pressure gauge was used to determine the gas flow at the ejector intake. According to the results of differential manometer calibrations, a new formula was obtained which reduces its inaccuracy to 1% at pressures below 0.6 MPa. In addition, according to the pressure-energy diagrams, it was determined that the gas injection with excess pressures in the ejector suction chamber significantly increases the efficiency of the pumping-ejector system overall.

Keywords: Liquid-gas ejector · Pump-ejector system · Liquid-gas mixture · SWAG

1 Introduction

Nowadays, a global trend of oil fields with increased residual oil saturation is observed (Drozdov and Drozdov 2012). There is also problem with associated petroleum gas (APG) utilization, which is not used rationally and is directed to gas flares. Currently, one of the most promising areas in oil production is the simultaneous water and gas injection (SWAG) technology. The definite advantage of the SWAG is the ability to use APG as a liquid-gas mixture compound. This method makes it possible to mount simple maintaining, reliable and efficient equipment, while providing a significant reduction in power consumption and increasing the SWAG efficiency (Drozdov et al. 2012).

2 Methods and Approaches

The investigation was conducted based on bench tests. The electric centrifugal pump and the liquid-gas ejector were used, which represent a single bench-model of the pump-ejector system. This stand is designed to investigate the characteristics of model ejectors, multistage centrifugal pumps and pump-ejector systems with liquid-gas mixtures using fresh water as a liquid, and air as a gas. The differential pressure gauge was used to determine the gas flow, with allowance for the excess pressure created in

© The Author(s) 2019
S. Glagolev (Ed.): ICAM 2019, SPEES, pp. 219–220, 2019.
https://doi.org/10.1007/978-3-030-22974-0_52

the suction chamber of the ejector. For operations at low pressures, the differential pressure gauge was pre-calibrated before.

3 Results and Discussion

Calibration of the differential manometer made it possible to clarify the existing formula for determining the gas flow rate at the ejector suction chamber at pressures below 0.6 MPa. The inaccuracy of the calculated values obtained by this method was less than 1%. According to the results of the conducted research, the pressure-energy characteristics of the ejectors were constructed during gas suction by the liquid. There occurs a significant expansion of liquid-gas ejector working range by the injection of gas with excess pressure in the ejectors suction chamber as well as the areas of maximum values of injection coefficients and efficiency rates are shifted to higher region of operating pressure.

4 Conclusions

Further study of the pressure-energy characteristics behavior and the causes affecting this change will facilitate the increase of efficiency of technological processes which use the pumping-ejector systems.

References

Drozdov AN, Drozdov NA (2012) Laboratory researches of the heavy oil displacement from Russkoye field's core models at the SWAG injection and development of technological schemes of pump-ejecting systems for the water-gas mixtures delivering. In: SPE 157819, presented at the SPE heavy oil conference canada held in Calgary, Alberta, Canada, pp 12–14

Drozdov NA, Drozdov AN, Malyavko EA (2012) Investigation of SWAG injection and prospects of its implementation with the usage of pump-ejecting systems at existing oil-field infrastructure. In: SPE 160687, presented at the SPE Russian oil and gas exploration and production technical conference and exhibition held in Moscow, Russia, pp 16–18

Integrated Use of Oil and Salt Layers at Oil Field Development

V. Malyukov and K. Vorobyev[✉]

Department of Mineral Developing and Oil & Gas Engineering,
Engineering Academy, RUDN University, Moscow, Russia
k.vorobyev98@mail.ru

Abstract. Combined development of oil and salt layers of the oil field allows to obtain a mineralized solution to intensify the extraction of oil from the reservoir and create an underground reservoir in the salt reservoir for underground storage of hydrocarbons, including the creation of an underground gas storage (UGS) in rock salt.

Keywords: Oil reservoir · Salt formation · Field · Combined use

1 Introduction

The complex development of deposits and the most complete use of the resource potential of the developed deposits, as well as the multifunctional use of the waste space of mineral deposits, is one of the main tasks of mining science (Vorob'ev et al. 2017).

In a number of oil fields, the screen is a salt layer of various capacities located above the productive oil reservoir, which can be used to produce mineralized water with subsequent injection of mineralized water into the productive oil reservoir, and the production-capacity (underground reservoir) formed in the salt reservoir can serve as a storage of hydrocarbons extracted from the productive reservoir (natural gas, associated oil gas, oil).

2 Methods and Approaches

To obtain mineralized water, it is advisable to drill wells on the salt formation and conduct underground dissolution of rock salt to obtain a solution of rock salt of a certain concentration for injection through injection wells into the productive formation and extraction of oil from the productive formation through producing wells.

Pumping more fresh water into the clay-containing collector than the reservoir water reduces the permeability of the collector and makes it a low-permeable collector. Controlling the mineralization of injected water and the properties of clays in productive formations can significantly increase the oil recovery rate.

© The Author(s) 2019
S. Glagolev (Ed.): ICAM 2019, SPEES, pp. 221–222, 2019.
https://doi.org/10.1007/978-3-030-22974-0_53

3 Results and Discussion

The use of mineralized water can reduce the hydration of formation clays, but it is desirable to select the composition of water that is most compatible with the formation components of the productive formation.

The use of mineralized water obtained by dissolving the salt layer of the oil field can largely solve the problem of compatibility of pumped mineralized water with reservoir water and mineral composition of the reservoir (Lyashenko et al. 2018).

When developing a salt formation with a capacity of several tens of meters with the supply of solvent through the drilling well, according to the technological regulations, an underground production is created-a container in rock salt (vertical or horizontal) for storage of petroleum gas along the way.

4 Conclusions

Complex use of oil and salt layers of the oil field allows to obtain a mineralized solution to intensify the extraction of oil from the reservoir and to create an underground reservoir in the salt reservoir for underground storage of associated petroleum gas (APG), i.e. to create an underground gas storage (UGS) in rock salt.

At Talakan oil and gas condensate field (Republic of Sakha) the implementation of combined development of oil and salt layers of oil and gas condensate field with the creation of an underground gas storage in rock salt and the use of mineralized solution for the intensification of oil extraction was started.

References

Lyashenko V, Vorob'ev A, Nebohin V, Vorob'ev K (2018) Improving the efficiency of blasting operations in mines with the help of emulsion explosives. Min Miner Deposits 1:95–102
Vorob'ev A, Chekushina T, Vorob'ev K (2017) Russian national technological initiative in the sphere of mineral resource usage. Rudarsko Geolosko Naftni Zbornik. 2:1–8

Oil and Gas Reservoirs in the Lower Triassic Deposits in the Arctic Regions of the Timan-Pechora Province

N. Timonina[(✉)]

Institute of Geology Komi SC UB RAS, Syktyvkar, Russia
nntimonina@geo.komisc.ru

Abstract. The Lower Triassic sediments of the Timan-Pechora oil and gas bearing province were studied by the complex of lithological, petrophysical and geochemical methods. It is established that productive deposits are represented by various-grained sandstones that formed in arid climates in the vast alluvial-lacustrine plain. The research allowed to study and characterize the structure features of rock and mineral aggregates, forming a pore space. The high heterogeneity of the composition and structure of the cement minerals of the reservoir caused by local facies-paleogeographic sedimentation conditions, were the cause of the significant variability of the pore space. Rocks reservoirs are complex, with high content of clay component, an effective development of which requires special methods of stimulation.

Keywords: Oil and gas bearing province · Lithological types · Reservoir · Clay minerals · Pore space · Porosity · Permeability

1 Introduction

In recent years, researches were intensified in the field of the formation of natural reservoirs. The basis for sedimentological reconstructions lies in the idea that the morphology and filtration characteristics of natural reservoirs are largely predetermined by ancient sedimentation situations, which are closely associated with the tectonic history of the territories. The Timan-Pechora province occupies the region of the northeastern Russian platform. Hydrocarbon accumulation occurred in Triassic rocks largely in the northern part of the basin, where pools were found in the rocks in the Varandey, Toravey, Labagan fields of the Sorokin swell, Kumzhinskoe, Korovinskoe – in Denisov Depression. Triassic deposits are distributed almost throughout the Timan-Pechora oil and gas bearing province, with the exception of the axial zones of large positive structures. The Lower Triassic includes strata of the Charkabozhsky suite, which thickness varies from the first meters in the southwest (in the Seduyahinsky swell) to 380 m in the central part of the Kolvinsky megaswell, the Khoreyver depression, the average thickness of the suite is 150–250 m (Morakhovskaya 2000). The sediments are represented by rhythmic alternation of red-brown clays, greenish-gray siltstones and gray sandstones with conglomerates and gravelites. Sandstones are characterized by a variety of granulometric composition from fine to coarse-grained, as well as a wide range of textures: massive, diagonal and horizontal bedding.

S. Glagolev (Ed.): ICAM 2019, SPEES, pp. 223–226, 2019.
https://doi.org/10.1007/978-3-030-22974-0_54

2 Methods and Approaches

Oil and gas reservoirs are geological bodies consisting of reservoir beds, lenses, and reservoirs of weakly and impermeable rocks of intra-reservoir tires, forming a single hydrodynamic system, bounded below and above by inter-reservoir tires. Accumulation of hydrocarbons in the reservoir and their safety are determined by the quality of each of the elements. The main reason for the differentiation of natural reservoirs by properties is their formation. The structural features of sedimentary layers determine the patterns of distribution of collectors and seals in them, their interrelationship, and ultimately predetermine the morphology and properties of natural reservoirs.

This paper proposes a conceptual model of the formation of a natural reservoir confined to the deposits of the Charkabozhskaya suite. The construction of working models was carried out using the results of complex processing of all available information, including well-logging, and study of core material.

The results of this work are based on combination of analytical tools, including polarizing microscope, scanning electron microscope, X-ray diffraction and electron microprobe, were used in order to identify and estimate authigenic mineral type sand to determine paragenetic sequences.

3 Results and Discussion

At the first stage of creating a geological model of a terrigeneous reservoir, separate strata were correlated. The selection and tracing of the layers was carried out by logging diagrams; local surfaces associated with homogeneous rocks, more or less sustained over the area, were taken as main reference points; within individual areas, additional benchmarks were used, which were characterized by stable geophysical characteristics. Two productive formations were confined to the Lower Triassic deposits. To maximally take into account the features of the structure of productive layers at the first stage of modeling, the study of microscopic in homogeneities of productive sediments was carried out on the basis of lithofacies and sedimentation models.

The main reason for the heterogeneity of natural reservoirs in terms of the properties of the reservoirs and tires that form them is the conditions of their formation. Detailed lithofacies analysis of Lower Triassic deposits was carried out, based on a set of investigations by both domestic (Muromtsev 1984; Bruzhes 2010; Morozov 2013) and foreign researchers (Celli 1989; Hellem 1983). The analysis of geological and geophysical information, the study of well cores, and the interpretation of well logging data allowed reconstructing conditions for the formation of lower Triassic natural reservoirs. The formation of these sediments took place in the continental conditions of the alluvial plain.

According to classification by A.G. Kossovskaya and M.I. Tuchkova (Kossovskaya, Tuchkova 1988) sandstones fall into the field of polymictic (SiO_2 content 62–78%) and volcanic (SiO_2 content 54–64%). According to classification of Pettijohn (1976) points of the composition of sandstones are localized in the fields of graywacke.

Dominant cement minerals in Triassic sandstones include calcite, kaolinite, smectite, illite, and chlorite. Carbonate cement include pure calcite and siderite. The clear transparency and delicate crystalline habit of most of the clay, as revealed by microscope and SEM analysis, leads us to conclusion that the most clays in these sandstones are authigenic. Kaolinite occurs mostly as pore-lining and pore-filling cement, some of them appear as visible alteration of detrital feldspars. The distribution of kaolinite is uneven both in the section and in area. In the transition from coarse to fine-grained sediments, a decrease in the content of kaolinite is noted. Chlorite is common as pore-filling and lining cement. Smectite occurs both as grain-coating and filling pores. The studied deposits are characterized by a wide distribution of smectite minerals with increasing content upward the section. In the fine-grained sandstones of the basal stratum, its content does not exceed 50–60%, whereas in the upper part of the section its amount increases to 80–90%. The distribution and amount of clay minerals in sandstone cement is determined by both conditions of sedimentation and post-sedimentation transformations. The most widespread collectors are of III–V classes according to the A.A. Khanin classification (Khanin 1976). Class VI reservoirs are characteristic of floodplain formations and represented by aleurolites and fine-grained sandstones, in which large pore channels are practically absent. Class V collectors are represented by fine-grained sandstones with pore cement predominantly of smectite composition, low values of filtration properties are due to the insignificant content of large pore channels (less than 5%) and increasing number of non-filtered pores. These formations were deposited in floodplain conditions. The class IV reservoirs include fine- and medium-grained sandstones with polymineral cement: clayey pore and carbonate clot-pore type.

Collectors of II–III classes are represented by coarse-grained and medium-grained poorly sorted sandstones with pore-type cement, formed in an environment with a relatively quiet hydrodynamic regime.

4 Conclusions

The analysis of the reservoir properties of sandstones shows that sandstones formed under channel conditions are characterized by high median and average values of porosity and permeability, respectively 24% and 20.8% (porosity), $56 \cdot 10^{-12}$ and $40 \cdot 10^{-12}$ m^2, the lowest values characterize sediments formed in floodplain conditions: the median values of porosity and permeability do not exceed 11% and $0.6 \cdot 10^{-12}$ m^2, arithmetic averages reach 14% and $1.6 \cdot 10^{-12}$ m^2. Analysis of the graphs of the dependence of porosity-permeability for productive layers showed that the highest values of filtration properties were typical to deposits of the basal layer, where I–II class reservoirs were identified.

A reliable seal of subregional distribution is the deposits of the Upper charkabojskaya sub-suite, having high quality and thickness. Minerals of the smectite dominate in the mineral composition, chlorite and illite are also present. Middle-Upper Triassic clays are floodplain and lake originated, in their composition chlorite and illite predominate, which reduces the quality of the foamed seal compared to the Charkabozhskaya seal.

The differences in the composition and type of cement requires an individual approach, a balanced choice of technologies in determining the development strategy of fields and careful selection of a set of methods aimed at increasing oil recovery for different sections of fields.

Acknowledgements. The article was created under partial financial support by project of Ural Branch of Russian Academy of Sciences № 18-5-5-13 "The Geological Models, Environmental Conditions and Prospects of Oil and Gas bearing of Phanerozoic deposits in Arctic regions of Timan-Pechora province".

References

Bruzhes LN, Izotov VG, Sitdikova LM (2010) Litofacial conditions of formation of the horizon of the Yu1 Tevlinsko-Russkinskoye field of the West Siberian oil and gas province. Georesources 2(34):6–9

Celli RCh (1989) Ancient sedimentation. Nedra, Moscow, p 294

Hellem E (1983) Interpretation of facies and stratigraphic sequence. Mir, Moscow, p 328

Khanin AA (1976) Petrophysics of oil and gas reservoirs. Nedra, Moscow, p 259

Morakhovskaya ED (2000) Trias of the timan-ural region (reference sections, stratigraphy, correlation). Biochronology and correlation of the Phanerozoic oil and gas basins of Russia, vol 1, p 80. SPb: VNIGRI

Morozov VP, Shmyrina VA (2013) Influence of secondary changes in reservoir rocks on reservoir properties of the БС111 and ЮС11 productive layers of the Kustovoye deposit. Uchenye zapiski Kazan University, Kazan, vol 155, pp 95–98

Muromtsev VS (1984) Electrometric geology of sand bodies - lithological traps of oil and gas. Nedra, p 260

Pettijohn FJ, Potter PE, Siever R (1976) Sand and sandstone. Mir, Moscow, p 536

Associated Petroleum Gas Flaring: The Problem and Possible Solution

A. Vorobev[1,2(✉)] and E. Shchesnyak[2]

[1] Atyrau University of Oil and Gas, Atyrau, Kazakhstan
fogel_al@mail.ru
[2] Peoples' Friendship University of Russia (RUDN University),
Moscow, Russia

Abstract. The article analyzes the current state and prospects for utilization of a hydrocarbon component dissolved in oil and released during its extraction and preparation - associated petroleum gas (APG). The authors studied the properties, characteristics and component composition of APG.

The analysis of the APG use at the international and regional levels is carried out. The main causes of flaring were discussed and the shortage of production capacities for APG processing in the Russian Federation was noted as one of the main factors in the high level of APG flaring in the country.

The paper notes possible ways of utilization of associated petroleum gas, which depend on oil production conditions, such as field characteristics, oil/gas ratio (gas-oil factor), and market opportunities for recovered gas. An overview of all APG utilization methods are presented, which focuses on unit costs, economic benefits and environmental impact reduction. The authors analyzed the innovative experience of effective APG use in the USA and Canada. Special attention is paid to the need to solve the problem of the effective use of APG in the Russian Federation, especially the reduction of its burning in flare plants.

Keywords: Associated petroleum gas · Associated petroleum gas utilization · APG flaring · Environmental pollution

1 Introduction

Associated petroleum gas (APG) is a kind of natural gas that is in oil deposits, either dissolved in oil or as a free "gas cap" over oil into the deposits. Regardless of the source, as soon as it separates from crude oil, it usually exists in mixtures of other hydrocarbons such as ethane, propane, butane and pentane; in addition, APG contains water vapor, hydrogen sulfide (H_2S) and carbon dioxide (CO_2), nitrogen (N_2) and other mixtures. Associated petroleum gas containing such impurities so it can not be transported and used without purification, as it is extracted in the process of oil production (Kartamysheva et al. 2017).

The volume and composition of APG depends on the area of production and on the specific properties of the field. In the process of extraction and separation of one ton of oil, it is possible to obtain from 25 to 800 m^3 of associated gas. Some of this gas is used or stored, because governments and oil companies have made significant investments

S. Glagolev (Ed.): ICAM 2019, SPEES, pp. 227–230, 2019.
https://doi.org/10.1007/978-3-030-22974-0_55

for its extraction. However, individual companies burn APG because of technical, regulatory or economic constraints. As a result, thousands of flare stacks from more than 17,000 oil production facilities around the world burn about 140 billion cubic meters of natural gas per year, resulting in more than 350 million tons of CO_2 and a large variety of pollutants, including very dangerous.

The overall increase in global flaring compared to previous years is largely due to negative developments in only a few countries: Iran, Russia and Iraq. Satellite data show an increase in flaring in Iran by more than 4 billion m^3, in Russia by almost 3 billion m^3 and more than 1 billion m^3 in Iraq. Flaring in Russia is close to the global average compared to oil production; in the other two countries, flaring intensity is higher. This is the expenditure of a valuable energy resource that can be used to promote the sustainable development of producing countries. Thus, 149 billion m^3 of associated petroleum gas, burned in 2018, could turn into 750 billion kWh of electricity, which exceeds its total annual consumption by all countries of the African continent.

According to official data, the volume of extracted APG in Russia increased by more than 7% - to 65 billion m^3 in 2010 and over 70 billion m^3 in 2018. The immediate impact on the increase in the volume of recoverable APG was made by the growth of oil production in new areas, including the East Siberian fields.

2 Methods and Approaches

For a long time, oil companies simply burned this unwanted by-product. Its flaring requires a significant portion of the security system.

The term "gas flaring" indicates a gas combustion (without energy recovery) in an open flame, which is continuously lit on top of the flare stacks in the field of oil production (Knizhnikov et al. 2017).

Flaring occurs for three main reasons:

- emergencies: limited incineration for reasons of safety for short periods of time can always be necessary even after connecting the gas gathering pipeline;
- deficiency of gas utilization capacity - isolated well flaring: if the well starts to produce oil and gas without connecting to gas gathering pipeline or other gas utilization technology, the gas can be shut off;
- lack of gas utilization capacity - well burning through the pipeline: if the well is connected to gas gathering pipeline, but these systems can not process all gas from the well (due to lack of power or compression), some or all of the associated gas from the well can be flared.

Billions of cubic meters of natural gas are burned at oil production sites around the world. Gas combustion is a costly energy resource that can be used to support economic growth and progress (Vorob'ev et al. 2017).

Since 2012, the US National Oceanic and Atmospheric Administration and the Global Gas Flaring Reduction Partnership have begun to apply a method for estimating the volume of APG flared. This method consists in the use of satellite observational data in the visible and near infrared ranges.

The conclusion drawn from the research results is that the volume of flared APG in the world increased to 147 billion m^3 in 2015, compared with 145 billion m^3 in 2014 and 141 billion m^3 in 2013. According to data for 2015, Russia led this "anti-rating", burning 24 billion m^3 of APG, followed by Iraq (17.5 billion m^3), Iran (16 billion m^3) and the United States (8 billion m^3). Russia is also a "leader" (in third place after the USA and Canada) with 1,814 flare stacks, which burn APG.

At present, there are other possible ways of utilization of associated gas, alternative to flaring. Among them it is necessary to single out the following:

- re-injection of APG into oil reservoirs to maintain pressure and increase oil recovery (as a method of increasing oil recovery), or for possible conservation of it as a resource and use in the future;
- the use of gas as an energy source for production site or at oil producing facilities in the vicinity;
- the most effective way of utilization of associated petroleum gas is its processing at gas processing plants to produce dry stripped gas (SOH), a wide fraction of light hydrocarbons (LH), liquefied natural gas (LNG) and stable gasoline (SG).

Below is an overview of all APG utilization methods, which focus on unit costs, economic benefits and environmental impacts.

3 Results and Discussion

The indicator of its useful use has remained stable since the 2000s, within 73–79% of the total amount of extracted APG in the country. Only in 2014–2017, according to the public accounts of companies, it rose to 85–86%. According to the statement of representatives of government organizations, the indicators of productive processing of APG amounted to 90% in 2018.

The amendments to the law "On environmental protection" (№ 219) adopted in July 2014 caused such as significant reduction in the share of associated gas combustion, according to these amendments, the company is obliged to establish its technological standards at the level of application of the best available technologies. The total investment in increasing the useful use of APG was estimated at 200 billion rubles. According to the Ministry of energy of the Russian Federation, it is expected that the target of 95% of the associated gas will be used by the end of 2020 (Vorobyev et al. 2018).

Foreign experience of utilization of APG shows that flaring of gas in torches has decreased slightly over the past two years, and oil production has also declined. In particular, Nigeria reduced volume APG combustion to 8 billion m^3 of nearly 18% compared with 2013 year. The volume of associated gas flaring in the USA decreased from 11 billion m^3 in 2016 to less than 9 billion m^3 in 2018 due to the use of a number of innovative low-volume technologies.

One of the innovative technologies for the production of liquefied natural gas with small volumes of associated petroleum gas is LNG Production (Production Natural Gas Liquids, LH-Pro). The "LH Pro" process combines dehydration, compression, cooling and conditioning, eliminating the need for expensive glycol and refrigeration systems.

Hydrate formation is excluded due to the thermal integration system. The technology was developed by ASPEN and is used in the US and Canada.

4 Conclusions

Thus, the most rational ways of utilization of associated petroleum gas in Russia, depending on the volumes of its extraction are:

- at small volumes - covering own energy needs;
- with increased volumes - electricity generation and primary processing of APG to produce lean dry gas (LDG) as fuel for the boiler room and light hydrocarbons (LH) for disposal to the oil collector;
- at resources from 50 to 150 million m^3/year - processing with obtaining LDG, as well as LH and electricity;
- with the amount of APG in excess of 150 million m^3/year, processing of LDG, and NGL is recommended.

References

Kartamysheva YeS, Ivanchenko DS (2017) Associated petroleum gas and the problem of its utilization. Molodoy uchenyy. № 25, pp 120–124

Knizhnikov AYU, Il'in AM (2017) Problems and prospects of associated petroleum gas use in Russia. 2017 WWF Rossii, Moscow, p 34

Vorob'ev A, Chekushina T, Vorob'ev K (2017) Russian national technological initiative in the sphere of mineral resource usage. Rudarsko Geolosko Naftni Zbornik 2:1–8

Vorobyev AE, Zhang L (2018) Analysis of production and consumption of associated petroleum gas in China. Bulletin of Atyrau Institute of oil and gas № 2(46), pp 137–142

Innovative Technology of Using Anti-sand Filters at Wells of the Vankor Oil and Gas Field

K. Vorobyev$^{(\boxtimes)}$ and A. Gomes

Department of Mineral Developing and Oil & Gas Engineering,
Engineering Academy, RUDN University, Moscow, Russia
k.vorobyev98@mail.ru

Abstract. In this article, the authors considered an innovative technology for restricting sands to reduce complications and watering during the development of the Vankor gas and oil field.

Keywords: Sand control filters · Waterflood · Well · Borehole zone · Formation pressure drop

1 Introduction

The development of oil and gas fields in complex mining and geological conditions is associated with various complications - a drop in reservoir pressure, an increase in the water content of the crude product and the amount of mechanical impurities, destruction of the integrity of the well bottom zone. Analysis of the sources (Vorob'ev et al. 2017) shows that sand manifestations are a multifactorial and multi-element sophisticated technical system.

The task of managing the sand formation processes includes such elements as sand forecasting and effective methods of influencing the sand manifestation phases in order to reduce negative effects.

2 Methods and Approaches

Studies confirm that during the removal of rock particles from the reservoir in the operation of wells in the bottom zone form the high permeability channels of various widths and lengths of fractures along and bedding planes, by which the bulk of gas and produced water is filtered (Ozhogina et al. 2017).

When considering the performance of geological and technical measures at the stockwell of OJSC Gazprom (Lyashenko et al. 2018) in the direction of technology for the elimination of sand production and the removal of wells from inactivity, the causes and factors of sand production are given: poorly cemented formation; the viscosity of the formation fluid; the velocity of fluid particles in the reservoir.

© The Author(s) 2019
S. Glagolev (Ed.): ICAM 2019, SPEES, pp. 231–232, 2019.
https://doi.org/10.1007/978-3-030-22974-0_56

3   Results and Discussion

Pilot tests of filters of various designs at the Vankor oil and gas field. During 2018, pilot industrial tests of filters of various designs were carried out: slotted and multi-layer mesh.

The filter descends on an unremovable packer into the perforated interval zone and is placed opposite the entire interval in order to minimize the skin effect. The well is operated using a high-rate installation of an electric submersible pump.

At the base of a 168 mm column, filters with a total length of about 200 m are lowered into the perforation intervals. Standard filter life is 24 months. Behind the filter, another is formed - a natural filter.

The organization of the reservoir pressure maintenance system at the Vankor oil and gas field involves the use of both produced water and water from water wells.

4   Conclusions

According to the results of the analysis of the geological and technical conditions of water wells and the characteristics of the aquifers of the Vankor oil and gas field, it was decided to use downhole anti-sand filters to reduce the effect of removal of mechanical impurities. The results of the pilot-scale industrial tests were considered successful, which gave rise to the decision to implement filters on an industrial scale. As of January 2019, all water wells in the Vankor oil and gas field are equipped with anti-sand mesh and slot-type filters

References

Lyashenko V, Vorob'ev A, Nebohin V, Vorob'ev K (2018) Improving the efficiency of blasting operations in mines with the help of emulsion explosives. Min Mineral Deposits. 1:95–102

Ozhogina EG, Shadrunova IV, Chekushina TV (2017) Mineralogical rationale for solving environmental problems of mining regions. Gornyi zhurnal. 11:105–110

Vorob'ev A, Chekushina T, Vorob'ev K (2017) Russian national technological initiative in the sphere of mineral resource usage. Rudarsko Geolosko Naftni Zbornik. 2:1–8

Analytical Methods, Instrumentation
and Automation

Thermometry of Apatite Saturation (The Kozhym Massif, The Subpolar Urals)

Y. Denisova[✉], A. Vikhot, O. Grakova, and N. Uljasheva

Institute of Geology of the Komi SC UB of RAS,
Syktyvkar, Komi Republic, Russia
yulden777@yandex.ru

Abstract. The results of the study of accessory apatite from the Kozhym massif rocks have been presented in this paper. Apatites of the same morphological type were found in granites. The Kozhym massif granites formation temperatures by apatite were determined by the Watson and Bea saturation thermometry. These temperatures were compared with the previously obtained ones for accessory zircon of the same massif.

Keywords: Apatite · Granite · The Kozhym massif · The Subpolar Urals · Watson · Bea

1 Introduction

The Kozhym massif is located in the northeastern part of the Subpolar Urals on the left and right banks of the Kozhym River in the Oseu and Ponyu Rivers basins (Fig. 1) among deposits of the Puyvinian Middle Riphean Formation. This massif is the second by area among the geobodies composing the Kozhym intrusion which includes the Kuzpuayu granite massif. The most fully preserved Kozhym massif granites are medium-grained pink greenish-gray rocks. They (the rocks) are characterized by a massive coarse-platy texture with well-defined tectonic gneissiness. The studied massif belongs to the A-type according to B. Chappel classification (Fishman et al. 1968).

The accessory massif apatite is represented by yellow mat elongated crystals of a hexagonal dipyramidal-prismatic habitus. The crystals size is 0.1–0.4 mm, the elongation coefficient is 1.5–3. The mineral shape is represented by a combination of a prism (1010) and a dipyramid (1011). The faces surface is fractured. Characteristic inclusions are quartz, plagioclase, zircon.

2 Methods

Apatite is increasingly used as a geothermometer. The E. Watson saturation thermometry was used to determine the mineral crystallization temperatures. This method allowed determining the apatite and rock formation temperature by the distribution of the phosphorus oxide content between apatite and the rock containing the mineral. The level of phosphorus saturation necessary for the apatite crystallization depends on the

S. Glagolev (Ed.): ICAM 2019, SPEES, pp. 235–238, 2019.
https://doi.org/10.1007/978-3-030-22974-0_57

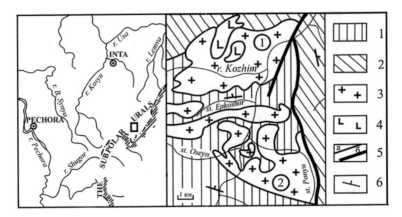

Fig. 1. The geological Kozhym granite massif map. 1 – mica-quartz schists, green orthoschists, quartzits; 2 – mica-quartz schists, porphyries, porphyrites, marbles and quartzits interlayers; 3 – granites; 4 – gabbro; 5 – contact lines: a – stratigraphic and magmatic, b – tectonic; 6 – planar structured bedding elements. Massifs (numbers in circles): 1 – the Kuzpuayu; 2 – the Kozhym.

silica content in the meta-aluminic rock (A/CNK < 1) and the temperature according to the Harrison and Watson calculations (Harrison et al. 1984):

$$InDp = (8400 + 26400\,(SiO_2 - 0,5))/T - 3,1 - 12,4\,(SiO_2 - 0,5),$$
$$P_2O_5(HW) = 42/Dp,$$

where Dp is the ratio of P concentration in apatite and melt, P_2O_5; SiO_2 is the weight fraction of the phosphorus oxide, silica in the melt, wt. %, T is the temperature, Kelvin.

Bea (Bea et al. 1992) proposed the following addition to the Watson's formula for peraluminum rocks (A/CNK > 1):

$$P_2O_5(Bea) = P_2O_5(HW) * P_2O_5(HW) * e^{\frac{6429(A/CNK-1)}{T-273,15}}.$$

3   Results and Discussion

The apatite saturation temperatures based on the data of the chemical granites composition of the Kozhym massif (Table 1) were calculated by the author according to Watson and Bea (Table 2).

The Kozhym massif rocks formation temperatures distribution histograms by apatite were compiled on the obtained temperatures for each calculation method (Fig. 2).

Table 1. Chemical granites composition of the Kozhym massif

Component, wt. %	Sample number									
	K-1	K-2	K-3	K-4	K-5	K-6	K-7	K-8	K-9	K-10
SiO_2	77.78	76.89	75.95	75.89	76.49	78.12	77.54	76.26	77.48	76.95
TiO_2	0.16	0.22	0.48	0.52	0.48	0.11	0.24	0.42	0.31	0.59
Al_2O_3	11.88	11.95	12.69	12.52	10.05	11.34	11.78	12.22	11.09	10.92
FeO	1.72	1.29	1.15	1.24	0.56	0.50	0.59	0.62	0.61	1.03
Fe_2O_3	0.84	1.12	0.52	1.05	0.92	1.21	0.87	1.02	0.89	0.56
MnO	0.02	0.00	0.00	0.01	0.02	0.01	0.02	0.04	0.03	0.03
MgO	0.16	0.25	0.17	0.33	0.38	0.39	0.18	0.29	0.19	0.18
CaO	0.31	0.29	0.22	0.38	0.59	0.28	0.45	0.42	0.37	0.51
Na_2O	3.65	3.33	4.22	4.02	3.08	3.15	3.22	4.51	3.01	3.89
K_2O	3.88	4.51	4.09	3.89	4.15	5.17	4.99	3.78	4.65	3.28
P_2O_5	0.01	0.03	0.02	0.02	0.03	0.02	0.01	0.01	0.03	0.02
ппп	0.05	0.29	0.59	0.15	0.75	0.62	0.39	0.98	1.02	1.23
$\sum$	100.46	100.17	100.10	100.02	97.50	100.92	100.28	100.57	99.68	99.19

Note. The chemical composition was obtained using the silicate method in CUC ≪Science≫ of Institute of geology of Komi SC UB RAS (analyst Koksharova O.V.)

Table 2. Saturation temperatures for the Kozhymmassif apatite

Temperature, °C	Sample number									
	K-1	K-2	K-3	K-4	K-5	K-6	K-7	K-8	K-9	K-10
According to Watson	784	860	819	818	856	838	782	770	865	828
According to Bea	722	798	770	764	856	836	768	770	840	826

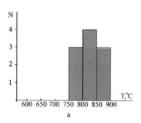
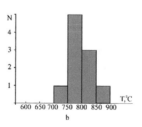

Fig. 2. Formation temperatures distribution histograms for the Kozhym massif granites. a - according to Watson, b - according to Bea

The presented histograms show that the studied granites are high-grade rocks. The Kozhym massif formation occurred at temperatures from 770 °C to 865 °C and averaged 822 °C according to Watson. The application of Bea refinements allowed concluding that the massif rocks formation occurred at temperatures from 722 °C to 856 °C and an average of 795 °C.

4 Conclusions

The Kozhym massif rocks formation occurred at high temperatures. The similar massif granites formation temperature ranges: 770–865 °C according to Watson, 722–856 °C according to Bea were obtained using the Watson saturation thermometry by apatite and the Watson formula adjustment for aluminous rocks according to Bea. The data confirmed the temperatures obtained earlier by the author using the Watson saturation thermometry (749–816 °C) and the classical evolutionary-morphological Pupin analysis (700–900 °C) for zircon of the same massif (Denisova 2016; 2018a; b). It can be affirmed that the Watson saturation thermometry and the refined Bea formula for apatite provide the same information about regime temperature evolution during the granites formation such as on the Watson saturation thermometry and the Pupin and Tyurko evolutionary-crystallomorphological analysis for zircons.

Acknowledgements. The work was supported by the Basic Research Program of RAS № 18-5-5-19.

References

Bea F, Fershtater GB, Corretgé LG (1992) The geochemistry of phosphorus in granite rocks and the effects of aluminium. Lithos 48:43–56

Denisova UV (2018a) Apatite of the Nikolaishor granite massif (the Subpolar Urals). Vestnik of Institute of geology of Komi SC UB RAS, Syktyvkar, № 9, pp 24–29

Denisova UV (2018b) Crystal morphology of zircon in solving problems of the Kozhimsky massif granites genesis (the Subpolar Urals). Trends in the development of science and education, №35, Part 3. Samara, pp 45–48. https://doi.org/10.18411/lj-28-02-2018-51

Denisova UV (2016) Temperature survey of zircon from the granitoids of the Subpolar Urals. Vestnikof Institute of geology of Komi SC UB RAS, Syktyvkar, №12, pp 37–44

Fishman MV, Yushkin NP, Goldin BA, Kalinin EP (1968) Mineralogy, typomorphism and genesis of accessory igneous rocks minerals of the Urals and Timan north. M.-L.: Sci 252

Harrison TM, Watson EB (1984) The behavior of apatite during crustal anatexis: equilibrium and kinetic considerations. Geochim Cosmochim Acta 48:1467–1477

Studies of Structural Changes in Surface and Deep Layers in Magnetite Crystals After High Pressure Pressing

P. Matyukhin$^{(\boxtimes)}$

Belgorod State Technical University named after V.G. Shukhov,
Belgorod, Russia
mpvbgtu@mail.ru

Abstract. The article introduces the study of structural changes in surface and deep layers in magnetite crystals, on samples under high pressure pressing. Magnetite (magnetite iron-ore concentrate) is widely used as a filling compound of new composites which are planned to be used in nuclear-construction. These compounds are based on the aluminum containing matrix with a filling compound. This modern composite material can be used in construction structures able to resist significant loads, operate in such extreme situations as abrupt dynamic loads, fires with further alternating temperature oscillations.

Keywords: Magnetite · Pressure · Crystal · Structure · Layer · Material

1 Introduction

Nowadays many scientists develop new kinds of materials, including composites which can be used in nearly all spheres of human life. Composite materials based on organic and inorganic components with different additives and fillers are created. These components can be the base for different matrixes. Fillers with different qualities are introduced into these matrixes depending on the purpose, scope of operation, conditions of service of the designed composite. These matrixes can have metal, ceramics, concrete, polymer and other bases, and the fillers may be cast iron shot, barites, metal processing remains, mining company products (for example, iron containing rocks) and of other enterprises (Grishina and Korolev 2016; Barabash et al. 2017; Matyukhin 2018; Gulbin et al. 2018; Kruglova 2009; Laptev et al. 2015; Gulbin et al. 2016; Matyukhin et al. 2011, Samoshin et al. 2017).

Composite materials based on metal matrixes can be used in bearing constructions production. Such constructions can resist high mechanical impacts, intensive ionizing radiation and alternating temperatures, that can widely be used in nuclear construction industry, and foremost, for biological protection on sites with power supplies of different nature. Nowadays, the issue of aluminum containing matrixes and iron containing fillers utilization in designing new types of radiation-protective construction composite materials becomes acute. One of promising fillers for new composite

S. Glagolev (Ed.): ICAM 2019, SPEES, pp. 239–243, 2019.
https://doi.org/10.1007/978-3-030-22974-0_58

materials is magnetite. The introduction of modern composite materials of natural iron ore raw materials will give them additional mechanical strength, increase their radiation shielding characteristics. This filler is widely found in nature, is relatively cheap, and composites on its base will meet ecological requirements as they are a part of the ecosystem. Here a possibility of combining magnetite iron ore concentrate with aluminum containing matrix and receiving composite material by high pressure pressing is of great scientific interest. In this research it is necessary to solve issues about maximum possible filler compacting in its composite material matrix, studying composite filler behaviour at high pressing pressures, in particular studying structural changes in surface and deep layers of magnetite crystals at pressing pressures when such a composite material is produced, and compact-space position of its particles relative to each other.

2   Methods and Approaches

As an object of research we used a highly-dispersive enriched magnetite iron-ore concentrate from the Lebedinsk deposit of the Kursk Magnetic Anomaly with density 4950 kg/m^3, Mohs hardness – 6, black colour; the magnetite is in form of irregular needle-shaped grains, octahedral crystals, shell-like fracture; the mineral composition is introduced by magnetite with inclusions of carbonate (chalk-stone) and siliceous (quartz) admixtures. After cleaning and chemical treatment of the magnetite iron ore concentrate per 99,8% its mineral composition is introduced by magnetite ($FeO.Fe_2O_3$) with fractional composition of particles 0,05–15 mcm. In the research we used a modern electronic microscope "TESCAN MIRA 3 LMU".

3   Results and Discussion

For experiment fairness the magnetite samples in the form of a high dispersive powder with particles 0,05–15 mcm were used, without adding it into the aluminium containing matrix. Figure 1 shows the structure of surface and deep layers of magnetite particles without pressure treatment (Fig. 1a) and after pressure pressing 5000 MPa (Fig. 1b) with magnifying power 10 mcm. The analysis of the received data showed that the surface and deep layers of magnetite crystal particles after pressure pressing at 5000 MPa (Fig. 1b), in comparison with the micro photo of magnetite crystal particles not pressed (Fig. 1a), there are spaces with loose distribution of its particles along the whole sample volume: there are numerous empty spaces; particles have irregular form with hard aggregation and rough edges. Figure 2 shows the structure of surface and deep magnetite particle layers, after pressure 10000 MPa and 20000 MPa with magnifying power 10 mcm.

a)   *b)*

Fig. 1. Magnetite surface with magnifying power 10 mcm: (*a*) without pressure pressing treatment, (*b*) after pressure treatment 5000 MPa

The received data analysis showed that at pressure pressing increase up to 10000 MPa (Fig. 2a) magnetite crystal particles were distributed more compactly within the whole volume of the studied sample: also there are numerous empty spaces but with a slight reduction of their geometrical sizes; the magnetite particles structure changed insignificantly, as before they had irregular form with strong aggregation and rough edges, but here particles of smaller fraction appeared, that evidenced the process of partial mechanical destruction of its particles.

 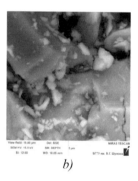

a)   *b)*

Fig. 2. Magnetite surface with magnifying power 10 mcm: (*a*) after pressure pressing 10000 MPa, (*b*) after pressure pressing 20000 MPa

When the pressure increases up to 20000 MPa on the magnetite crystal particles (Fig. 2b), there is a more compact visual magnetite particle packing both in surface and deep layers of the studied sample; there is a significant increase of zones with good compaction. There is a greater degree of destruction of magnetite crystal particles surface in the total amount of material samples in comparison with previous pressure pressing; somewhere surface aggregation and hematite particles edges is smoothed.

It evidences qualitative distribution of hematite particles of all fraction composition along the whole sample volume, but micro photos show remaining zones with "insufficient compaction", where there is a lack of fine-fractioned magnetite particles.

4 Conclusions

Based on the studies of structure changes in surface and deep layers of magnetite crystals, after high pressure pressing and studied with electron microscope, we can make an assumption that the structure of the composite may be denser by adding magnetite into its filler 15–25% (by volume) of its particles, but smaller fraction in comparison the studied one. It is theoretically possible that after adding such an amount of magnetite fine-fraction a higher degree of sample material compacting will be achieved. Or to conduct the studies with adding the same percentage ratio into the volume of studied composite material matrix samples in the form of aluminum powder, as it has particles less in size than magnetite particles. We think that these studies have high theoretical and practical significance and they should be taken into account when developing new composite materials based on magnetite filler and different metal matrixes.

Acknowledgements.   The work is realized in the framework of the Program of flagship university development on the base of the Belgorod State Technological University named after V.G. Shukhov, using equipment of High Technology Center at BSTU named after V.G. Shukhov.

References

Barabash DE, Barabash AD, Potapov YuB, Panfilov DV, Perekalskiy OE (2017) Radiation-resistant composite for biological shield of personnel. In: IOP conference series: earth and environmental science. C. 012085

Grishina AN, Korolev EV (2016) New radiation-protective binder for special-purpose composites. Key Eng Mater 683:318–324

Gulbin VN, Kolpakov NS, Gorkavenko VV, Boikov AA (2018) Research of the structure and properties of radio and radiation-protective polymer nanocomposites. J Electro-magnetic Waves Electron Syst 23(1):4–11

Gulbin VN, Martsenuk AV, Gorkavenko VV, Cherdyntsev VV (2016) Polymeric composites for radio and radio active protection. Sci Intensive Technol 17(10):7–12

Kruglova AN (2009) Radiation protective materials based on industrial wastes: physic-mechanical properties. Reg Archit Const 1:53–56

Laptev GA, Potapov Y, Yerofeev VT (2015) Development of manufacturing technology metalloconcretes. Build Reconst 1(57):123–129

Matyukhin PV (2018) The choice of iron-containing filling for composite radioprotective material. In: IOP conference series: materials science and engineering 11. International conference on mechanical engineering, automation and control systems 2017 - material science in mechanical engineering. C. 032036

Matyukhin PV, Pavlenko VI, Yastrebinskiy RN, Bondarenko YuM, (2011) Prospects of creating modern highly constructive radiation-protective metalocomposites. Bulletin of BSTU named after V.G. Shukhov. 2, 97

Ochkina NA (2018) Heat stability of radio-protective composite based on aluminous cement and polymineral industrial waste. Pridneprov Sci Bull 3(2):007–010

Samoshin AP, Korolev YV, Samoshina YN (2017) Internal stresses at metal concrete structure formation for protection from radiation. Bull. BSTU Named After V.G. Shukhov 6:13–17

The Potential of Lacquer Peel Profiles and Hyperspectral Analysis for Exploration of Tailings Deposits

W. Nikonow[✉] and D. Rammlmair

Federal Institute for Geosciences and Natural Resources (BGR),
Hanover, Germany
wilhelm.nikonow@bgr.de

Abstract. Lacquer peel profiles are a valuable technique to preserve and study sedimentary structures and depositional features outside the field. The use of the lacquer fluid Mowiol® has shown to work well as a simple and rapid preparation technique for tailings material from a tailings heap in Copiapó, Chile. The combination of lacquer peels with 2D mapping techniques such as Hyperspectral Imaging (HSI) can provide important sedimentary information such as particle size distribution on a large scale with little effort, which becomes quantifiable due to continuous 2D information and modern image analysis. The presented relationship of element or mineral distribution with particle size can serve as a tool for targeted and more focused sampling for mineral exploration in tailings and tailings analysis in general, considering future selective mining for economic or environmental reasons.

Keywords: Lacquer profile · Sedimentology · Hyperspectral Imaging ·
Tailings exploration

1 Introduction

Creation of lacquer peels is an old, but rarely recognized technique enabling preservation of sedimentary structures of granular materials. The profiles are also known as sedimentary peels, sediment plates or lacquer profiles and are similar to the early description of soil monoliths by e.g. Voigt (1936). As a relatively simple and inexpensive technique with great possibilities in various geoscientific fields it represents a valuable method to study and preserve complex depositional processes and sedimentary textures on a scale from millimeters up to several square meters.

The general procedure is to apply the lacquer fluid on the target area, then let the fluid dry, solidify, remove the peel and fixate it on a board for transport and display. A detailed technical description of the preparation process can be found e.g. in Van Baren and Bomer (1979). The impregnation takes place due to gravity and capillary forces, which makes it possible to create vertical profiles.

The advantage of lacquer peel preparation is the possibility to preserve and transport features of interest for additional analysis on the site or in the laboratory. The combination of lacquer peels with emerging analytical techniques for 2D mapping such

© The Author(s) 2019
S. Glagolev (Ed.): ICAM 2019, SPEES, pp. 244–247, 2019.
https://doi.org/10.1007/978-3-030-22974-0_59

as Hyperspectral Imaging (HSI) can provide non-destructively valuable information from e.g. textural, sedimentological and geochemical analysis.

HSI is commonly used in remote sensing and finds its path into other fields of geoscience. HSI measures the reflectance of the samples in a wavelength range of, in this work, the visible and near infrared region (VIS-NIR: 400–2500 nm). It utilizes molecule - light interaction such as vibrational processes or electron energy level transitions which produce characteristic absorption features at certain wavelengths (Hunt and Salisbury 1970).

2 Methods and Approaches

The lacquer profiles were taken from a tailings heap in the city of Copiapó in northern Chile. According to historic aerial images it was deposited ca. 1993 on an area of ca. 230×130 m with an approximated mass of 400.000 t in 2018. There is only little information about the origin of the heap, but the material is most probably from a nearby processing plant using flotation or leaching to process ore from one of the many mines in the Punta del Cobre district around Copiapó. Since the heap is not covered, it is a target for erosion by wind and rain affecting the adjacent area including the Copiapó River.

Three lacquer profiles were taken from the steep parts of the eastern and western sides, where the internal layering becomes visible and accessible. To create the profiles the polyvinyl alcohol Mowiol® 4-88 was used. It is dissolved in hot water a day before preparation. At the site, the selected surface is scraped even and the Mowiol is applied on the surface by a spray bottle or a brush. For stabilization, several layers of Mowiol and gauze are laid on the back of the profile. Then the profiles are left to dry for about 24 h. Finally, they can be removed with a knife and, wrapped in cloth, can be transported easily. Preparation of one profile of about 30×50 cm takes about 30 min excluding the time for drying. One profile was taken from the eastern side (LP1) and two profiles from the western side (LP2, LP3). Additional samples from each layer were taken next to the profiles for geochemical, particle size analysis and verification of the HSI analyses.

The dry profiles were analyzed in the laboratory with the SisuRock system from Specim. The images were taken with two push broom cameras covering the wavelength range from 400 to 2500 nm (VNIR and SWIR) at a spatial resolution of 875 μm (SWIR) and 280 μm (VNIR). The data were processed using vertical de-striping and minimum noise fraction transform.

3 Results and Discussion

From the tailings impoundment three lacquer profiles were prepared. The profiles were left to dry for about 24 h to minimize the risk of destroying structures during removal and transport. When the profiles were removed from the tailings wall, only small areas at the edges or in clay layers were lost, but overall, the texture was well preserved.

Hyperspectral images of all profiles were obtained in the laboratory with the SisuRock system within minutes. The data show clearly the layered structure, which seems to be mostly an effect of particle size. Figure 1 shows the optical and the hyperspectral image of LP2. Layers that were differentiated optically in the field are indicated by capital letters. In the hyperspectral image areas of coarse particle size appear dark, while layers of fine particle size appear light grey or white. The data show a regression coefficient of $R^2 = 0.76$ at 860 nm between particle size and reflectance.

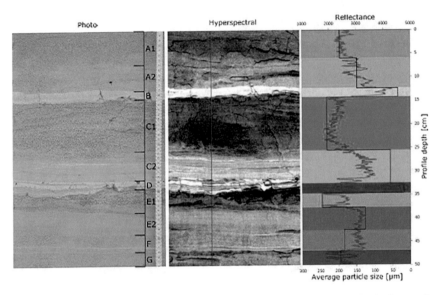

Fig. 1. Left: True color photo of the lacquer peel LP2 with layers selected for bulk sampling. Middle: Hyperspectral scan of LP2 at 860 nm. The red line indicates the position of the reflectance profile in the right graph. Right: Distribution of reflectance (red line) at 860 nm and average particle size (black line) at lacquer peel LP2.

The effect of particle size on reflectance in the visible and infrared wavelength region has been studied in the literature and was described as a negative correlation (Hunt and Vincent 1968; Okin and Painter 2004; Salisbury and Hunt 1968). However, the application on and implications for tailings evaluation still have to be described. The particle size distribution is an important information for the evaluation of the economic value of tailings deposits. The extractability of e.g. Cu depends largely on the particle size of the Cu minerals, since flotation or heap leaching depend on the access to free mineral surfaces. The geochemical data from the additional samples analyzed by XRF show a possible trend of increasing Cu concentration with increasing particle size above 100 μm. These areas of interesting particle size can be identified, located and quantified with HSI. After establishing the methods in the laboratory, it could be expanded to drill cores or in-situ measurements in the field. Therefore, the combination of creating lacquer profiles and HSI and image analysis serves as a rapid and simple tool to acquire important information for targeted and more selective sampling and even evaluation of tailings material for possible selective mining.

4 Conclusions

The proposed method of creating lacquer profiles has worked well on tailings material. It is a simple and rapid method to preserve sedimentary features for display and further analysis. The combination with 2D mapping techniques such as HSI provides rapidly continuous 2D information on sedimentary structures and particle size distribution on a scale from micrometer to meter. The information is quantifiable by image analysis and can be used for various geoscientific applications including sedimentology, sequential analysis and tailings exploration.

Acknowledgements. The results of this work are part of research that is funded by the German Federal Ministry of Education and Research (BMBF) within the projects SecMinStratEl (033R118B) and SecMinTec (033R186B). We thank Dominic Göricke for preparation of the lacquer profiles and Dr. Martin C. Schodlok for support with the hyperspectral data processing. We are very grateful to the reviewers for their helpful comments.

References

Hunt GR, Salisbury JW (1970) Visible and near-infrared spectra of minerals and rocks: I silicate minerals. Mod Geol 1:283–300
Hunt GR, Vincent RK (1968) The behavior of spectral features in the infrared emission from particulate surfaces of various grain sizes. J Geophys Res 73:6039–6046
Okin GS, Painter TH (2004) Effect of grain size on remotely sensed spectral reflectance of sandy desert surfaces. Remote Sens Environ 89:272–280
Salisbury JW, Hunt GR (1968) Martian surface materials: effect of particle size on spectral behavior. Science 161:365–366
Van Baren J, Bomer W (1979) Procedures for the collection and preservation of soil profiles. International soil museum
Voigt E (1936) Die Lackfilmmethode, ihre Bedeutung und Anwendung in der Paläontologie, Sedimentpetrographie und Bodenkunde. Zeitschrift der deutschen geologischen Gesellschaft 88:272–292

Methods of Extraction of Micro- and Nanoparticles of Metal Compounds from Fine Fractions of Rocks, Ores and Processing Products

A. Smetannikov[✉] and D. Onosov

PFIC UB RAS "GI UB RAS", Perm, Russia
smetannikov@bk.ru

Abstract. The presented extraction methods are based on the features of the state of liquids in the capillary space in the form of a weak electrolyte. These methods make possible to extract micro- and nanoparticles adsorbed in matrix minerals from a suspension placed in the graphite substrate into a capillary solution. After the particle deposition in the substrate due to evaporative concentration the microprobe analysis is performed. The method is known as the capillary method of extracting micro and nanoparticles.

The described methods was used as a prototype for extracting micro- and nanoparticles from suspensions associated with the use of an external electric field. The field is created by connecting the electrodes to a graphite substrate and applied suspension with a direct current source using the voltage of $4 \div 6$ V. The micro- and nanoparticles adsorbed in matrix minerals are extracted into the capillary solution. The deposition of micro- and nanoparticles in a capillary solution is made by the method of evaporative concentration. The application of an external electric field intensify extraction of micro- and nanoparticles.

Keywords: Nanoparticles · Capillaries · Electrolyte · Graphite ·
Electric potential · Adsorption

1 Introduction

The following methods of extraction of minerals trace from rocks, ores and products of their processing are well known.

1. In heavy liquids with subsequent magnetic and electromagnetic separation of heavy fractions (Chueva 1950; Mitrofanov et al. 1974). The disadvantage of this method is the aggregation of nanoparticles with minerals with a particle size less than 0.045 mm and their absence in the heavy fraction.
2. Extraction of rhenium from ores of black shale formations (Oleynikova et al. 2012. The method for the extraction of micro- and nanoparticles of metal compounds is not calculated.
3. Extraction of nanoparticles from disperse systems (Zhabreev 1997) using the electric field created by electromagnets. The method does not provide for the extraction of all classes of metal compounds.

© The Author(s) 2019
S. Glagolev (Ed.): ICAM 2019, SPEES, pp. 248–251, 2019.
https://doi.org/10.1007/978-3-030-22974-0_60

4. Method of capillary extraction of nanoparticles (Smetannikov 2014). The material of fine fractions (<0.25 mm) mixing with a liquid forms of suspension, placed in the form of a hemisphere in a graphite substrate. The distance between minerals is comparable to their size and forms a capillary space. The liquid acquires structure, charge and becomes electrolyte (Deryagin et al. 1989). Nanoparticles adsorbed in matrix minerals are extracted into the capillary solution.

Due to the evaporation concentration, the nanoparticles migrate to the base of the hemisphere. After drying (1 ÷ 1.5 h), the hemisphere is removed, and the nanoparticles on the substrate are examined under a microprobe (Fig. 1). More than 50 minerals have been identified: intermetallic compounds and solid solutions of Au, Cu, Ag, Zn, Pb, Ni, Sn, Cu, Fe, Cr, Ti zircon, Sn minerals, monazite, which are trace substances in the water-insoluble salt residues of the Verkhnekamsk deposit (Smetannikov and Filippov 2010).

The disadvantage of this method is incomplete extraction of nanoparticles. This method served as a prototype for developing method for extracting particles using the methods of intensifying the natural properties of capillary systems by an external electric field.

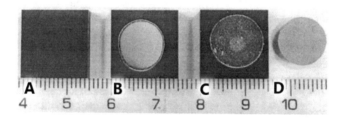

Fig. 1. A - clean graphite plate; B - a plate with a dried hemisphere of the investigated suspension; C- a plate with a hemisphere removed; D - Dried hemisphere (flat side up)

2   Methods and Approaches

The suspension is prepared from the material of fine fractions using distilled water and adding salt. The suspension is applied to a graphite plate, forming the hemisphere. An electrode with a "plus" sign (forming an anodic part of the system with a graphite plate) is connected to the underside of the plate. The second electrode is immersed in the suspension (without touching the graphite substrate) with a minus sign serves as a cathode. Then there is inclusion of the direct current voltage of 4 ÷ 6 V. Further there is an electrolytic extraction of nanoparticles and deposition on the substrate. The dried hemisphere is removed and the sediment is analyzed.

There were conducted three experiences. The first is capillary sedimentation. The second experiment is anodic deposition, with a graphite plate serves as an anode. The third experiment is with pole reversal, when the upper (positive) electrode serves as the anode and the graphite plate as the cathode. Figure 2 posted three photos of deposition traces in three experiments.

In the first experiment, there were matrix minerals practically absent, the number of nanoparticles is minimal (Fig. 2A). In the second experiment, the number of matrix minerals and nanoparticles is 1–2 orders of magnitude higher than in the first experiment (Fig. 2B). In the third experiment, an intermediate result (Fig. 2C). The maximum of nanoparticles is fixed in the second experiment. Here, the deposition process is enhanced by heating the graphite plate.

Experiments have shown the natural properties of water in a capillary space change created by a constant current source under the influence of an electric field. Capillary solution acquires the properties of an electrolyte. The salt concentration in the suspension is 0.5–1%.

Fig. 2. A - sediment after capillary leaching; B - sediment after electrolytic anodic leaching; C - sediment after cathodic leaching

3   Results and Discussion

Extraction of micro- and nanoparticles from insoluble residues of salts, rocks, ores and products of their processing is provided by creating an electric field. The field creates the extracting properties of the electrolyte in suspension from the material of fine fractions and salt solution.

This factor ensures maximum extraction of micro- and nanoparticles and their fixation on a graphite substrate for subsequent analysis is by the microprobe method. Moreover, it is possible to solve the direct problem - the extraction of micro- and nanoparticles from insoluble residues of salts, ores and technogenic products, as well as the inverse problem - "cleansing" any small fraction of the material from impurities of minerals and metal particles.

While using the capillary-electrolytic method of extracting micro- and nanoparticles, the leaching effect (extraction) is achieved by 1–2 orders of magnitude higher than the results of the application of the prototype. The main goal is to obtain information of the form of finding elements-microimpurities. This result has been achieved. Researches are confirmed by the patent (Smetannikov and Onosov 2018).

4   Conclusions

1. The method relates to the methods for extracting micro- and nanoparticles of metal compounds from various media under the influence of direct electric current in a suspension of the material under study, water and salt, placed in a graphite substrate.

2. The new method is different from the prototype by the use of an external electric field, the intensity of extraction of nanoparticles and the completeness of extraction.

References

Chueva MN (1950) Mineralogical analysis of schlich and ore concentrates. Gosgeolizdat, Moscow, p 179

Deryagin BV, Churaev NV, Ovcharenko FD et al (1989) Water in disperse systems. Chemistry, 288 p

Mitrofanov SI et al (1974) Mineral studies for enrichment. Nedra, Moscow, p 352

Oleynikova GA et al (2012) A nanotechnological method for extracting rhenium from rocks and ores of black shale formations. Patent No. 245237 RU, St. Petersburg State University (RU), Claims 12 June 2010. Publ. 10 July 2012, Bul. No. 19

Smetannikov AF (2014) Capillary method of extracting micro- and nanoparticles of minerals from fine fractions for subsequent microprobe analysis. Yushkinsky readings 2014, Materials of the mineralogical seminar with international participation, Syktyvkar, pp 177, 178

Smetannikov AF, Filippov VN (2010) Some features of the mineral composition of salt rocks and products of their processing (for example, Verkhnekamsk salt deposit). Scientific readings of the memory of P.N. Chirvinsky: Sat. articles, Perm, Issue 13, pp 99–113

Smetannikov AF, Onosov DV (2018) Capillary electrolytic method of extracting micro- and nanoparticles of metal compounds from fine fractions of rocks, ores and industrial products. Patent №2659871, PFIC UB RAS (RU), Declared. 20 December 2016, Published 04 July 2018

Zhabreev VS (1997) Installation for the extraction of substances and particles from suspensions and solutions. Patent 32098193 (RU), Chelyabinsk State Technical University, Appl. 26 July 1995, Publ. 10 December 1997

Advanced Materials with Improved Characteristics, Including Technical Ceramics and Glass

Efficiency Evaluation for Titanium Dioxide-Based Advanced Materials in Water Treatment

M. Harja[1], O. Kotova[2], S. Sun[3], A. Ponaryadov[2(✉)],
and T. Shchemelinina[4]

[1] Gheorghe Asachi Technical University of Iasi, Iași, Romania
[2] Institute of Geology Komi SC UB RAS, Syktyvkar, Russia
avponaryadov@geo.komisc.ru
[3] Institute of Non-metallic Minerals, Department of Geological Engineering,
School of Environment and Resource, Southwest University of Science
and Technology, Mianyang, People's Republic of China
[4] Institute of Biology Komi SC UB RAS, Syktyvkar, Russia

Abstract. We present a comparative evaluation of efficiency of titanium dioxide polymorphs as an active photocatalyst (commercially available DegussaP25, anatase (Sigma Aldrich), natural leucoxene concentrate (Pizhemskoe deposit, Russia) and titanium dioxide nanotubes based on it). The materials obtained on the basis of relatively inexpensive and affordable ilmenite-leucoxene ore have the same efficiency as more expensive commercial products.

Keywords: Anatase · Ilmenite-leucoxene ores · Nanotube · Water treatment

1 Introduction

In recent years, advanced oxidation processes have been proposed as alternative methods for eliminating toxic organic pollutants from aquatic systems. Semiconductor heterogeneous photocatalysis is one of the most promising and effective method. This method is environment friendly, because the reaction products of the oxidation of organic pollutants are carbon dioxide and water. A comparative analysis of the economic efficiency of water purification showed that the photocatalytic method was the cheapest (Duduman et al. 2018, Kotova et al. 2016b).

The treatment of water from phenols, in particular, containing chlorine (2, 4, 6 - trichlorophenol, TCP), is an important public health task because of their estrogenic, mutagenic or carcinogenic effects. Their toxicity depends on the degree of chlorination and the position of chlorine atoms in relation to the hydroxyl group. Removing these compounds from the water is necessary to protect both human health and the environment. To produce semiconductor photocatalyst based on titanium dioxide, multi-stage synthesis methods are most often used, using orthotitanium acid or titanium tetrachloride as precursors.

S. Glagolev (Ed.): ICAM 2019, SPEES, pp. 255–258, 2019.
https://doi.org/10.1007/978-3-030-22974-0_61

The aim of the work is the comparative evaluation of efficiency of commercially available titanium dioxide (Degussa P25, Anatase Sigma Aldrich), natural (leucoxene concentrate, Pizhemskoe deposit, Russia) and titanium dioxide nanotubes based on natural leucoxene as active photocatalysts (TiNT).

2 Methods and Approaches

Titanium dioxide. Degussa P25 (80% anatase, 20% rutile; Sigma Aldrich, France) was used as photocatalyst without any purification. It has a BET surface (average) of 50 m^2/g, a particle size of 20–50 nm. Anatase (Sigma Aldrich) was used as photocatalyst without any purification. It has a BET surface (average) of 80 m^2/g, elongated particles with a size of 15–30 nm.

Leucoxene concentrate (LC) was obtained from the Pizhemskoe deposit (Russia). Chemical composition (wt%): TiO_2 – 42.12, SiO_2 – 46.57, Fe_2O_3 - 1.04, Al_2O_3 – 7.57, K_2O - 1.61, MnO - 0.06, CaO - 0.13, MgO - 0.37, SO_3 - 0.06, P_2O_5 - 0.17, ZrO_2 - 0.05, NbO - 0.11. The particle size is about 20–40 mcm.

Titanium dioxide nanotubes (TiNT) were obtained using a hydrothermal treat ment procedure. The detailed description is given elsewhere (Kotova et al. 2016a).

The photocatalytic activity of the samples was studied using a test reaction of decomposition of trichlorophenol in Hereaus circular reactor of a volume of 350 cm^3. Vertically to the reactor axis the TQ150 Z2 mercury lamp (150 W, 352–540 nm) was located. The control solutions were analyzed by liquid chromatography (Hypersil C18 reverse phase HPLC column). The solvent was a solution of acetonnitrile in water in a ratio of 3:2. The solvent flow was 0.5 ml/min.

3 Results and Discussion

The initial leucoxen (Fig. 1A) is a mixture of two phases: rutile and quartz. The peaks are clear, which indicates a high crystallinity of these phases. There are weak reflexes of ilmenite and anatase. Leucoxen is a rutile microcrystalline matrix, saturated with the finest inclusions of quartz (Ponaryadov 2017). The synthesized sample (Fig. 1B) is a mixture of two phases: quartz and sodium titanate $Na_2Ti_6O_{13}$. The chemical composition (semi-quantitative): TiO_2 – 74.68%, SiO_2 – 12.64%, Fe_2O_3 – 5.44%, Al_2O_3 – 4.71%, K_2O – 0.93%.

The structural rearrangement at the nanoscale level – formation of titanium di-oxide nanotubes – leads to decreasing band gap: anatase – 3.1, LC –2.8, TiNT – 2.4 eV. Another important parameter is the specific surface area. During formation of titanium dioxide nanotubes we observed increasing specific surface, which is associated with formation of external and internal surfaces. For the studied samples, the specific surface area was: anatase – 80, LC – 13, TiNT – 230 g/m^2.

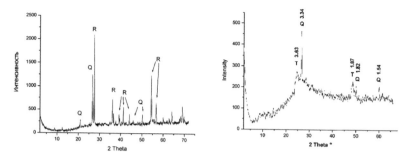

Fig. 1. XRD patterns for leucoxene concentrate, heavy fraction (A) and as-synthesized TiNT (B)

Kinetics of heterogeneous photooxidation reaction in liquid medium in the presence of a catalyst is described by the Langmuir-Hinshelwood model. For the reaction of decomposition of trichlorophenol, the time dependence $\ln(C_0/C)$ is linear, at that the slope ratio gives a constant k_{app}. The time dependence graphs for the studied samples are presented in Fig. 2.

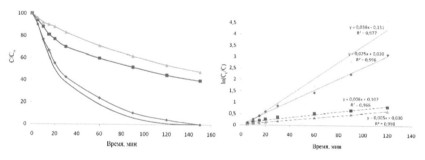

Fig. 2. Curves of decomposition of trichlorophenol in water medium (▲ - leucoxene concentrate, ■ - anatase, ♦ - DegussaP25, × - titanium dioxide nanotubes)

The adsorption and decomposition reaction on surface occur simultaneously, most likely, they do not determine the reaction rate. In the initial period of time (0–10 min), trichlorophenol is adsorbed on the sample surface and the reaction rate increases. Upon reaching the full coverage of the surface with adsorbate, the reaction rate is maximal and does not change in the future. Based on the received data, the values k_{app} of reaction constants were calculated: 0.005 for leucoxene concentrate, 0.006 for anatase, 0.025 for Degussa P25, 0.036 for titanium dioxide nanotubes.

Thus, TiNT, produced by the hydrothermal method from ilmenite-leucoxene ore, are competitive photocatalysts in water treatment from organically contaminants in comparison to the above stated synthetic analogues.

4 Conclusions

We studied the dependence of the kinetics of photoinduced decomposition of trichlorophenol in water solutions in the presence of various types of catalysts based on titanium dioxide: commercially available DegussaP25 and anatase (Aldrich), leucoxene concentrate (Pizhemskoe deposit), titanium dioxide nanotubes. We calculated reaction constants of the photoinduced decomposition of trichlorophenol. It is shown that advanced materials on the basis of relatively inexpensive and affordable ilmenite-leucoxene ore have the same efficiency as expensive commercial products.

Acknowledgements. This research was supported by UB RAS project № 15-18-5-44 and project AAAA-A17-117121270037-4 "Scientific basis for effective development and use of the mineral resource base, development and implementation of innovative technologies, geological and economic zoning of the Timan-North Ural region".

References

Duduman CN, de Salazar y Caso de Los Cobos JMG, Harja M, Barrena Pérez MI, Gómez de Castro C, Lutic D, Kotova O, Cretescu I (2018) Preparation and characterisation of nanocomposite materials based on TiO_2-Ag for environmental applications. Environ Eng Manag J 17(4):2813–2821

Kotova O, Ozhogina E, Ponaryadov A, Golubeva I (2016b) Titanium minerals for new materials. In: IOP conference series: materials science and engineering, p 012025. https://doi.org/10.1088/1757-899x/123/1/012025

Kotova OB, Ponaryadov AV, Gömze LA (2016a) Hydrothermal synthesis of TiO2 nanotubes from concentrate of titanium ore Pizhemskoe deposit (Russia). Vestnik IG Komi SC UB RAS 1:34–36

Ponaryadov AV (2017) Mineralogical and technological features of ilmenite-leucoxene ores of Pizhemskoe deposit, Middle Timan. Vestn Inst Geol Komi SC UB RAS 1:29–36. https://doi.org/10.19110/2221-1381-2017-1-29-36 (in Russian)

The Use of Karelia's High-Mg Rocks for the Production of Building Materials, Ceramics and Other Materials with Improved Properties

V. Ilyina[(✉)]

Institute of Geology KarRC RAS, Petrozavodsk, Russia
Ilyina@igkrc.ru

Abstract. The possible use of high-Mg host rocks, such as serpentinite and pyroxenite, from the Aganozero chromium ore deposit and serpentinite from the Ozerki soapstone deposit, Republic of Karelia, for the production of heat-insulating building materials, ceramic pigments and filters for purification of technogenous solutions is assessed. The results of analysis of the mineralogical compositions of serpentinites and pyroxenite, as well as the physico-mechanical properties (strength, heat conductivity, shrinkage upon roasting, moisture resistance, etc.) and structural characteristics of the ceramic and heat-insulating materials produced on their basis are reported.

Keywords: Serpentinite · Pyroxenite · Ceramics · Heat conductivity · Mechanical strength · Facing material

1 Introduction

Non-conventional high-Mg rocks can be used for the production of ceramic, building and other materials because they are widespread but mainly because of the chemical, mineral and structural characteristics of their mineral constituents: periclase, forsterite, diopside, augite, enstatite and serpentine. The main high-Mg mineral constituents of ceramics suffer phase transformations upon heating, as a result of the disintegration and recrystallization of their lattice (Deere et al. 1965), forming crystalline phases that improve the physic-mechanical properties of ceramics. As forsteritic ceramics suffers no polymorphic transformations, it does not age and is mechanically strong. It is used for the production of dielectrics, heat-insulating materials, facing ceramics and filters for water purification. The aim of the present project is to study serpentinites, kemistites (rocks of hydrotalcite-serpentine composition) and pyroxenites that host chromium ores at the Aganozero deposit and serpentinites at the Ozerki soapstone deposit and to use them for the production of ceramics with a dominant forsteritic crystalline phase, filters for purification of technogenic solutions and ceramic pigments.

© The Author(s) 2019
S. Glagolev (Ed.): ICAM 2019, SPEES, pp. 259–262, 2019.
https://doi.org/10.1007/978-3-030-22974-0_62

2 Methods and Approaches

The mineral composition of the analyzed samples was studied in the IG KarRC RAS by optical microscopy methods, X-ray phase analysis (XPA) and thermal analysis (TA). Rock-forming minerals were studied by Vega II LSH scanning electron microscope with INCA Energy 350 energy dispersion analyzer. X-ray phase analysis was performed by ARL X'TRA diffractometer with CuKl radiation. The physico-mechanical properties of the materials and ceramics were assessed in accordance with State All-Russia standards.

3 Results and Discussion

Differences in the mineral composition of serpentinites affect their chemical composition. Aganozero kemistites and serpentinites are the richest in magnesium (36–38%) and contain minor quantities of impurities (0.1–0.5% Al_2O_3 and 0.24–0.5% CaO) and elevated quantities of crystallization water (loss on ignition is 15–18.5%). Ozerki serpentinite is rich in MgO (36.92%) and contains Al_2O_3 (2.2) and CaO (0.22%) as impurities. All the samples are iron-rich (3.46–10.02% Fe_2O_3, 1.72–3.9% FeO). The mineralogo-analytical study (RPA, TA) of serpentinites showed that Ozerki serpentinites consisted of 89.1% fine-grained lamellar antigorite aggregate and that Aganozero serpentinites contained 78% lizardite. Ore minerals are represented by magnetite and ilmenite. The main minerals of natural pyroxenite (wt%) are augite (67, 2), forsterite (4, 3), enstatite (23, 7) and serpentine (4, 8). Kemistite-based porous heatinsulating ceramics was developed and its properties were studied (Patent no. 2497774, 2013). Electron microscopy study has shown that forsterite, produced by serpentine recrystallization, is the main crystal-line phase of heat-insulating ceramics. The properties of heat-insulating ceramics are shown in Fig. 1.

Pyroxenite (20–70%) and hydromica-based facing ceramics (Il'ina et al. 2017) displays water absorption of 13–15.8% at a roasting temperature of 900–1100 °C, which is consistent with the current standards. At 1200 °C, the water absorption of all the masses decreases rapidly from 0 to 1.2%. Their bending strength is 9–10 MPa.

Pyroxenite-based ceramic pigment has been developed. Unlike the well-known pigment, it can be used to obtain a stable color after roasting over a wide temperature of 750–1250 °C.

The results obtained (Ilyina et al. 2018) show that Ozerki serpentinite can be used for the production of an Mg-silicate reagent for the removal of heavy metals from solutions, e.g. the removal of heavy metal compounds from highly polluted technogenous solutions by filtration through loading from a granulated reagent.

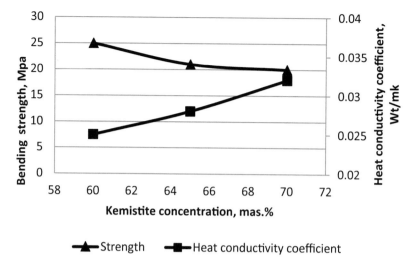

Fig. 1. Dependence of variations in the properties of ceramics on kemistite concentration

4   Conclusions

The materials developed display a porous structure and, consequently, low density, low heat conductivity and high strength. Hence, they can be widely used in industrial and civil engineering for the insulation of buildings, for the heat insulation of the hot surfaces of equipment (furnaces and pipelines), for intensifying high temperature processes and for fuel saving.

References

Deere WA, Haui RA, Zoosman J (1965) Rock-forming minerals. Chain silicated. vol 2, p 405

Il'ina VP, Inina IS, Frolov PV (2017) Ceramic mix based on pyroxenite and low-melting clay. Glass and Ceramics pp 1–4

Ilyina VP, Kremenetskaya IP, Gurevich BI, Klimovskaya EE, Ivashevskaya SN (2018) The study of serpentinized ultramafics from the Kareli-an-Kola Region and the production of a Mg-rich-silicate reagent on their basis for the removal of heavy metals from solutions. In: 18th International Multidisciplinary Scientific GeoConference SGEM2018: Conference proceedings. STEF 92 Technology Ltd., 51 "Alexander Malinov" Blvd., 1712 Sofia, Bulgaria, Energy and Clean Technologies Issue: 4, 2. 2 July – 8 July 2018. Albena, Bulgaria, vol 18 (13), pp 207–213

Ilyina VP, Shchiptsov VV, Frolov PV (2013) Raw mixture for the pro-duction of a porous heat-insulating material, Bull no 31 Patent no 2497774 RF, MPC SO4B 33/132

Kinetic Features of Formation of Supramolecular Matrices on the Basis of Silica Monodisperse Spherical Particles

D. Kamashev[✉]

Institute of Geology Komi SC UB RAS, Syktyvkar, Russia
kamashev@geo.komisc.ru

Abstract. We have established that under such conditions, when the formation of a supramolecular structure from spherical silica particles 220–320 nm in diameter is limited by the rate of introduction of particles into the sedimentation zone (sedimentation deposition in a constant cross section tube), the particle deposition rate, as well as the formation rate of the supramolecular structure, is strictly linear. At the same time, under conditions of excess of disperse phase in the zone of formation of the supramolecular structure (sedimentation deposition in a tube with a modified cross section), the deposition rate is also linear, but there is some delay in formation of the supramolecular structure in time, the larger the smaller is the particle size of the disperse phase. After the formation of the supramolecular structure is completed, a region with an increased concentration of the disperse phase remains, the height of which is greater, the smaller is the particle size. It is shown that a certain concentration of the disperse phase is necessary to begin the formation of a supramolecular structure, below which the formation of a supramolecularly ordered structure does not occur. The concentration of the disperse phase, necessary for beginning of formation of a supramolecular structure, is a constant value that does not vary with time and depends only on the size of the particles.

Keywords: Supramolecular structure · Monodisperse spherical silica particles

1 Introduction

The supramolecularly ordered structures, based on monodisperse spherical silica particles, generated interest relatively long ago, back in the 70s of the last century, in connection with attempts to synthesize artificial analogs of noble opal on their basis (Stober et al. 1968). The overwhelming majority of studies of that time were aimed at developing conditions for synthesis of spherical silica particles and supramolecular structures based on them, and ended with the development of methodological bases for production of synthetic noble opals (Deniskina et al. 1987). However at present the supramolecular silica particles are more often considered as promising objects for synthesis of new composition materials, photonic crystals and nanostructured materials based on them. The 3D ordered closest packing of monodisperse silica spheres is an ideal matrix for creating a wide class of new nanostructured materials, but it greatly increases requirements for the monodispersity of particles, their size and the defects of

S. Glagolev (Ed.): ICAM 2019, SPEES, pp. 263–266, 2019.
https://doi.org/10.1007/978-3-030-22974-0_63

derived supramolecular structures (Kamashev 2012). With this aim and within the task to develop the basis of synthesis of supramolecularly ordered matrices we carried out experimental studies of the rate of precipitation of silica particles and formation of a supramolecular structure on their basis in various conditions (growth modes).

2 Methods and Approaches

Monodisperse spherical silica particles with radius 109, 138 and 158 nm were synthesized by Stober-Fink method (Stober et al. 1968). The sizes of the obtained particles were determined by Photocor Complex dynamic light scattering spectrometer at a laser wavelength 661 nm, a scattering angle 90°, and a correlation function accumulation time 20 min. The initial concentration of the dispersed phase particles was about 2 wt.%. The deposition rate of the silica particles was measured in glass tubes 750 mm high and 20 mm in diameter, with the control of deposition boundary advancing every 30 days with accuracy 0.5 mm. The average daily temperature was also taken into account to calculate the temperature correction associated with the expansion of the dispersion environment. The following values were obtained for the silica particles of different radius (Fig. 1).

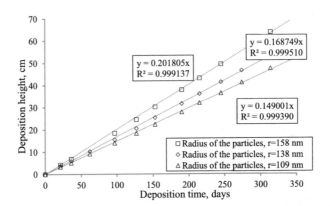

Fig. 1. Dependence of height of deposition of silica particles of different sizes over time. The coefficient before x (0.201805, 0.168749 and 0.149001) in the equation of approximating curve represents the rate of particle deposition, cm/day.

To measure the rate of formation of the supramolecular structure, the deposition of particles was carried out in glass tubes with a narrowing in the lower part, which reduced the area (concentration of suspension) by about 80 times. This mode of formation of the supramolecular structure is characterized by a constant increased content of the dispersed phase, unlike deposition in tubes with a constant cross section, where the formation of region with a high content of particles is limited by their deposition rate. The obtained data on the rate of formation of a supramolecularly ordered structure are presented in Fig. 2. At such rates one monolayer of silica particles with radius 158 nm is formed approximately within 90 s.

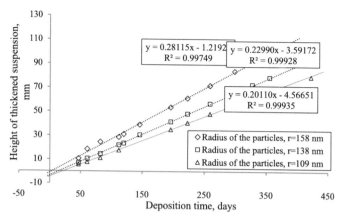

Fig. 2. Dependence of height of a supramolecular structure formed from silica particles of different size over time. The coefficient before "x" (0.28115, 0.22990 and 0.20110) in the equation of approximating curve is a growth rate of the supramolecular structure in conditions of high concentration (excess) of the dispersed phase, mm/day.

3 Results and Discussion

Our data testify to that the particle deposition rate (Fig. 1) is linear and starts at the origin and strictly obeys the Stokes equation throughout the time. In turn, the analysis of data on the formation rate of the supramolecular structure (Fig. 2) shows that the formation rate of the ordered structure is also linear, however, the process of supramolecular crystallization does not begin immediately, but after some time, due to creation of "supersaturation" or some increased concentration of particles of dispersed phase in the bottom region. This time depends on size of particles and increases with their decrease.

As a result of deposition of monodisperse silica particles and formation of a supramolecular structure we noted that initially a zone with an increased concentration of dispersed phase was formed in the bottom region. The boundaries of this zone were clearly expressed, it possessed a constant height, both in the process of the beginning of deposition of particles, and after its completion. The height of the condensed

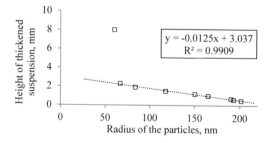

Fig. 3. Dependence of thickened suspension height on the particle size of the disperse phase after completion of deposition.

suspension zone after deposition of particles was constant and depended only on size of dispersed particles (Fig. 3) with the exception of particles of radius less than 60 nm. In this case a similar zone with clearly defined boundaries was not formed, and the supramolecular structure was not formed.

4 Conclusions

The obtained experimental data on deposition rate of monodisperse spherical silica particles with formation of a supramolecular structure showed that its formation was the result of a kind of second-order phase transition (self-organization of particles). This transition was associated with a threshold concentration of particles of dispersed phase in the area of deposition, below which the formation of the supramolecular structure did not occur. When this value was exceeded, the process of formation of a supramolecular structure started, and the rate of its formation was strictly linear and independent of the concentration of the dispersed phase. If for some reason the threshold concentration of particles was not reached (for example, very small sizes, low density of particles or high viscosity of the dispersion medium), then the formation of a supramolecular structure did not occur.

Acknowledgements. The work was accomplished with partial financial support by UB RAS program No. 15-18-5-44, and RFBR No. 19-05-00460a.

References

Deniskina ND, Kalinin DV, Kazantseva LK (1987) Noble opals. Novosibirsk, Nauka, Sib. otd. (Trudy institute geologii I geofiziki) 693, p 184
Kamashev DV (2012) Synthesis, features and model of formation of supramolecular silica structures. Phys Chem Glass 38(3):69–80
Stober W, Fink A, Bohn E (1968) Controlled growth of monodisperse silica spheres in the micron size range. J Colloid Interface Sci 26:62–69

Three-Cation Scandium Borates $R_xLa_{1-x}Sc_3(BO_3)_4$ (R = Sm, Tb): Synthesis, Structure, Crystal Growth and Luminescent Properties

A. Kokh[1(✉)], A. Kuznetsov[1], K. Kokh[1,2], N. Kononova[1],
V. Shevchenko[1], B. Uralbekov[3], A. Bolatov[3], and V. Svetlichnyi[4]

[1] Sobolev Institute of Geology and Mineralogy SB RAS, Novosibirsk, Russia
a.e.kokh@gmail.com
[2] Novosibirsk State University, Novosibirsk, Russia
[3] Al-Farabi Kazakh National University, Almaty, Kazakhstan
[4] Tomsk State University, Tomsk, Russia

Abstract. Complex ortohoborates of rare earth metals with the general chemical formula $R_xLa_{1-x}Sc_3(BO_3)_4$ (R = Sm, Tb) have been obtained by solid state synthesis and spontaneous crystallization. These crystals belong to the huntite family with the space group R32 and for x = 0.5 have unit cell parameters a = 9.823(6), c = 7.975(3) (SLSB) and a = 9.803(3), c = 7.960(4) Å (TLSB).

Keywords: Crystal · Borate · Structure · Huntite · Growth · Luminescence

1 Introduction

Orthoborates with the general formula $RX_3(BO_3)_4$, where R = Y, Ln; X = Al, Ga, Sc, Cr, Fe are practically important and interesting from the point of view of crystal chemistry objects for research. One of the important properties of these compounds is the ability to form a non-centrosymmetric structure, which is called huntite-like. Such a structure causes, for example, non-linear optical properties.

To understand the formation of the huntite-like structure of three-cation scandoborates, we consider the lanthanum – scandium borate $LaSc_3(BO_3)_4$. The authors (He et al. 1999) distinguish three modifications of this crystal: high-temperature monoclinic with the C2/c space group, medium temperature trigonal with the R32 space group (huntite-like) and low-temperature monoclinic with the Cc space group. As a result of our research (Fedorova et al. 2013) identity of the X-ray patterns of polymorphic modifications high and low was shown.

The stabilization of the huntite-like structure can occur if an additional isomorphic cation is introduced into the $LaSc_3(BO_3)_4$ structure, that was confirmed in (Li et al. 2001) who initiated the new three-cation scandoborate with the huntite-like structure $Nd_xLa_{1-x}Sc_3(BO_3)_4$. Further, in a number of works by adding a third cation

© The Author(s) 2019
S. Glagolev (Ed.): ICAM 2019, SPEES, pp. 267–271, 2019.
https://doi.org/10.1007/978-3-030-22974-0_64

$R_xLa_ySc_z(BO_3)_4$ nonlinear optical crystals with a stable huntite-like structure were obtained with R = Gd (Xu et al., 2011); Y (Ye et al. 2005) and Lu (Li et al. 2007).

The existence of a huntite-like structure for the boundary members of the REE series suggests the stability of such a structure with the rest of the REE. This paper presents data on the huntite-like structures SLSB and TLSB for systems $R_xLa_{1-x}Sc_3(BO_3)_4$ (R = Sm, Tb).

2 Methods and Approaches

Polycrystalline sample of $R_xLa_{1-x}Sc_3(BO_3)_4$ (x = 0–0.5)were prepared by the method of two stage solid state synthesis in a Pt crucible. The stoichiometric mixtures of pure raw La_2O_3, Sc_2O_3, H_3BO_3 and R_2O_3(R = Sm, Tb) reactants were heated at 800 °C for 5 h to decompose H_3BO_3. At the second stage, the mixtures were grinded in an agate mortar and heated again at 1300 °C for 12 h until the powder X-ray method showed no peaks of initial compounds (Fig. 1).

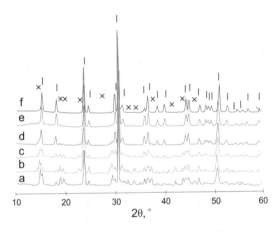

Fig. 1. X-ray pattern of $Sm_xLa_{1-x}Sc_3(BO_3)_4$, where x = 0(a), 0.2(b), 0.3(c), 0.4(d), 0.5(e), 1(f). I- huntite (R32), X- monoclinic (C2/c) structure.

Spontaneous crystals of $R_xLa_{1-x}Sc_3(BO_3)_4$with dimensions 30 × 30 × 10 mm with a transparent area of 5 × 5 × 5 mm were grown from LiBO$_2$- LiF flux Fig. 2. A Pt crucible containing $R_{0.5}La_{0.5}Sc_3(BO_3)_4$, Li_2CO_3, H_3BO_3 and LiF in the molar ratio of 1:1,5:1,5:3 was heated to 1000 °C. The charge was held in a melted state for a day to achieve homogenization. After this stage a platinum wire with a loop was placed in the center of the melt surface and the temperature was decreased to 900 °C. Then the melt was cooled with the 2 °C/day to 850 °C and following cooling at the rate of

15 °C/day to room temperature. The crystal was chosen for x-ray analysis. Powder diffraction patterns were refined using the Rietveld method within the GSAS- II program.

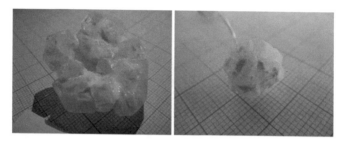

Fig. 2. Crystals grown from $LiBO_2$-LiF flux: TLSB (left) and SLSB (right)

The chemical composition of obtained crystals was measured by X-ray fluorescent analysis using XRF 1800 (Shimadzu, Japan). The results of the analysis are conformed with the formula obtained after crystal structure refinement: (Table 1)

Table 1. Composition of TLSB and SLSB based on X-ray fluorescent elemental analysis

Composition of the $Tb_xLa_ySc_z(BO_3)_4$		Ratio of Tb/La
Starting melt	$Tb_{0.5}La_{0.5}Sc_3(BO_3)_4$	1
Center	$Tb_{0.22}La_{0.78}Sc_3(BO_3)_4$	0.28
Edge	$Tb_{0.24}La_{0.75}Sc_{2.99}(BO_3)_4$	0.32
Composition of the $Sm_xLa_ySc_z(BO_3)_4$		Ratio of Sm/La
Starting melt	$Sm_{0.5}La_{0.5}Sc_3(BO_3)_4$	1
Center	$Sm_{0.32}La_{0.69}Sc_{2.98}(BO_3)_4$	0.46
Edge	$Sm_{0.35}La_{0.68}Sc_{2.97}(BO_3)_4$	0.52

3   Results and Discussion

Structure. According to Rietveld refinement both SLSB and TLSB crystalize in the trigonal space group R32 with unit cell parameters: a = 9.823(6), c = 7.975(3) (SLSB) and a = 9.803(3), c = 7.960(4)Å (TLSB). The structure framework is composed of the R, La atoms, Sc atoms and B atoms occupy trigonal prisms, octahedra and planar triangle of oxygen, respectively. The isolated $(R, La)O_6$ trigonal prisms alternate along the c-axis with BO3 triangle that are perpendicular to the c-axis. ScO_6octahedra link to each other along the edge and form twisted chain along c, which separate $(R, La)O_6$ prisms as well. The discrepancies between refined diffraction spectra with model calculations can be explained by crystal cleavage along {202} and {113}.

Luminescence. Typical excitation and luminescence spectra of SLSB are shown on Fig. 3(a). The strongest excitation peak of samarium crystal corresponds to $^6H_{5/2} \rightarrow {}^4F_{7/2}$ transition located at 407 nm. Whereas luminescent spectrum of SLSB has some peaks corresponding to $^4G_{5/2} \rightarrow {}^6H_J$ (J = 5/2, 7/2, 9/2 и 11/2) and located at 566, 602, 645 and 708 nm.

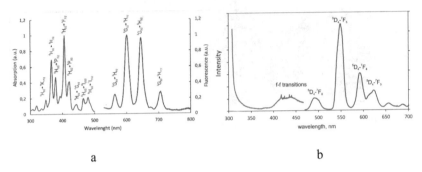

a   b

Fig. 3. Luminescent properties of (a) SLSB, (b) TLSB

On Fig. 3(b) TLSB excitation and luminescence spectra are shown with a wide strip at 300 nm corresponding to ^4F - ^5D transition. Luminescent spectra consist of 5 peaks at 490, 505, 585 and 640 corresponding to $^5D_4$ - ^7F_J (J = 6, 5, 4, 3) transitions.

4   Conclusions

The formation of a huntite structure in systems $R_xLa_{1-x}Sc_3(BO_3)_4$, (R = Sm, Tb), as well as the dependence of the compositions stable in the required structure depending on the production method is shown. The spectral characteristics confirm the potential of using crystals as luminescent materials.

Acknowledgements. This work is supported by RFBR project#19-05-00198a, state contract of IGM SB RAS and partially by Project GF MES RK IRN AP05130794.

References

Fedorova MV, Kononova NG, Kokh AE, Shevchenko VS (2013) Growth of MBO3 (M = La, Y, Sc) and LaSc3(BO3)4 Crystals from LiBO2–LiF Fluxes. Inorg Mater 49:482–486

He M, Wang G, Lin Z et al (1999) Structure of medium temperature phase β-LaSc3(BO3)4 crystal. Mater Res Innov 2(6):345–348

Li W, Huang L, Zhang G, Ye N (2007) Growth and characterization of nonlinear optical crystal Lu0.66La0.95Sc2.39(BO3)4. J Cryst Growth 307:405–409

Li Y, Aka G, Kahn-Harari A, Vivien D (2001) Phase transition, growth, and optical properties of NdxLa1−xSc3(BO3)4 crystals. J Mater Res 16:38–44

Xu X, Ye N (2011) $Gd_xLa_{1-x}Sc_3(BO_3)_4$: a new nonlinear optical crystal. J Cryst Growth 324:304–308

Ye N, Stone-Sundberg JL, Hruschka MA et al (2005) Nonlinear Optical Crystal $Y_xLa_ySc_z(BO_3)_4$ $(x + y + z = 4)$. Chem Mater 17:2687–2692

Rational Usage of Amorphous Varieties of Silicon Dioxide in Dry Mixtures of Glass with Specific Light Transmittance

N. Min'ko[✉] and O. Dobrinskaya

Belgorod State Technological University named after V G Shukhov,
Belgorod, Russia
minjko_n_i@mail.ru

Abstract. The paper studied high-silicon amorphous rocks from the perspective of their application for glass production of different purpose. The results contained data of calculation of dry mixtures for producing heat-protective glass using amorphous varieties of silicon dioxide. The obtained glass specimens were melted and studied for spectral characteristics.

Keywords: Amorphous silicon dioxide · Light transmittance · Dry mixture · Heat-protection glass

1 Introduction

High-silicon amorphous rocks as raw materials have a wide range of valuable features. Primarily, this is the amorphous (metastable) state of silicon dioxide (Kondrashov and Kondrashov 2013). Moreover, one can distinguish some peculiarities of amorphous varieties that can be regarded as drawbacks (Manevich et al. 2012):

- instability of chemical composition;
- SiO_2 is accompanied by other components (up to 40%) that can play the role of auxiliary raw materials;
- increased content of aluminum oxides and iron.

One of the main glass spectral characteristics is light transmittance. The main components that affect light transmittance of glass products and that should be strictly controlled are oxides of coloring metals that are encountered in conventional raw materials (dolomite, feldspar concentrate, sands). These compounds include iron oxides; their content in glass is strictly regulated:

- sheet glass – 0.09–0.12%;
- heat-protective – 0.6–0.7%;
- clear container glass – $0.1 \pm 0.01\%$;
- brown container glass – $0.8 \pm 0.1\%$;
- green container glass – not regulated.

S. Glagolev (Ed.): ICAM 2019, SPEES, pp. 272–276, 2019.
https://doi.org/10.1007/978-3-030-22974-0_65

The composition of amorphous varieties of silicon dioxide (Distantov 1976) is characterized by increased content of iron oxides (Table 1), which impedes their wide application in production of glass articles.

Table 1. Variation of chemical composition of amorphous silicic raw materials

Rock	Content of oxides [wt.%]							
	SiO_2	Al_2O_3	Fe_2O_3	CaO	MgO	K_2O	Na_2O	TiO_2
Diatomites	73.0–90.0	3.3–7.5	2.0–5.2	less than 0.6	0.6–1.7	less than 1.0	less than 0.5	less than 0.3
Opokas	52.1–91.4	2.5–15.4	1.0–5.0	0.43–17.1	less than 2.48	0.6–4.0	0.1–1.0	less than 0.2
Pearlites	68.5–75.3	11.2–16.3	less than 3.0	0.5–2.0	less than 1.0	1.5–4.0	2.0–6.2	0.1–0.5
Tripolites	35.3–86.7	2.5–11.6	0.3–3.4	0.4–31.2	0.2–1.6	0.85–2.1	less than 0.5	less than 0.2

Taking into account increased content of iron and aluminum oxides, these rocks can be used as aluminum- or iron-containing raw materials in production of heat-protection glass or dark-glass containers.

2 Methods and Approaches

Current work assesses amorphous silicon dioxide (ASD) as aluminum-containing raw material that can partially replace quartz sand and other conventional raw materials (Table 2).

Table 2. Composition of average samples of amorphous silicon dioxide varieties

Rock	Content of oxides [wt.%]							
	SiO_2	Al_2O_3	Fe_2O_3	CaO	MgO	K_2O	Na_2O	TiO_2
Diatomite	87.22	6.79	2.22	0.431	1.25	1.08	0.245	0.265
Opoka	92.47	3.38	1.32	0.506	0.712	0.88	–	0.174
Pearlite	72.0	16.45	1.06	0.863	0.422	4.30	4.21	0.145
Tripolite	79.93	10.99	3.17	0.838	1.81	1.99	0.273	0.687

To conduct experimental studies, the dry mixtures for heat protective glass with ASD were calculated. The glass was melted from the dry mixtures using pearlite. The content of iron oxides in pearlite is insufficient for production of heat-protective glass, which necessitates the introduction of iron containing material such as magnetite.

The glass was melted in an electric kiln with silicon carbide heating elements at maximum melting temperature of 1420 °C. Another batch of glass was melted without pearlite.

Spectral light transmittance in the visible range was measured automatically on SF-56 spectrophotometer (Russia). The specimens were prepared by mechanical grinding and polishing on laboratory setups.

3 Results

The results showed that the application of ASD for preparation of dry mixtures allowed reducing number of conventional materials for glass melting (Table 3).

Table 3. Economy of conventional raw materials after replacement by ASD, %

Rock	Raw material			
	Sand	Soda	Creta	Dolomite
Opoka	**72.7**	2.44	1.61	19.71
Pearlite	43.8	**20.15**	**38.7**	**40.38**
Diatomite	41.7	6.5	4.08	25.14
Tripolite	27.6	7.79	7.66	27.06

The glass produced in the laboratory differed in color (Fig. 1), which depends on the shift of equilibrium $Fe^{2+} \leftrightarrow Fe^{3+}$.

Fig. 1. Specimens of sheet glass: 1 – with pearlite; 2 – with pearlite and magnetite; 3 – with pearlite and magnetite (+ coal)

We studied spectral characteristics of the glass specimens: sheet glass melted from conventional components; sheet glass with addition of pearlite as aluminum-containing raw material; sheet glass with addition of magnetite (magnetite was reduced by coal) (Fig. 2).

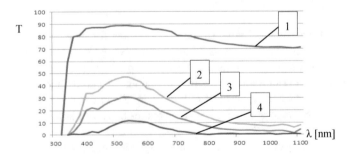

Fig. 2. Spectral light transmittance of sheet glass: 1 – conventional raw materials; 2 – with pearlite; 3 – with pearlite and magnetite; 4 – with pearlite and magnetite (+ coal)

The data from Table 4 demonstrated that in the visible spectrum, the light transmittance of the glass with application of pearlite differed from glass melted with conventional raw materials. The highest light transmittance was shown by sheet glass melted from conventional raw materials, since the content of iron in quartz sand was minimal (0.022 wt%).

Table 4. Light transmittance of sheet glass specimens calculated for the thickness of 10 mm

Specimen	Light transmittance T [%]				Diathermancy index = 10^{-1}
	$\lambda = 570$ nm		$\lambda = 1100$ nm		T_{1100}
	As per equation	As per nomogram*	As per equation	As per nomogram*	
1	88.6	88.1	67.4	67.7	6.7
2	48.4	48.8	8.4	8.2	0.8
3	30.8	30.5	5.1	4.8	0.4
4	12.3	12.0	3.7	3.2	0.3

* - Amosov's nomogram

The heat-protective characteristics are assessed by light transmittance at $\lambda = 1100$ nm. In the infrared range of the spectrum, the light transmittance of specimens 2, 3 and 4 reduces due to the presence of impurities of Fe^{2+}, which provisions the heat-protective characteristics.

In IR-range, the light transmittance of studied glasses is different (from 3 to 67%), i.e. high heat-protective properties are possessed by specimens 3 and 4, the IR-light transmittance is 3–5%. However, the production of glass with such content of FeO is unreasonable due to low light transmittance in the visible spectrum. Such dry mixtures can be used for special glass.

4 Conclusions

The calculations allowed determining the amorphous varieties of silicon dioxide that could replace most conventional materials in dry mixtures. They were: opoka (as a replacement for quartz sand) for both aluminum- and iron-containing raw materials, and pearlite capable of replacing the biggest amount of soda, creta and dolomite when using it as an iron-containing material.

Additional studies of glass properties produced from dry mixtures with pearlite were required (spectral characteristics, crystallization capacity and impact of redox conditions of melting). Nevertheless, the work showed that pearlite could be used in the technology of heat-protective glass.

Acknowledgments. The work is realized in the framework of the Program of flagship university development on the base of Belgorod State Technological University named after V.G. Shukhov, using the equipment of High Technology Center at BSTU named after V.G. Shukhov.

References

Distantov AG (1976) Silicon rocks of USSR. Tatar Press, Kazan

Kondrashov VI, Kondrashov DV (2013) Perspectives of synthesis of industrial compositions of float-glass on the basis of crystalline and amorphous silicon dioxide. GlassRussia. 3:31–33

Manevich VE, Subbotin RK, Nikiforov EA, Senik NA, Meshkov AV (2012) Diatomite as silica-containing material for glass industry. Glass Ceram 5:34–39

Peculiarities of Phase Formation in Artificial Ceramic Binders for White-Ware Compositions

I. Moreva[(⊠)], E. Evtushenko, O. Sysa, and V. Bedina

Belgorod State Technological University named after V G Shukhov,
Belgorod, Russia
moreva_bstu@mail.ru

Abstract. The production of sanitary white wares traditionally uses multi-component mixes, which is necessitated by a whole complex of properties: high density and low humidity of molding slurries with low thixotropic strengthening and good filterability of the slurry. However, modern understanding of the structure and properties of materials open opportunities for optimization of technological process and production of higher-quality articles. The implementation of activation technologies and replacement of traditional molding slurries by artificial ceramic binders will reduce the number of components in mixes and optimize the production of white wares. Since the achievement of working performance of ceramic materials substantially depends on the phase formation processes in a material during sintering. Current work analyses the phase transformations occurring at all stages of white ware production as per proposed technology.

Keywords: Artificial ceramic binders · Phase conversions · Porcelain ·
Pottery · Ceramic slurry

1 Introduction

Artificial ceramic binders (ACBs) are molding suspensions produced by the technology of high-concentration ceramic binding suspensions (Pivinskii 2003) from nonplastic materials (quartz sand, quartz glass, fire clay, etc.). The binding properties of such suspensions are determined not by clay minerals (as in the case of traditional ceramic slurries), but by a colloid component that forms during milling and which content can reach up to 5%. Previous studies have shown the possible production of ACBs on the basis of clay materials after preliminary thermal activation. The technology allows producing high-density molding suspensions with record-low moisture content: W_{susp} = 16%, ρ = 1950 kg/m^3 (for traditional slurry 35% and 1740 kg/m^3, respectively). This technology is interesting for the production of white wares, because it allows unifying the properties of implemented raw materials and reduce the number of components in the paste from more than 7 down to 3. This work studies the peculiarities of ACB formation on the basis of thermally activated mixes of typical white ware composition: 50% of clay (kaolin from Prosyanoye deposit), 25% of fluxes (Vishnevogorsk spar) and 25% of filling aggregates (Ziborovsk sand).

© The Author(s) 2019
S. Glagolev (Ed.): ICAM 2019, SPEES, pp. 277–280, 2019.
https://doi.org/10.1007/978-3-030-22974-0_66

2   Methods and Approaches

Initial materials were broken, mixed and wetted to 15–20% with consequent formation of bricks. After drying, the bricks were burnt at the temperature of 950–1000 °C with shock cooling for increased system activity. The thermally activated mix was milled as per the technology of highly concentrated ceramic binders for suspensions with introduction of electrolytes and liquid phase at the loading of the first part of the material. The XRD analysis of the investigated materials was made on DRON-3 diffractometer. The diffractograms were recorded with filtered CuKα-radiation (Ni filter); the tube voltage was 20 kV; tube anode current was 20 mA; the measurement limit was 10,000–4,000 pulses per second; the detector rotation speed was 2.4 min^{-1}; the angular mark was 10. The phases were identified using JCPDF cards. The samples with the dimensions of 30 × 30 × 30 mm from obtained suspensions and slurries were formed by molding into gypsum molds. The dried samples were burnt in a periodic kiln with silicon carbide heating elements at the temperature of 800–1200 °C with holding at maximum temperature for 5–10 min.

3   Results and Discussion

The preliminary thermal treatment of the studied raw materials allowed weakening crystal lattices of the minerals, creating the structural non-stability due to polymorph transformations of quartz (d/n, Å – 4.27; 3.35; 2.46; 2.29) and dehydration of clay and hydromicaceous minerals (d/n, Å – 7.23; 3.36; 4.48; d/n, Å – 10.16; 3.30; 2.91) (Fig. 1). After such a thermal treatment, a part of material passes into active state (Evtushenko et al. 2007).

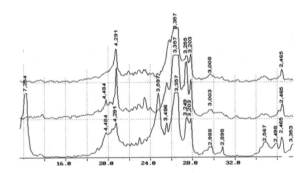

Fig. 1. Changed phase composition of studied mixes after activation and consequent milling: (a) initial untreated mix; (b) mix after thermal activation at 950 °C; (c) after wet milling.

The preliminary thermal activation of the material, consequent wet milling and high density of the samples facilitates the intensification of burning, in particular, appearance of mullite seeds (d/n, Å – 3.39; 5.42; 2.71; 2.55) beginning from 950 °C. In traditional compounds, the formation of mullite begins only at 1100 °C. A part of crystalline phases of the ACB in the white ware composition transits into the melt, which is witnessed by the appearance of amorphous phase and decrease in the intensity of reflections that are typical for quartz (d/n, Å – 3.35; 4.27; 2.46) and feldspars (d/n, Å – 3.24; 3.22) (Fig. 2).

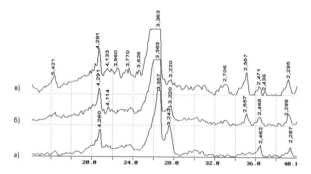

Fig. 2. X-ray diagrams of burnt samples of ACB in white ware composition: (1) 950 °C; (2) 1100 °C; (3) 1200 °C

Probably, a denser structure of the molding leads to an intense amorphization and sintering at temperatures close to 1100 °C, while at 1200 °C close-packed structures start to crystallize and stabilize. The high intensity of sintering and formation of a stronger structure of ACB samples is testified by gradual increase in the coefficient of crystallinity (Cc) with increased burning temperature: from 58.5% at 600 °C up to 91% at 1200 °C, while for cement plant slurry it is 66.5% at the final burning temperature. The high values of Cc for samples of ACB are probably connected with more intense processes of mullite and cristobalite formation.

4 Conclusions

Thus, the revealed peculiarities of phase formation in ACBs of white ware compositions demonstrate that the material has high reactivity in a wide temperature range. The preliminary thermal activation of raw materials with consequent sintering of the material facilitates earlier beginning of mullite and melt formation. The mentioned processes are determining for the formation of the strong structure of the material and high operation characteristics of the ceramic products.

Acknowledgements. The work is realized in the framework of the Program of flagship university development on the base of Belgorod State Technological University named after V.G. Shoukhov, using equipment of High Technology Center at BSTU named after V.G. Shoukhov.

References

Evtushenko EI, Sysa OK, Moreva IYu (2007) Controlling the properties of raw materials, casting slips, and pastes in fine-ceramic technology. Stroit Mater 8:16–17

Pivinskii YuE (2003) Highly concentrated ceramic binding suspensions (HCBS) and ceramic castables. Stages in research and development. Refract Ind Ceram 44(3):152–160

Experimental Modeling of Biogeosorbents

T. Shchemelinina[1], O. Kotova[2], E. Anchugova[1], D. Shushkov[2(✉)],
G. Ignatyev[2], and M. Markarova[2]

[1] Laboratory of Biochemistry and Biotechnology, Institute of Biology,
Komi Science Center, Ural Branch of the Russian Academy of Sciences,
Syktyvkar, Russia
[2] Laboratory of Technology of Mineral Raw, Institute of Geology named after
Academician N.P. Yushkin Komi Science Center of the Ural Branch
of the Russian Academy of Sciences, Syktyvkar, Russia
dashushkov@geo.komisc.ru

Abstract. A new trend in the modeling algorithm oil sorption materials is the
adsorptive immobilization of strains of microorganisms on mineral sorbent. The
objects of study were clay and zeolite raw and biogeosorbents with oil-oxidizing
microorganisms from Biotrin preparation, immobilized on them. During our
work we modeled biogeosorbents and estimated their sorption and destructive
properties in reference to petroleum hydrocarbons.

Keywords: Biogeosorbents · Biotrin · Petroleum hydrocarbons · Sorption ·
Zeolites · Clays

1 Introduction

One of the most promising ways to solve problem of oil pollution is to use biogeosorbents, which are sorbent-carriers with oil-oxidizing microorganisms immobilized on their surface (Shchemelinina et al. 2017). It is possible to use zeolite and clay
raw as mineral carriers, which have high ion-exchange and sorption properties (Kotova
et al. 2017). The purpose of this work is to study sorption and destructive properties of
biogeosorbents based on clay and zeolite raw as mineral carriers for Biotrin
biopreparation.

2 Methods and Approaches

The objects of study were:

1. Mineral carriers based on analcime-bearing rocks from Koinskaya zeolite area
   (Shushkov et al. 2006), clinoptilolite-bearing clays and glauconite-bearing rocks
   from Chim-Loptyugskoe oil shale deposit (Saldin et al. 2013; Simakova 2016),
   located in Komi Republic (Russia). For comparison, Ionsorb ™ quartz-glauconite
   sand from Bondarskoe deposit of Tambov region was taken as a control.

© The Author(s) 2019
S. Glagolev (Ed.): ICAM 2019, SPEES, pp. 281–285, 2019.
https://doi.org/10.1007/978-3-030-22974-0_67

2. Strains of microorganisms in the composition of Biotrin biopreparation (Conclusion … 2017): bacteria *Pseudomonas yamanorum* VKM B-3033D, isolated from heavily soiled railway bed near the city of Syktyvkar (Patent 2615458 RU); yeast *Rhodotorula glutinis* VKM Y-2998D (Patent 2658134 RU); microalgae *Chlorella vulgaris* Beijer. f. *globosa* V. Andr. A1123. Microorganisms (cell titer 10^9) were cultivated according to standard methods. Immobilization of the biopreparation on mineral carriers was carried out in the ratio of 1 part of the biopreparation to 6 parts of the sorbent. Initial sorbents (without Biotrin) and biogeosorbents were added to oil-contaminated water, aerated for 4 days, and total petroleum hydrocarbons (TPH) content in water samples, filtered initial sorbents and biogeosorbents was measured (Method … 1998).

3 Results and Discussion

Norms for maximum permissible concentration (MPC) of TPH in water of fishery value are 0.05 mg/dm^3 (Order … 2016). The TPH content in the control water sample is 2.4 times higher than MPC (Table 1).

Experiments showed that samples of the initial analcime-bearing rocks (551, 56403, 1/83) presented adsorption activity in relation to oil products. As a result of the introduction of these samples into oil-polluted water, the content of pollutant in water decreases 2.5–3 times in 4 days, to the MPC. When biogeosorbents are applied to contaminated water (551-B, 56403-B, 1/83-B), the efficiency of water purification decreases and does not reach MPC standards, which indicates decreasing sorption properties after immobilization of microorganisms on the mineral carriers. This is probably due to decreasing surface area of the mineral carriers covered by microorganisms. During the study of initial samples and Biotrin treated samples for destructive properties we revealed that the efficiency of oxidation of oil products in samples 551-B, 56403-B, 1/83-B increases in 4.4, 3.5 and 1.14 times, respectively.

The sorption properties of clinoptilolite-bearing clays are most attractive in sample 541-31. However, taking into account a highly destructive activity of microorganisms in 538-35-B biogeosorbent, it is preferred for remediation of oil contaminated water.

Samples of initial glauconite-bearing rocks have high sorption properties (539-40, 531-56, 315-10, TG). The TPH content in the experimental water is reduced by 3.4–5 times in 4 days relative to a control sample. Biodegradation of hydrocarbons in samples of Biotrin immobilized glauconite-bearing rocks (539-40-B, 531-56-B, 315-10-B, TG-B) ranges from 62 to 76%.

Table 1. Change in the concentration of oil products in water in the presence of initial mineral carriers and biogeosorbents

Initial samples	TPH content*	Biotrin treated samples	TPH content*
Analcime-bearing rocks			
551	$\dfrac{0.04 \pm 0.014}{250 \pm 60}$	551-B	$\dfrac{0.11 \pm 0.04}{57 \pm 23}$
56403	$\dfrac{0.046 \pm 0.016}{130 \pm 50}$	56403-B	$\dfrac{0.061 \pm 0.021}{37 \pm 15}$
1/83	$\dfrac{0.048 \pm 0.017}{250 \pm 60}$	1/83-B	$\dfrac{0.071 \pm 0.025}{220 \pm 90}$
58603	$\dfrac{0.071 \pm 0.025}{250 \pm 60}$	58603-B	$\dfrac{0.064 \pm 0.022}{90 \pm 40}$
Clinoptilolite-bearing clays			
538-35	$\dfrac{0.085 \pm 0.030}{50 \pm 20}$	538-35-B	$\dfrac{0.037 \pm 0.013}{40 \pm 16}$
541-31	$\dfrac{0.035 \pm 0.012}{250 \pm 60}$	541-31-B	$\dfrac{0.058 \pm 0.021}{100 \pm 40}$
Glauconite-bearing rocks			
539-40	$\dfrac{0.024 \pm 0.009}{58 \pm 23}$	539-40-B	$\dfrac{0.072 \pm 0.025}{20 \pm 8}$
531-56	$\dfrac{0.027 \pm 0.009}{63 \pm 25}$	531-56-B	$\dfrac{**}{15 \pm 6}$
315-10	$\dfrac{0.09 \pm 0.03}{11 \pm 4}$	315-10-B	$\dfrac{0.021 \pm 0.007}{17 \pm 7}$
TG	$\dfrac{0.035 \pm 0.012}{90 \pm 40}$	TG-B	$\dfrac{0.045 \pm 0.016}{34 \pm 14}$
Oil-contaminated water (control)	0.12 ± 0.041		

Note: * – in the numerator, TPH content in the experimental water, mg/dm^3, in the denominator – TPH content in the initial sorbents and biogeosorbents after the experiment, mg/g. ** – no data

4   Conclusions

Our experiments resulted in modeling of biogeosorbents based on clay and zeolite raw and oil-oxidizing microorganisms from Biotrin biopreparation immobilized on them. We determined that samples of initial sorbents showed a high adsorption activity with respect to oil products. TPH content in water was reduced by 2.5–5 times, up to or substantially below MPC. We revealed that microorganism cells could reduce sorption properties of mineral carriers, at the same time providing oil destruction. Biodestruction of oil products with biogeosorbents for 4 days was 12–77%.

Acknowledgments. The authors express their gratitude to the Center for Collective Use "Geonauka", ecoanalytical laboratory of the Institute of Biology of the Komi Science Center UB RAS for their assistance in analytical work. The work was carried out with the partial financial support of UB RAS Programs (project 18-5-5-44), UMNIK project (12412GU/2017), of the State task "Development of biocatalytic systems based on enzymes, microorganisms and plant cells, their immobilized forms and associations for the processing of plant raw, production of biologically active substances, biofuels, remediation of contaminated soils and wastewater treatment" No. AAAA-A17-117121270025-1, "Scientific basis for effective subsoil use, development and exploration of mineral resource base, development and implementation of innovating technologies and economic zoning of the Timan-Nothern Ural region" No. AAAA-A17-117121270037-4.

References

Certificate on the toxicological and hygienic assessment of "BIOTRIN" consortium of strains of oil-oxidizing microorganisms. State federal enterprise for science Research center for toxicology and hygienic regulation of biopreparations at Federal medico-biological agency, Serpukhov, 28 September 2017 (in Russian)

(1998) Method for performing measurements of the mass fraction of petroleum products in soil samples on a Fluorat-02 analyzer. Institute of Biology of Komi Science Center of the Ural Branch of the Russian Academy of Sciences, 16.1.21-98, Moscow, 15 p (in Russian)

Patent of the Russian Federation No. 2615458

Patent of the Russian Federation No. 2658134

Order of the Ministry of Agriculture of the Russian Federation of December 13, 2016 No. 552. On approval of water quality standards for water bodies of fishery importance, including standards for maximum permissible concentrations of harmful substances in the waters of water bodies of fish-economic importance (in Russian)

Saldin VA, Burtsev IN, Mashin DO, Shebolkin DN, Inkina NS (2013) Marking horizons in the Upper Jurassic deposits of the Yarengsky shale region (north-east of the Russian plate), no 11, pp 26–29. Vestnik of the Institute of Geology of the Komi Science Center UB RAS (in Russian)

Simakova YuS (2016) Features of globular layered silicates of the Chim-Loptyugsky oil shale deposit, № 9–10, pp 52–57. Vestnik of the Institute of Geology of the Komi Science Center UB RAS (in Russian)

Shushkov DA, Kotova OB, Kapitanov VM, Ignatiev AN (2006) Analcime rocks of Timan as a promising type of minerals. Syktyvkar, 40 p (Scientific recommendations to the national economy/Komi Science Center UB RAS, issue 123) (in Russian)

Shchemelinina TN, Kotova OB, Harja M, Anchugova EM, Pelovski I, Kretesku I (2017) New trends in the mechanisms of increasing the productivity of materials on a mineral basis, no 6, pp 40–42. Vestnik of the Institute of Geology of the Komi Science Center UB RAS (in Russian)

Kotova OB, Harja M, Cretescu I, Noli F, Pelovski Y, Shushkov DA (2017) Zeolites in technologies of pollution prevention and remediation of aquatic systems, no 5, pp 49–53. Vestnik of the Institute of Geology of the Komi Science Center UB RAS

Heating Rate and Liquid Glass Content Influence on Cement Brick Dehydration

V. Strokova[(✉)] and D. Bondarenko

Department of Material Science and Material Technology, Belgorod State
Technological University named after V.G. Shukhov, Belgorod, Russia
vvstrokova@gmail.com

Abstract. Peculiarities of Portland cement dehydration in different hydrate
phases with sodium water glass have been given. Three endoeffects were
determined during non-isothermal heating, connected with ettringite dehydration
and water extraction at temperature ranges 98.7–110.0 °C, calcium hydroxide
decomposition at temperature ranges 439.4–450.7 °C and secondary carbonates
decomposition at temperature ranges 657.4–669.3 °C. We experimentally
proved that the rates of dehydration of hydrated Portland cement was significantly influenced by the liquid glass concentration. Optimum liquid glass content was grounded in the protective layer of composite finishing material,
modified with low-temperature plasma.

Keywords: Plasma-chemical modification · Dehydration · Cement brick ·
Soda water glass

1 Introduction

Plasma-chemical modification is one of the promising technologies of creating protective–decorative coatings in the manufacture of finishing construction materials for
building and construction faces (Bondarenko et al. 2018a, b; Bessmertny et al. 2018;
Bondarenko et al. 2016; Volokitin et al. 2016). Dehydration, melt formation and
accumulation during plasma melting take second fractions, and the surface is heated up
to 2000 °C. As a result of high temperature impact hydro silicate dehydration in the
surface layers can result in micro cracking and protective-decorative coating softening,
as well as coating adhesion strength reduction and lowering of cold resisting properties.

Insufficient technology elaboration on reducing heat impact consequences and
dehydration minimizing plasma melting of cement concrete does not allow wide
application of these materials on the national market. That is why the main task in
developing treatment technologies for materials based on cement matrix is composition
development for protective coating which eliminate these processes.

© The Author(s) 2019
S. Glagolev (Ed.): ICAM 2019, SPEES, pp. 286–289, 2019.
https://doi.org/10.1007/978-3-030-22974-0_68

2   Methods and Approaches

To prove the efficiency of Portland cement and liquid glass application in the protective coating during manufacture of composite finishing material with plasma surface treatment the samples were prepared at water\concrete ratio 0.3 from pure Portland cement (CEM I 42,5 H) and with 5 and 10% of soda water glass ($\rho = 1,4$ g/sm^3, silica modulus 2.8) of water of mixing. After hardening at normal conditions during 28 days, the samples were exposed to differential-thermal analysis.

Plasma-chemical surface modification is done in non-isothermal conditions at heating rate 3000 °C/min. It makes impossible to study dehydration in real conditions of plasma heating. This process was simulated with simultaneous thermal analysis device Netzsch STA 449 F3 Jupiter at heating rates 5 °C/min and 10 °C/min with maximum heating rate 1000 °C.

3   Results and Discussion

The thermograph of pure hydrated Portland cement shows three endoeffects (Table 1). The first endoeffects, in the temperature range 98.7–110.0 °C in the low temperature area, is connected with ettringite dehydration ($Ca_6Al_2(SO_4)_3(OH)_{12}\cdot26H_2O$) and water extraction. Endoeffects of these two processes superimpose each other. The second is connected with calcium hydroxide dehydration ($Ca(OH)_2$) and happens at temperature ranges 439.4–450.7 °C. The third endoeffects (657.4–669.3 °C) is connected with the secondary hydro carbonates dehydration ($CaCO_3$). Complete water extraction is at 900 °C.

Table 1. Changing of endoeffects with the introduction of liquid glass and heating range 5 and 10 °C/min

Endoeffect producer	Pure hydrated Portland cement		Portland cement with 5% liquid glass		Portland cement with 10% liquid glass	
	Heating range, °C/min					
	5	10	5	10	5	10
Ettringite and physically-coupled water	98.7	110.0	92.6	106.4	92.8	108.0
Calcium hydroxide	439.4	450.7	437.8	450.6	438.0	452.6
Secondary carbonate and hydro silicate	657.4	669.3	662.0	683.1	663.0	693.6

Similar results were received with the hydrated Portland cement after adding 5 and 10% liquid glass (Table 1).

A positive effect of liquid glass adding on secondary carbonate and hydro silicate endoeffects, which are responsible for cement brick softening and micro cracking at higher temperature range, can be explained by effect of encapsulation of hydrate phases with coating of liquid glass.

Adding sodium silicate solute into Portland cement in the amount 5 and 10% takes down mass loss (TG) in ettringite dehydration area (Fig. 1). But in high temperature area dehydration intensity increases up to 2–3%, it is especially notable with 10% liquid glass.

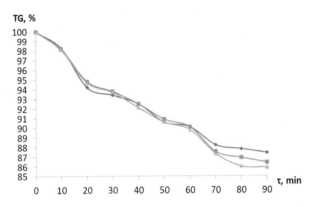

Fig. 1. Dependence of water loss on time at heating range 10 °C/min: —— Portland cement; —— Portland cement with 5% liquid glass; —— Portland cement with 10% liquid glass

The highest dehydration rate is in low temperature area (Fig. 2), which is caused by ettringite dehydration (first climax). The second and third climaxes are connected with dehydration of calcium hydroxide, secondary carbonate and different hydro silicates, are below the first climax by magnitude.

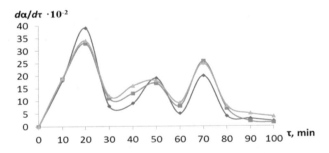

Fig. 2. Dependence of water loss on time at heating rates 10 °C/min: —— Portland cement; —— Portland cement with 5% liquid glass; —— Portland cement with 10% liquid glass

Dehydration rate decrease during the first and the second climax for cement brick with 5 and 10% of liquid glass has a significant impact on micro cracking minimization in the surface layer of protective–decorative coating of composite finishing material at plasma-chemical modification. Dehydration intensification can result in microcracks increase and reduce to zero the positive influence of liquid glass in the coating. This effect takes place at adding 10% liquid glass into Portland cement. Thus, the received laws analysis of mass loss and dehydration rate of the studied compositions drew to a conclusion that the optimal is liquid glass component in the ratio 5% of mixing water.

4 Conclusions

The influence pattern character has been determined of liquid glass content on ettringite dehydration rate, calcium hydroxide and secondary carbonate, which is in endoeffects shift in low temperature area to lower temperatures, and in high temperature area to the area of high temperatures. This minimizes dehydration in low temperature area, cement brick softening and micro cracking and as a result provides adhesion strength improvement of protective-decorative coating with the concrete layer.

Acknowledgements. The work is realized with the financial support of the Grant of the President for scientific schools, No. NSh-2724.2018.8, using equipment of High Technology Center at BSTU named after V.G. Shukhov.

References

Bessmertny VS, Puchka OV, Bondarenko DO, Antropova IA, Bragina LL (2018) Plasmochemical modification of wall building materials. Constr Mater Prod 1(2):11–18

Bondarenko DO, Bondarenko NI, Bessmertnyi VS, Strokova VV (2018a) Plasma-chemical modification of concrete. Adv Eng Res 157:105–110

Bondarenko NI, Bessmertnyi VS, Borisov IN, Tymoshenko TI, Burshina NA (2016a) The concrete with protective and decorative coverings on the basis of alyuminatny cements which are melted off by the plasma stream. Bulletin of Belgorod State Technological University named after V.G. Shukhov, vol 2, pp 6–12

Bondarenko NI, Bondarenko DO, Burlakov NM, Bragina LL (2018b) Investigation of influence of plasmochemical modification on macro- and microstructure of surface layer of autoclave wall materials. Constr Mater Prod 1(2):4–10

Volokitin O, Volokitin G, Skripnikova N, Shekhovtsov V (2016) Plasma technology for creation of protective and decorative coatings for building materials. In: AIP conference proceedings, vol 1698, p 070022

Structure and Surface Reactivity Mediated Enzymatic Performances of Clay-Based Nanobiocatalyst

S. Sun$^{(\boxtimes)}$, K. Wang, F. Dong, B. Ma, T. Huo, Y. Zhao, H. Yu, Y. Huang, and J. Huang

Institute of Non-metallic Minerals, Department of Geological Engineering, School of Environment and Resource, Southwest University of Science and Technology, Mianyang, People's Republic of China
shysun@swust.edu.cn

Abstract. The nanobiocatalyst is an emerging innovation that synergistically integrates advanced nanotechnology with biotechnology for improving enzyme activity, stability, capability and engineering performances. Enzymes and clay minerals are two important essential substances in the earth. As the important nanostructured minerals, clay minerals have played an important role in enzyme-related biochemical processes such as the origin and evolution of life, soil ecological environment, pollution remediation and element geochemical cycling, and even some clay minerals have intrinsic enzymatic activities. Based on the study of clay minerals-based enzyme-like activities and enzymatic catalysis in confined environment by clay minerals, lipase and 3-aminopropyltriethoxysilane (APTES) functionalized montmorillonite (Mt) were selected to biomimetic construct lipase-Mt nanobiocatalyst. Experimental results indicated that lipase-Mt activity was 40.65 U/mg, which was nearly 4-fold higher than that of free lipase under optimal conditions. The kinetic parameters of K_m and V_{max} for lipase-Mt were 3.406 mM and 312.5 mM/L min, respectively, indicating a higher affinity of the immobilized lipase in nanobiocatalyst towards the substrate compared with free lipase. The present study has provided a promising way for screening, optimizing and rational design of efficient nanomineral-based enzymatic nanobiocatalyst. The present work was supported by the National Natural Science Foundation of China (41672039). The Longshan Academic Talent Research Support Program of the Southwest University of Science and Technology (18LZX405).

Keywords: Surface reactivity · Clay minerals · Enzyme · Surface modification · Nanostructured minerals · Nanobiocatalyst

© The Author(s) 2019
S. Glagolev (Ed.): ICAM 2019, SPEES, pp. 290–291, 2019.
https://doi.org/10.1007/978-3-030-22974-0_69

Structural-Phase Stabilization of Clay Materials in Hydrothermal Conditions

O. Sysa[⊠], E. Evtushenko, I. Moreva, and V. Loktionov

Institute of Chemical Technology, BSTU named after V.G. Shukhov,
Belgorod, Russia
sysa1975@inbox.ru

Abstract. The results of thermal and X-ray phase analysis of hydrothermal stabilized kaolin clay are given. An evaluation method has been suggested of crystalline structure order according to the strength degree of elementary contacts in clay suspensions. It has been noted that hydrothermal stabilization may result in crystallohydrate crystalline structure perfection, in saturation, extraction or rearrangement of water molecule in kaolinite clay structure, and influence new phase generation during ceramic material baking.

Keywords: Kaolin clay · Hydrothermal processing · Degree of sophistication · Crystalline · Structural-phase changes · Ceramics

1 Introduction

Structure imperfection often defines material quality and properties. The structure perfection for layered silicates is determined by the structure of aluminosilicate layers themselves and their positioning within crystallite. It is known that kaolinite perfect crystals are hexagonal plane particles of the regular shape which allows free sliding relative each other and providing plasticity, liquescence and close arrangement goods formation. Crystallite distortion results in distortion of kaolinite plates due to lineal and point defects. It deteriorates rheological behaviour of kaolinite suspensions, liquescency instability (Kukovskiy et al. 1966).

It is possible to affect clay material structure by several methods (natural, mechanical, chemical, biological and thermal) (Evtushenko et al. 2009). Hydrothermal clay treatment can be the most effective when it is accompanied by Rehbinder effect (adsorptive plastification) (Rebinder et al. 1972), which accelerates restructurisation and minimizes crystal defects in clay minerals for a short period of time (Evtushenko et al. 2006).

2 Methods and Approaches

X-ray phase material analysis has been done with a diffractometer "DRON-3". XRD-pattern was shot with CuKα filtered radiation, – radiation (Ni – filter); voltage across the tube is 20 kw; anode tube current is 20 mA; measurement range is 10000–4000 counts per second; detectors rotation rate is 2,4 rot/min; angular mark is 10. Card index

© The Author(s) 2019
S. Glagolev (Ed.): ICAM 2019, SPEES, pp. 292–295, 2019.
https://doi.org/10.1007/978-3-030-22974-0_70

JCPDF was used for phase identification of elementary contacts durability formed without particles in clay slurry. The calculation was done according to Ur'ev model (1972, 2002) for low aggregate suspensions (Zubehin et al. 1995):

$$F_1 = \frac{\gamma_m \eta_{min} d^2}{6,4}, \tag{1}$$

where γ_m is share rate corresponding to complete structural destruction, s-1; η_{min} is effective viscosity corresponding to complete structural destruction, Pa s; d is average particle diameter, m.

Differential-thermal analysis was done with a derivatograph OD-102.

3   Results and Discussion

Kaolin clay of local deposits has been studied, which show structural changes of clay mineral after steaming in a pressure vessel at pressure from 1 up to 4 MPa.

With X-ray phase analysis we determined constant phase clay composition before and after treatment, with significant deviations of kaolinite crystalline structure. Hydrothermal modification at saturated steam pressure up to 4 MPa drives up crystallinity index according to Hinckley and intensivity of main diffractional kaolinite reflections (Table 1), that testifies to a greater mineral structure order in the treated raw material (Shlykov et al. 2006).

But "Hinckley crystallinity index" C_h decrease depending on defects in layer composition. This parameter is not always applicable to index correlation of index characteristics with its real structure.

Basing on the said above the authors developed an estimation method of material structure sophistication degree according to the strength magnitude of elementary contact, formed between particles in clay suspensions.

Table 1. Change in "Hinckley crystallinity index" (Ch) of kaolinite after hydrothermal treatment

Kaoline clay deposits	Initial kaoline	Kaoline after pressure treatment 1,6 MPa	Kaoline after pressure treatment 4,0 MPa
Zhuravliniy Log	0,20	0,22	0,46
Gluhovetskoe	0,77	0,83	1,27
Kyshtymskoye	0,55	0,47	0,61
Prosyanovskoye	0,51	0,63	0,63
Novoselitskoye	0,26	0,35	0,31

It has been found that elementary contact strength decreases at hydrothermal treatment pressure and temperature increase by more than an order. Ranging by crystalline structure defects these kaolines can be placed as follows: Zhuravliniy Log > Glukhovetskiy > Kyshtymskiy > Prosyanovskiy and Novoselitskiy.

Some differences in endo-and exothermal processes during baking in kaoline samples depending on hydrothermal modification have been studied (Gorshkov et al. 1988). It has been determined that maximum endothermic effect of kaoline dehydration shifts towards higher temperatures that testifies to crystalline structure sophistication. Restructurisation results in additional hydration and lessening connection energy of crystal water in kaoline, facilitates water molecule extraction penetrating into basis of tetrahedral kaoline layers.

Thermal capacity change dependence of the studied materials during baking has been studied. In the temperature range 950–1020 °C there are several exothermic extreme values testifying to crystallization possibility of a variety of phase (β-crystobalite, mullite, sillimanite, γ-alumina). Structural change of the initial material causes temperature shifts at the initial crystallization phase, as in clays with clearly seen crystalline structure mullite is formed at lower temperature and in greater quantity than from disordered structure minerals.

4   Conclusions

Hence it has been determined that hydrothermal treatment improves significantly stabilization of clay minerals structure as well as ceramic material baking processes, that has been proved by several research methods. Due to crystallohydrate structure change, saturation, water molecule distribution in kaolinite structure dehydration parameters change, processes take place which cause temperature shift of new phase chilling point in the interval of lower temperatures.

Acknowledgement. The work is realized in the framework of the Program of Flagship University development on the base of the Belgorod State Technological University named after V.G. Shukhov, using equipment of High Technology Center at BSTU named after V.G. Shukhov.

References

Evtushenko EI, Sysa OK (2006) Structural modification of clay raw material in hydrothermal conditions. University news. North-Caucasian region. Technical sciences series, no 2, pp 82–86

Evtushenko EI, Sysa OK, Moreva IYu, Bedina VI, Trunov Y (2009) Raw material preparation refining during activation processes in ceramics technology. Glass Ceram 1:15–16

Gorshkov VS, Saveliev VG, Fedorov NF (1988) Physical chemistry of silicates and other high-melting compounds. High School, Moscow

Kukovskiy Y (1966) Structure peculiarities and physic-chemical properties of clay minerals. Naukova Dumka, Kiev

Rebinder PA, Ur'ev NB, Shukin Y (1972) Physic-chemical mechanics in chemical technology of disperse systems. Theor Bases Chem Technol 6:16–24

Shlykov VG (2006) X-ray analysis of disperse soils mineral composition. GEOS, Moscow

Zubehin AP, Strakhov VI, Chekhovskiy VG (1995) Physico-chemical research methods of high-melting non-metal and silicate materials. Synthesis, Saint-Petersburg

Phase Changes in Radiation Protection Composite Materials Based on Bismuth Oxide

S. Yashkina[✉], V. Doroganov, E. Evtushenko, O. Gavshina,
and E. Sysa

Department of Technology of Glass and Ceramics, Belgorod State Technological
University named after V.G. Shukhov, Belgorod, Russia
yashkina_asp@mail.ru

Abstract. XRF method was used to study the mineralogical compounds of the radiation protection ceramic materials based on a high-alumina binder and a "heavy" aggregate, bismuth oxide. The content of Bi_2O_3 in the test samples was kept in the range of 38.5–75 wt%. Along with the bismuth oxide, the aggregate was aluminum oxide (Al_2O_3). The binder synthesis followed the principle of obtaining ceramic concretes based on artificial ceramic binders (ACB).

The paper establishes the specifics of sintering of the composites under study, fired under different temperatures.

Keywords: Bismuth oxide · Artificial ceramic binders · X-ray fluorescence · Radiation protection ceramic composites

1 Introduction

Bismuth oxide is a widely used bismuth compound due to its chemical stability, low toxicity, and its unique physical and chemical properties (Gulbin et al. 2014). Its sufficiently high density (8.9 g/cm^3) allows to use it in the synthesis of "heavy" materials employed in the protection from gamma rays. As it is known, in this case, the leading role in characterizing the protective properties is played by the density of the material. In the way of exercising high structural properties, the most efficient were the ceramic matrices (Correya et al. 2018).

This paper studies the phase transformations of the radiation protection materials based on the high-alumina ACBs, bismuth oxide, and aluminum oxide sintered at different temperatures.

2 Methods and Approaches

Obtaining the artificial ceramic binder (ACB) was carried out through the technique of wet grinding in a periodic action ball mill with a stepwise loading of material. This principle of obtaining ceramic concretes allows to obtain matrices with the desired phase composition, improved physical and mathematical properties and ensures the broad opportunities for using various aggregates.

S. Glagolev (Ed.): ICAM 2019, SPEES, pp. 296–299, 2019.
https://doi.org/10.1007/978-3-030-22974-0_71

The suspension based on high-alumina grog possesses the thixotropic dilatant flow (the destruction of the initial thixotropic structure and the following dilatant formation of structure). The bismuth oxide powder used as aggregate represents spherical particles sized no more than 35 μm.

The source materials were mixed in the proportions presented in Table 1. Then, the technique of static pressing was used with the specific pressure of 100 MPa to form the bar. After drying, the samples were subjected to burning within the interval of 200° to 1000 °C with holding at the maximum temperature for one hour.

Table 1. Compositions of radiation protection materials

Number of the compound	Content, wt%		
	ACB	Al_2O_3	Bi_2O_3
1	38.5	23.0	38.5
2	19.0	23.0	58.0
3	–	25.0	75.0

The X-ray fluorescence analysis was performed using the DRON-3 diffractometer. XRD patterns were recorded using the CuKα radiation (Ni filter); tube voltage: 20 kV; tube anode current: 20 mA; measuring range: 10000–4000 PPS; detector turn rate: 2.4 rpm; elevation angle: 10. The JCPDF database was used for identification.

3   Results and Discussion

The analysis of the transformations of the samples with the admixture Bi_2O_3 at varied temperatures (Figs. 1, 2 and 3) indicates that heating within the interval of 100–500 °C, all compounds are mineralogically constant and exhibit phases of mullite, corundum, and Bi_2O_3.

When increasing the burning temperature to 600 °C, starts the formation of bismuth silicates $2Bi_2O_3 \cdot 3SiO_2$ and $12Bi_2O_3 \cdot 2SiO_2$. Further increase of the burning temperature to 700 °C leads to a more intensive formation of bismuth silicates, indicated by the increase in the intensity of reflections for these compounds. In the process, the entire Bi_2O_3 turns into silicates, while the interplanar reflections of bismuth oxide totally disappear. There occurs a sharp decrease in the intensity of mullite peaks, while in the composition No. 3 this compound is totally absent.

When the temperature is increased up to 800 °C and further, all compositions display further formation of bismuth silicates, and in the compositions No. 2–3 there were identified two phases of α-Al_2O_3 и $12Bi_2O_3 \cdot 2SiO_2$, while in the composition No. 1, in addition to the indicated two, there also forms $2Bi_2O_3 \cdot 3SiO_2$ (Figs. 1, 2 and 3).

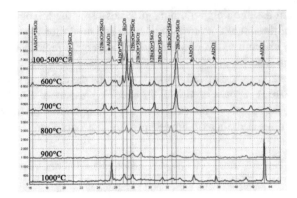

Fig. 1. X-ray patterns of the samples of the composition of radiation protection composites No. 1

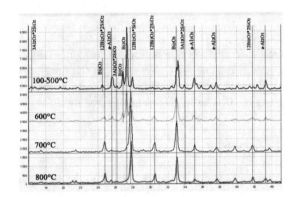

Fig. 2. X-ray patterns of the samples of the composition of radiation protection composites No. 2

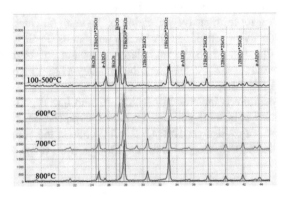

Fig. 3. X-ray patterns of the samples of the composition of radiation protection composites No. 3

It is possible to assume that compositions No. 1–2 show the formation of bismuth silicates as a result of mullite decomposition and binding of SiO_2 into the bismuth silicates, while in the composition No. 3 the silicates form due to the interaction of Bi_2O_3 and the nanosilica introduced as binder.

4   Conclusions

This paper studied the mineralogical compounds of the radiation protection ceramic materials based on a high-alumina ACB and bismuth oxide. It was identified that the higher the content of Bi_2O_3 in the source composition, the more intensive is the process of mullite decomposition. This way, at the content of Bi_2O_3 38,5%, mullite perseveres up to the temperature of 900 °C, and if bismuth oxide is increased to 58%, mullite disintegrates completely at the temperature of 700 °C, which is 200 °C lower than the previous composition.

Acknowledgement. The work is realized in the framework of the Program of Flagship University development on the base of the Belgorod State Technological University named after V.G. Shukhov, using the equipment of High Technology Center at BSTU named after V.G. Shukhov.

References

Correya AA, Mathew S, Nampoori VPN, Mujeeb A (2018) Structural and optical character-ization of hexagonal nanocrystalline bismuth-bismuth oxide core-shell structures synthesized at low temperature. Optik 175:930–935

Gulbin VN, Kolpakov NS, Polivkin VV (2014) Radio - i radiatcionno-zashchitnye kompozit-cionnye materialy s nanostrukturnymi napolniteliami [Radiation protection composite materials with nanostructure aggregates]. VSTU Bull 23:43–51

Development of Technology for Anti-corrosion Glass Enamel Coatings for Oil Pipelines

E. Yatsenko$^{(\boxtimes)}$, A. Ryabova, and L. Klimova

South Russian State Polytechnical University (NPI), Novocherkassk, Russia
e_yatsenko@mail.ru

Abstract. Among anticorrosive coatings for steel products, glass-enamel glass is the most reliable and versatile, based on aluminoborosilicate glasses of the SiO_2-Al_2O_3-B_2O_3-R_2O system. This anti-corrosion coating allows to increase the chemical resistance of the internal surface of pipelines to various groups of reagents. Therefore, in the course of the study, a previously developed composition was modified by introducing oxides SrO, ZrO_2, CaO, MoO_3, Li_2O and their acid and alkalinity of enamel frits and coatings based on them and it was found that the addition of strontium and zirconium oxide in the amount of 2% was optimal.

Keywords: Oil pipelines · Anti-corrosion coatings · Glass-enamel coatings · Steel protection · Chemical resistance · Acid and alkali resistance

1 Introduction

Currently, the oil industry of the Russian Federation is developing rapidly and efficiently. However, corrosion of equipment and facilities in the oil and gas industry is one of the main reasons for reducing their performance, causing huge economic losses and environmental damage. Presently, protective coatings (bitumen, epoxy, polyurethane, etc.) have become widely used to protect oil pipelines, among which glass-enamel coatings with high chemical resistance and thermomechanical properties, in particular, heat resistance, reaching 500 C, occupy a special place. Due to the fact that the silicate-enamel coating for steel pipelines is exposed to an aggressive environment containing hydrocarbons and formation water, in which chlorides, sulfates and organic acids are present, as well as up to 10% hydrogen sulfide and 10% carbon dioxide. The coating composition was based on an acid-resistant glass composition (Ryabova et al. 2018) with a high content of quartz and low boric anhydride and alkaline oxides, which will improve the chemical resistance of enamels and extend the range of their roasting (Yatsenko et al. 2018).

Also, when choosing the type of protective coating, the following factors should be taken into account: operating conditions, composition of the transported medium, temperature and pressure in the system, speed and nature of the flow movement, presence of abrasive particles in the fluid flow, composition and properties of associated petroleum gas (APG), presence of asphalt-resin-paraffin deposits (AFS), the manifestation of the life of microorganisms.

S. Glagolev (Ed.): ICAM 2019, SPEES, pp. 300–303, 2019.
https://doi.org/10.1007/978-3-030-22974-0_72

Therefore, the purpose of these studies was the development of anticorrosive glass-enamel coatings to protect steel pipelines from medium carbon steel 32G2S and the study of factors affecting the process of their defect-free formation and technical and operational properties.

2 Methods and Approaches

The technology of enameling the internal surface of pipelines includes the following technological stages: preliminary annealing of steel pipes at a temperature of ~ 750 °C in order to remove contaminants and decarburize the outer layers of steel; mechanical preparation of the inner surface of pipes using shot blasting using single or multiple shot blasting units to remove scale and roughen for better adhesion to the glass-enamel coating; preparation of enamel slip suspension on the basis of finely ground glass granulate; application of slip suspension on the inner surface of a horizontal pipe by sprinkling using an impeller; drying at a temperature of 100–120 °C and induction roasting at a temperature of 860–880 °C.

The aim of the work was to develop the composition and technology of applying a glass-enamel single-layer coating for medium-carbon steels, which has high rates of chemical and abrasive resistance and is capable of forming a defect-free smooth coating on the steel surface. To solve this problem, the aluminoborosilicate system SiO_2-Al_2O_3-B_2O_3-R_2O was chosen as the most acceptable in the single-layer enamelling technology, which was modified by introducing compounds such as SrO, ZrO_2, CaO, MoO_3, Li_2O, in order to improve chemical resistance and defect-free formation in the form of through pores. The introduction of enamels strontium, calcium and zirconium oxides that are insoluble in oxides helps to reduce the leaching of alkali and alkaline earth cations when exposed to acid coating. Lithium oxide together introduced with oxides of sodium and potassium contributes to the chemical resistance of enamel coatings due to the manifestation of polyalkalnochnogo and polycationic effects. The amount of additives introduced into the charge was 2 wt.%, Since the introduction of additives less than 1% slightly affects the properties of enamel coatings, and more than 2% can greatly affect the change in the technological properties of the melt.

Next, the compounded mixtures were boiled at a temperature of 1350 °C for 1 h in an electric oven in alundum crucibles, subjected to wet granulation and applied in the form of finely ground enamel slip to samples of 32G2S steel. After drying and firing at a temperature of 850 °C, the resulting enamel coatings were tested to study the effect of modifying oxides on the structure and properties of enamels.

To assess the corrosion resistance, tests were carried out to determine the acid resistance characterized by weight loss after exposure to 20% boiling hydrochloric acid and alkali resistance - weight loss after exposure to 8% sodium hydroxide solution. For frits, the tests were carried out by the grain method, and for enamel coatings, the impact on their surface.

3 Results and Discussion

For the developed modified frits and enamel coatings, chemical resistance to various groups of reagents for weight loss after boiling in acid and alkaline solutions was studied. The test results are presented in Table 1.

Table 1. Indicators of the properties of modified frits and enamel coatings depending on the composition

Name of enamel	Modifying additive	Chemical resistance			
		Frites, %		Coatings, mg/cm^2	
		20%-HCl	8%-NaOH	20%-HCl	8%-NaOH
10-0	Without additives	0,38	0,83	0,22	0,78
10-1	SrO	0,32	0,99	0,16	0,45
10-2	ZrO$_2$	0,40	1,20	0,20	0,60
10-3	CaO	0,38	0,75	0,21	0,63
10-4	MoO$_3$	0,54	0,98	0,22	0,68
10-5	Li$_2$O	0,42	0,87	0,23	0,63

The results obtained allow us to conclude that the introduction of modifying additives into the glass matrix composition has a significant impact on anti-corrosion properties.

For all compositions, the mass loss of frits after boiling is quite significant, since the specific surface of glass powders is much larger than the surface of the burned enamel coating. However, the composition of frits modified by strontium and zirconium oxides is characterized by less mass loss, which is caused by their positive effect on the increase in the packing density of the structural amorphous network, due to the large radius of these ions. Molybdenum oxide has almost no effect on chemical resistance, but it has a positive effect on the continuity of the coating contributing to a smoother formation and the absence of coating defects, due to a decrease in the surface tension of the enamel melt. Calcium oxide in such an amount does not affect the chemical resistance of frits and coatings. Lithium oxide increases chemical resistance and frit and coatings, although it is alkaline in itself, but it has an inhibitory effect due to the presence of several alkali cations (Na$_2$O and K$_2$O).

4 Conclusions

As a result of the research, the composition and technology of producing anticorrosive glass-enamel coatings for the internal protection of steel pipelines based on the SiO$_2$-Al$_2$O$_3$-B$_2$O$_3$-R$_2$O aluminoborosilicate system has been developed. The effect of various modifiers of the vitreous matrix on the acid and alkali resistance was studied, and it was found that the strontium and zirconium oxides in an amount of 2% are optimal.

Acknowledgements. The work was done with the financial support of the Russian Science Foundation under the agreement No. 18-19-00455 "Development of technology for the integrated protection of pipelines for oil and gas operated in the Far East of Russia" (headed by Yatsenko E. A.).

References

Ryabova AV, Yatsenko EA, Klimova LV, Goltsman BM, Fanda AY (2018) Protection of steel pipelines with glass-enamel coatings based on silica-containing raw materials of the far east of Russia. Int. J. Mech. Eng. Technol. 9(10):769–774

Yatsenko EA et al (2018) Investigation of chemical processes that ensure the adhesion strength of glass-enamel coating with steel pipelines. Butlerov Commun 56(11): 122–127

Building Materials

Optimization of Formulations of Cement Composites Modified by Calcined Clay Raw Material for Energy Efficient Building Constructions

A. Balykov[✉], T. Nizina, V. Volodin, and D. Korovkin

Department of Building Structures, Ogarev Mordovia State University,
Saransk, Russia
artbalrun@yandex.ru

Abstract. The paper presents the results of the study of the influence of formulation and process parameters of dehydrated raw material preparation based on polymineral clay rocks of the Republic of Mordovia used as independent mineral additives to cement composites. The possibility of increasing the studied physical and mechanical parameters of composites by optimizing the mode of clay raw material calcination and the content of the developed modifier is shown.

Keywords: Cement composite · Dehydration · Mineral additive · Clay

1 Introduction

Currently, Portland cement is the main binder in the construction industry. Introduction of fine-grained mineral additives of natural and man-made origin to Portland cement in order to improve the indicators of its physical and mechanical properties and partially replace clinker is one of the effective ways to ensure sustainable development in terms of resource conservation. In recent years, such mineral additives as microsilica and metakaolin have been increasingly used for more rational use of Portland cement and ensure the required level of cement composites characteristics. These modifiers help to increase the density of cement stone by controlling its phase composition and porosity, thereby allowing improvement of physical, mechanical and operational properties of cement composites at reduced cement consumption (Kirsanova et al. 2015; Nizina et al. 2017; Dvorkin et al. 2015).

However, the resources of the above additives do not meet the increasing needs of the construction industry. In this regard, researchers face the challenge of expanding the resource base for the production of mineral additives using available natural raw materials. One of the most promising in this respect are calcined clay rocks (Schulze et al. 2015). At the same time, according to the studies (Rakhimov et al. 2017; Fernandez et al. 2011) results, it was found that kaolinite, montmorillonite and muscovite/illite clays have the highest pozzolanic activity after heat treatment.

© The Author(s) 2019
S. Glagolev (Ed.): ICAM 2019, SPEES, pp. 307–310, 2019.
https://doi.org/10.1007/978-3-030-22974-0_73

The territory of Russia is rich in various types of clays. Ordinary (low-melting) clay in Russia is produced almost everywhere. For example, in the territory of the Republic of Mordovia there are more than fifty deposits of clay rocks, which allows classifying the development of active mineral additives based on clay raw materials as promising task of the construction industry, the solution of which a number of economic, technological and environmental problems of the cement industry both in the region and in country as a whole.

2 Methods and Approaches

Clay from Staroshaygovsky deposit (The Republic of Mordovia) was selected as a raw material for mineral additive development. To carry out the experimental studies, a plan was prepared, which includes 15 experiments allowing variation of the temperature and duration of calcination at three levels (400, 600 and 800 °C and 2, 3 and 4 h, respectively), and the content of mineral additive based on thermally-activated clay in the composition of cement composites on five levels – 2, 6, 10, 14 and 18% of the weight of Portland cement. Also, the additive-free composition (No. 16) was studied in addition to the 15 formulations included in the main block of the experiment plan. Manufacture of cement compositions was carried out at a fixed water-solid ratio of 0.3. The calcined clays were ground in a planetary mill for 1 h. The resulting fine powder was introduced into the cement binder based on Portland cement CEM I 42.5 N produced by Serebryakovcement PJSC. According to the results of the study, optimization of the modified cement binders was carried out and the most effective calcination modes were determined. Rational compositions were determined according to the analysis of an experimental statistical model describing the change in compression resistance of cement composites based on modified calcined clay raw materials:

$$
\begin{aligned}
y = 67,29 + 3,23 \cdot x_1 + 0,18 \cdot x_2 - 3,99 \cdot x_3 + 1,36 \cdot x_1 \cdot x_2 \\
-0,81 \cdot x_1 \cdot x_3 - 1,38 \cdot x_2 \cdot x_3 + 0,31 \cdot x_1 \cdot x_2 \cdot x_3 - 7,55 \cdot x_1^2 \\
-4,35 \cdot x_2^2 + 4,48 \cdot x_3^2 - 0,91 \cdot x_1^2 \cdot x_2 - 0,56 \cdot x_1 \cdot x_2^2 \\
-0,49 \cdot x_1^2 \cdot x_3 - 2,19 \cdot (x_1 \cdot x_2 \cdot x_3)^2
\end{aligned}
\tag{1}
$$

Identification of compromise solutions optimal areas for each factor separately was carried out using frequency ranges, which is one of the most descriptive ways to graphically represent the random variable probability density (Lyashenko et al. 2017).

3 Results and Discussion

According to the results of the conducted study, it was determined that a number of modified cement composites can achieve compression resistance equal to 70 ÷ 80 MPa, which is comparable with the control composition No. 16 (Fig. 1). The highest strength characteristics were achieved in compositions 2, 4, 6 and 13 at a content of calcined clay from 2 to 6% of the cement weight.

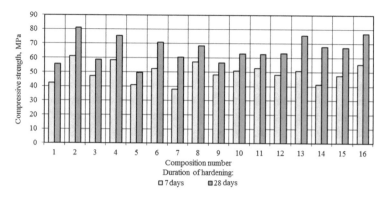

Fig. 1. Compression resistance of modified cement composites at the age of 7 and 28 days

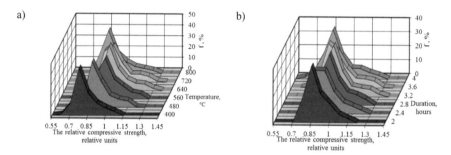

Fig. 2. The range of distribution of the compression resistance of modified cement composites at the age of 28 days: a – from calcination temperature, b – from calcination duration

The analysis of the ES model (1) based on frequency ranges (Fig. 2) showed that the compression resistance corresponding to the control composition can be provided for cement composites with a mineral additive at any studied temperature level and duration of calcination of the clay raw material. At the same time, for the accepted temperature and time intervals of the mineral additive calcination, the total proportion of compositions with enhanced or corresponding to the control composite character-istics varies from 22 to 41% depending on the duration and from 11 to 45% depending on the temperature of calcination. It was found that an increase in clay calcination time from 2 to 3 ÷ 4 h leads to an expansion of the relative values range of modified cement composites strength characteristics from 77.5 ÷ 115 to 62.5 ÷ 130%. Increasing the temperature of calcined clay rocks calcination from 400 to 720 °C allows changing the limit (achievable) range of compression resistance from 62.5 ÷ 107.5 to 85 ÷ 130%, a further increase in temperature leads to a certain decrease in the boundary values of the relative strength indicator to 77.5 (lower boundary) and 122.5% (upper boundary), respectively.

4 Conclusions

According to the results of the study, optimal formulation and process principles for the production of mineral additive based on clay raw materials were determined, which allow increasing compression resistance of modified cement composites in comparison with the additive-free composition. The most effective additives were obtained at calcination time from to 3 to 3.6 h at clay calcination temperature 640 ÷ 720 °C.

The data obtained indicate the prospects and relevance of the development of concrete with modifying additives based on thermally-activated polymineral clays, which allows expanding the range of modified cement composites produced today due to better use of local mineral resources base.

Acknowledgements. The reported study was funded by RFBR and Government of the Republic of Mordovia according to the research project № 18-43-130008.

References

Dvorkin LI, Zhitkovsky VV, Dvorkin OL, Razumovsky AR (2015) Metakaolin is effective mineral additive for concrete. Concrete Technol 9–10(110–111):21–24

Fernandez R, Martizena F, Scrivener KL (2011) The origin of the pozzolanic activity of calcined clay minerals: a comparison between kaolinite, illite and montmorrilonite. Cement Concrete Res 41:113–122

Kirsanova AA, Ionov YV, Orlova AA, Kramar LY (2015) Features of hydration and hardening of cement concretes with additives modifiers containing metakaolin. Cement Appl 2:130–135

Lyashenko TV, Voznesensky VA (2017) Composition-process fields methodology in computational building materials science. Astroprint, Odessa

Nizina TA, Balykov AS, Volodin VV, Korovkin DI (2017) Fiber fine-grained concretes with polyfunctional modifying additives. Mag Civil Eng 4:73–83. https://doi.org/10.18720/MCE.72.9

Rakhimov RZ, Rakhimova NR, Gaifullin AR, Morozov VP (2017) Properties of Portland cement paste incorporated with loamy clay. Geosyst Eng 6:318–325

Schulze SE, Pierkes R, Rickert J (2015) Optimization of cements with calcined clays as supplementary cementations materials. In: Proceedings of XIV international congress on the chemistry of cement, Beijing, China

Santa Maria Clays as Ceramic Raw Materials

Â. Cerqueira[1]([⊠]), C. Sequeira[1], D. Terroso[1], S. Moutinho[1], C. Costa[1], and F. Rocha[1,2]

[1] GeoBioTec, Geosciences Department, University of Aveiro,
3810-193 Aveiro, Portugal
angelamcerqueira@ua.pt
[2] RISCO, Civil Engineering Department, University of Aveiro,
3810-193 Aveiro, Portugal

Abstract. There are evidences and records concerning clay exploitation for pottery in the island of Santa Maria (Azores, Portugal) in the past. Nowadays this activity is almost extinct but this is the only island in the archipelago with abundant residual and sedimentary clay deposits. To evaluate the applicability of clays from this island for modern ceramics, a campaign made in May 2017 allowed collecting twenty samples in several outcrops all over the island. All samples were subsequently analyzed in terms of granulometry, mineralogy, chemical composition and physical properties. Results revealed to be interesting, namely concerning mineralogical composition, where phyllosilicates such as Kaolinite are in high percentages. Granulometry also revealed that most part of the samples is composed by fine grain size particles (<63 μm), which can be a good indicator of the existence of resources in great quantity.

Keywords: Santa Maria Island · Pottery · Physical characterization · Potentialities

1 Introduction

Santa Maria, the oldest and most oriental island from the archipelago, is very weathered, as consequence of intense volcanic activity alternated with sea level alterations and intense erosion episodes. There are residual and sedimentary clayey deposits in several parts of the island, one of which known as "Red Desert", corresponding to Feteiras Formation. In addition to the previous, Almagreira Formation is other important deposit of the island. These two in particular were more studied. The abundance of raw materials, gave this island an ancient tradition concerning exploitation of clays for pottery, being known as the "mother island" of clay. From this island, during centuries, white clays were extracted and exported to provide other islands. Recently a pottery oven was discovery "in the middle" of Vila do Porto and it date from the 18th century.

2 Methods and Approaches

Chemical composition was assessed by X-ray fluorescence, qualitative and semi-quantitative mineralogical analyses were carried out by X-ray diffraction and crystal-lochemistry analyses were carried on a Scanning Electron Microscope (SEM) Hitachi

© The Author(s) 2019
S. Glagolev (Ed.): ICAM 2019, SPEES, pp. 311–312, 2019.
https://doi.org/10.1007/978-3-030-22974-0_74

SU70 with Energy Dispersive X-Ray Spectroscopy (EDS) Brucker QUANTAX 400. Viscosity was assessed with a Haake Viscotester iQ. Plasticity Index was also computed from Atterberg Limits determination.

3 Results and Conclusions

Results are very positive since most part of samples are rich on phyllosilicates (69% to 98%, being kaolinite the most common), fine-grained, and showing adequate plasticity and viscosity. Geological formations, in particular Feteiras and Almagreira Formations, outcrops on a large part of the island, the residual ones showing always intensive alteration, therefore assuring the existence of good reserves.

Alkaline Activation of Rammed Earth Material – "New Generation of Adobes"

C. Costa[1(✉)], D. Arduin[1], C. Sequeira[1], D. Terroso[1], S. Moutinho[2], Â. Cerqueira[1], A. Velosa[1,2], and F. Rocha[1,2]

[1] GeoBioTec, Geosciences Department, University of Aveiro,
3810-193 Aveiro, Portugal
cristianacosta@ua.pt
[2] RISCO, Civil Engineering Department, University of Aveiro,
3810-193 Aveiro, Portugal

Abstract. Adobe is an extremely simple form of earth construction and with this technique the shrinkage associated with the construction of large structures is avoided. In Portugal, earthen materials have been used in load-bearing walls in the form of adobe or rammed earth for the construction of buildings especially in the southern and central coast. Most conventional consolidation treatments used in the past have not succeeded in providing a long-term solution because they did not tackle the main cause of degradation, the expansion and contraction of constituent clay minerals in response to humidity changes. Clay swelling could be reduced significantly by transforming clay minerals into non-expandable binding materials with cementing capacity using alkaline activation. in this study it was being developed adobes with water, and alkaline activated with NaOH and KOH. The obtained results allowed to conclude that the adobes with NaOH and KOH have an increase of its properties.

Keywords: Rammed-earth · Adobe construction · Alkaline activation · "New generation of adobes"

1 Introduction

The use of adobe construction in the Aveiro district reflected the properties of the existing raw materials available to be applied, namely sand, clay sediments and soils and lime (Silveira et al. 2012), and there is an evident heterogeneity of the adobes linked to the geographic distribution of the available resources. In Aveiro there was a semi-industrial production of adobe, some small companies employing 'adobeiros', for the manufacture of blocks of adobe, along with a domestic self-production (Costa et al. 2016). Most conventional consolidation treatments used in the past have not succeeded in providing a long-term solution because they did not tackle the main cause of degradation, the expansion and contraction of constituent clay minerals in response to humidity changes. Clay swelling could be reduced significantly by transforming clay minerals into non-expandable binding materials with cementing capacity using alkaline activation.

S. Glagolev (Ed.): ICAM 2019, SPEES, pp. 313–314, 2019.
https://doi.org/10.1007/978-3-030-22974-0_75

2 Methods and Approaches

The chemical composition of diatomite samples was assessed by X-ray fluorescence (XRF), qualitative mineralogical analyses were carried out by X- ray diffraction (XRD). Compressive strength tests use a universal mechanical compression testing machine (Shimadzu Autograph AG 25 TA).

3 Results and Conclusions

Mineralogically, samples are composed by quartz, feldspars and phyllosilicates, however, there is an evidence of the presence of amorphous alumino-silicate phases. Samples with higher mechanical strength are associated with higher specific surface areas, namely, those with NaOH. The alkaline activation promotes the increase of mechanical resistance. With the exception of one sample, all the samples present promising mechanical resistances for their application in rehabilitation works on adobe buildings, since these do not require high resistance, even for the sake of compatibility.

References

Costa CS, Rocha F, Velosa AL (2016) Sustainability in earthen heritage conservation. Sustainable use of traditional geomaterials in construction practice. Geol Soc London, Spec Publ 416. http://doi.org/10.1144/SP416.22

Silveira D, Varum H, Costa A, Martins T, Pereira H, Almeida J (2012) Mechanical properties of adobe bricks in ancient constructions. Constr Build Mater 28:36–44

Structurization of Composites When Using 3D-Additive Technologies in Construction

M. Elistratkin[1](✉), V. Lesovik[1], N. Chernysheva[1], E. Glagolev[2], and P. Hardaev[3]

[1] Department of Building Materials, Products and Designs,
Belgorod State Technological University named after V.G. Shukhov,
Belgorod, Russia
elistratkin.my@bstu.ru, mr.elistratkin@yandex.ru
[2] Department of Construction and Municipal Economy,
Belgorod State Technological University named after V.G. Shukhov,
Belgorod, Russia
[3] Department Industrial and Civil Engineering,
East Siberian State University of Technology and Management,
Ulan-Ude, Russia

Abstract. One of new and perspective lines of development in the field of construction technologies is the integration of additive production elements. The laboratory study of these issues revealed critical challenges that slightly slowed down the introduction of construction printing in daily construction practice. One of such problems is a big difference of structurization conditions of additive composites produced via traditional methods. The paper provides the analysis of factors exerting negative influence on structurization and considers the possibilities of solving such challenges.

Keywords: Construction printing · Mixing ratio ·
Structurization of composites · Composite binding agents · Mineral additives

1 Introduction

One of new and quite perspective development areas of construction technologies is the integration of additive production elements. The active development of construction printing launched a decade ago allowed creating some concepts demonstrating only a mere part of its potential and drawing public attention and investments to its development (De Schutter et al. 2018). However, it also caused some critical problems that slightly slowed down the introduction of construction printing in daily construction practice.

First of all, such problems include structural reinforcement. In the most cases traditional reinforcement methods (frames, rods, grids) do not correspond to the ideology of additive production, which implies that the construction robot installs the building structure without too much human involvement and use of various additional technical tools. Various methods of dispersed (Christ 2015) and textile reinforcement (Lesovik et al. 2017) may possibly solve the matter, and in the long-term perspective this problem may be solved by making the construction printer place or 'raise' the reinforcement according to the design project.

© The Author(s) 2019
S. Glagolev (Ed.): ICAM 2019, SPEES, pp. 315–318, 2019.
https://doi.org/10.1007/978-3-030-22974-0_76

Another problem, which is currently being tackled by scientists, is the production of efficient mixes for printing, which is considered controversial from the perspective of the traditional concrete technology and with regard to its sufficient properties. It is impossible to ensure competitiveness of construction additive technologies in relation to traditionally applied methods without the development of new principles of their creation that would integrate classic approaches and the latest achievements in construction materials science (Yi et al. 2017).

2   Methods and Approaches

The study used the traditional cement-sand mortar at the ratio of 1:4 at W/C = 0.45…0.5 as an extrusion printing mix. The following refer to special properties of mortar, which make its different from standard mixes:

- ability to easily pass through the extruder and a 20 × 20 mm nozzle without losing uniformity and sticking to walls;
- ability to hold its shape after extrusion and resist loading of at least 5 layers placed on top without intermediate curing.

The specified qualities were obtained due to combination of two Russian additives characterized by availability, low cost and good technological effectiveness. The used additives allow receiving various mixes for construction printing at the C:S ratio from 1:3 to 1:5 that preserve special properties within 25…30 min.

3   Results and Discussion

The printing with the developed mixes on a laboratory unit (Fig. 1) sets the task to obtain the flattest surface without post-processing.

Fig. 1. Printing with developed mixes on a laboratory unit

Table 1 shows the strength properties of the mix during processing in various conditions. The preparation of a molding compound is critical since this stage ensures the formation of traditional and special properties.

According to our experience in the study of construction printing, it is more preferable for industrial facilities to ensure continuous supply of the mix into a small bin feeder installed on a forming device.

Table 1. Strength of mixes (MPa) depending on curing conditions

Composition	1 day	3 day	7 day	28 day
1:3 (in water)	3.7	8.2	12.6	17.9
1:3 (in insulation)	4.2	8.2	11.3	14.3
1:3 (in air)	4	6.9	8.1	8.5
1:4 (in water)	2.9	6.3	9.4	12.9
1:4 (in insulation)	3	6.8	10.5	15.2
1:4 (in air)	3	5.4	6.6	6.8

The stage when the mix passes through the extruder is followed by its additional mixing, decrease in viscosity, and utilization of some amount of air. Printing with slight premolding of lower layers with the newly placed ones ensures the formation of stronger contact between them compared to free mix outflow, but at the same time imposes increased requirements on the ability of a mix to keep its shape after extrusion (Secrieru et al. 2017).

The third stage is characterized by the largest duration and ensures the formation of final indicators of the additive structure.

The curing conditions of composites in the printed structure significantly (for the worse) differ from traditional methods of concrete works (Zharikov et al. 2018). The wall assemblies made via outline printing have small effective sectional area at their quite big surface area, which leads to their fast dehydration. Table 1 shows the influence of curing conditions on strength accumulation velocity.

It is less likely possible to create favorable conditions for structurization of composites due to external moistening, therefore it is possible to define solution to this problem: reduction of a binder's demand in water and the maximum increase in early strength during the period until the main amount of liquid has not evaporated yet.

As it was noted in some works (Chernysheva et al. 2013; Sumskoy et al. 2018; Lesovik et al. 2014; Kuprina et al. 2014; Elistratkin et al. 2018), a good solution in such cases may be the replacement of portland cement with composite binding agents that contain mineral additives with developed microporosity (zeolites, utilization products of some ceramic construction materials). Such products are able to create some moisture stock, which maintains hydration for some time needed for the material to become strong.

4   Conclusions

Alongside with various processing technologies, quite often complicating the design of a construction printer, the essential positive effect may be achieved by using composite binding agents with required properties: required rheology, ability to quickly gain strength during solidification in the conditions of fast dehydration, reduced or zero shrinkage. Such measures will provide for a simpler solution to the task of creating favorable conditions for composite structurization in construction printing thus establishing a self-sufficient system not too much dependent on external factors.

Acknowledgements. The study is implemented in the framework of the RFBR according to the research project No. 18-03-00352, using equipment of High Technology Center at BSTU named after V.G. Shukhov.

References

Chernysheva NV, Ageeva MS, Elyan I, Drebezgova MYu (2013) The effect of mineral additives of different genesis on the microstructure of the gypsum-cement stone. Bull. BSTU named after V.G. Shukhov 4:12–18

Christ S (2015) Fiber reinforcement during 3D printing. Mater Lett 139:165–168

De Schutter G, Lesage K, Mechtcherine V, Nerella VN, Habert G, Agusti-Juan, I (2018) Vision of 3D printing with concrete – technical, economic and environmental potentials. Cement Concrete Res

Elistratkin MYu, Minakova AV, Jamil AN, Kukovitsky VV, Eleyan Issa Jamal Issa (2018) Composite binders for finishing compositions. Constr Mater Products 1(2):37–44

Kuprina AA, Lesovik VS, Zagorodnyk LH, Elistratkin MY (2014) Anisotropy of materials properties of natural and man-triggered origin. Res J Appl Sci 9(11):816–819

Lesovik VS, Chulkova IL, Zagordnyuk LK, Volodchenko AA, Yurievich PD (2014) The role of the law of affinity structures in the construction material science by performance of the restoration works. Res J Appl Sci 9(12):1100–1105

Lesovik VS, Popov DYu, Glagolev ES (2017) Textile-concrete is an effective reinforced composite of the future. Constr Mater 3:81–84

Secrieru E, Fataei S, Schröfl C, Mechtcherine V (2017) Study on concrete pumpability combining different laboratory tools and linkage to rheology. Constr Build Mater 144:451–461

Sumskoy DA, Zagorodnyuk LKh, Zhernovskiy IV (2018) Features of the formation of crystalline neoplasms in astringent compositions depending on the technology of their preparation. Bull BSTU named after V.G. Shukhov 6:71–78

Yi WDT, Biranchi P, Suvash CP, Nisar ANM, Ming JT, Kah FL (2017) 3D printing trends in building and construction industry: a review. Virtual Phys Prototyping 12(3):261–276

Zharikov IS, Laketich A, Laketich N (2018) Impact of concrete quality works on concrete strength of monolithic constructions. Constr Mater Products 1(1):51–58

Influence of Flow Blowing Agent on the Properties of Aerated Concrete Variable Density and Strength

V. Galdina[✉], E. Gurova, P. Deryabin, M. Rashchupkina, and I. Chulkova

Department of Building Structures, Ogarev Mordovia State University, Saransk, Russia
example@yandex.ru

Abstract. We presented the effect of gasifier flow on gas release kinetics and basic properties of aerated concrete with variable traverse density and strength. We found optimal consumption of gasifier for manufacturing of aerated concrete with variatropic properties from expanded clay sand, wherein we obtained aerated concrete with a strength of 1.42 MPa at an average density of 414 kg/m^3.

Keywords: Cellular concrete · Aerated concrete · Blowing consumption · Bloating · Gas release kinetics

1 Introduction

The scientists and technologists should develop a technology of new generation of cellular concrete with a higher strength and frost-resistance at a low average density.

2 Methods and Approaches

The production of concrete products in a closed mold can be related to new methods. Works by Chernov and Zavadsky are important at the present stage of production of aerated concrete. Their studies are mainly based on preparation of gas concrete mix in the mold with a hollow cap (without holes), or small holes in side and top faces of the mold (Chernov et al. 1983; Chernov 2003; Zavadsky 2005; Zavadsky 2001; Chernov 2002; Zavadsky, Kosach 1999).

The properties of aerated concrete with variable traverse density and strength of products are influenced by the following factors: area of surface of a cover; fluidity of mix; consumption and type of blowing agent; fill level of mix in the mold; type, flow rate and surface area of silica component and binder.

© The Author(s) 2019
S. Glagolev (Ed.): ICAM 2019, SPEES, pp. 319–322, 2019.
https://doi.org/10.1007/978-3-030-22974-0_77

3   Results and Discussion

The formation of a cellular structure and gas-concrete products in an individual mold was carried out as follows: gas-concrete mix was prepared, poured into a mold closed by a cover with a circular hole in the center of the mold (The method..., 2015).

The essence of formation of a cellular structure of concrete of variable traverse density and durability is as follows. Hydrogen releases when an alkaline component binds with aluminum, which results in blowing of viscoplastic mass. The blowing mix, having reached the internal surface of the cover with a circular hole, meets a barrier on the way and swells up on the way of the weakest resistance (through the hole), and on the periphery of the product there is a self-consolidation of aerated concrete due to overpressure. As a result, the pressing gas with more than 0.1 kgf/cm^2 and presence of the closed cover surface at the time of a blowing results in less dense and more porous products in the mid-range of the composite and denser, stronger – on the periphery.

The known formulations in production of gas-concrete products by traditional technology, in particular a gas developing agent consumption, are not absolutely correct for manufacturing of products in the closed mold since the raised gas developing agent consumption is necessary for effective self-consolidation of exemplars. In this regard it is necessary to reveal its influence on a kinetics of gas emission of the mix and main properties of aerated concrete.

V/T = 0.5 was applied to aerated concrete on the basis of expanded clay sand with mix spread at Southard viscosimeter 30 cm. Gazobetolyuks gas paste and earlier chosen area of the closed cover surface with the round section equal to 71% was applied as gas developing agent. The mix was poured up to 70% of the mold height.

The nature of a flatulence of the mix up to the cover in the period of time from 5 to 20 min at all gas-concrete mix was identical, different only by the amount of gas developing agent. Process of an aerogenesis in all mixtures practically finishes in 20–30 min (Fig. 1).

The highest limit of compression strength is observed for samples produced at a gasifier flow 1000 g per m^3 of the mix. As a result of the chemical reaction the released hydrogen swells the mix which reaches the edge of the mold, where it meets an obstacle of cover and is pressed on the way of the weakest resistance through a circular hole in the cover, whereby mostly form crusts and increase strength of samples during compression. When the flow reduces to 800 g the strength is reduced on average by 30%. This is a result of low flow of the gasifier, and the mix during swelling is not sufficient for self-sealing along the periphery, which is confirmed by gas release kinetics, which reaches the inner surface of the cover only in 20–22 min and the final height of swelling 35%. Since one of the main objectives is not only to obtain aerated concrete with a relatively high compressive strength, but also with reduced average density index, the optimum flow rate is from 1100 to 1300 g per m^3.

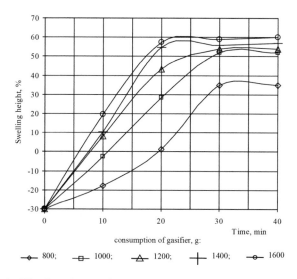

Fig. 1. Kinetics of gas emission of mix on the basis of expanded sand

Aerated concrete, prepared in the mold with a circular hole in the cover on expanded clay aggregate with the strength 1.42 MPa at an average density 414 kg/m^3 (Fig. 2), was obtained at a flow rate equal to 1,200 g of blowing agent per m3 of mix, which corresponds to D500 brand strength. The gassing process also takes about 25–30 min, after which the height of the crusts is 50–54% (Fig. 2).

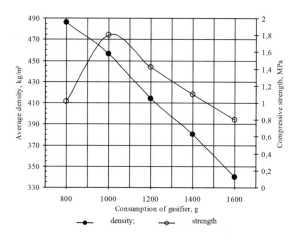

Fig. 2. The effect of the gasifier consumption on the average density and strength of aerated concrete based on expanded clay sand

4   Conclusions

We determined the optimal flow of gas agent on the basis of expanded clay sand equal to 1100–1300 g per m^3, at which the aerated concrete with a durability of 1.42 MPa at an average density of 414 kg/m^3 that corresponds to D500 brand was produced. At the same time the process of aerogenesis continues 25–30 min, after which the height of top crust is 50–54%.

Advantage of technology of aerated concrete with variatropic properties is a high stability of density of cellular concrete depending only on the accuracy of the gravimetric feeders measuring components. At the same time not only density, but also durability, deformability, heat conductivity and other properties are stabilized. The increased dispersion of operational parameters, inherent to all cellular concretes, is excluded.

Development of new and combination of already known methods of pore making in mass and also application of processing methods for products with variatropic properties at formation allow producing one and two-layer wall products of various configuration, sizes and required heat-shielding and operational properties.

References

Chernov AN (2003) Auto-frettage in aerated concrete technology. Constr Mater 11:22–23

Chernov AN (2002) Cellular concretes. Publishing house of SUSU, Chelyabinsk, 111 p

Chernov AN, Kozhevnikova LP, Khmelev SV, Tsarkov VV, Danilyuk MA, Moiseev EI, Stepanova ZA (1983) Technology of cellular concrete products with compacted surface layer. Constr Mater 8:12–13

The method of producing aerated concrete and the raw mix for its preparation. Patent of the Russian Federation P. Deryabin. No. 2560009 of July 20 2015

Zavadsky VF (1999) An integrated approach to solving the problem of thermal protection of walls of heated buildings. Constr Mater 2:7–8

Zavadsky VF (2001) Variants of wall constructions with the use of effective heaters - Novosibirsk: NGASU, 52 p

Zavadsky VF, Kosach AF (2005) Wall materials and products. SibADI Publishing House, Omsk, 254 p

Structuring Features of Mixed Cements on the Basis of Technogenic Products

M. Garkavi[1], A. Artamonov[1], E. Kolodezhnaya[2(✉)], A. Pursheva[1], and M. Akhmetzyanova[1]

[1] JSC "Ural-Omega", Magnitogorsk, Russia
[2] Department of Mining and Ecology, Institute of Comprehensive Exploitation of Mineral Resources, Russian Academy of Sciences, Moscow, Russia
gsm@uralomega.ru

Abstract. The structure of mixed cements with mineral additives of different nature enables to divide them into two types: with single-stage and multi-stage structure formation. Characterization of mixed cements is determined, mainly, by nature of the mineral admixture. The index of multi-stage structure formation of mixed cements has been suggested. This index is the value of instantaneous power of structure formation Wη. It has been shown that in case of the other equal conditions the nature of structure formation is determined by the ratio of the components and by milling method of the mixed cement. As a result of analysis of numerous experimental data it has been detected, that the strength of stone, which is formed during hardening of mixed cements, has an extreme nature depending on the content of the mineral admixture. Analysis of the thermodynamic stability of the structure of the stone, which is formed in the process of hardening of mixed cements, allows to divide mixed cements into three groups and identify field of their rational use.

Keywords: Mixed cement · Structure formation · Instantaneous power of structure formation

1 Introduction

Technogenic products have a low hydraulic activity. Therefore, the main condition is to increase their reaction capacity in making the mixed binders, which is achieved by increasing the degree of fineness and concentration of blemishes.

Milling is an energy intensive process operation. So, the energy reduction is achieved by using modern fineness of grind, of which vertical shaft impacted mill is the most effective.

The feature of VIS mill is their high-power rating (more than 10 kW/kg). This allows using them in mechanoactivation processing when solid materials increase capacity of reaction.

The simultaneous grinding and mechanoactivation in VIS mills provide to diversify types of cements (Garkavi et al. 2013).

S. Glagolev (Ed.): ICAM 2019, SPEES, pp. 323–326, 2019.
https://doi.org/10.1007/978-3-030-22974-0_78

2   Research Methods

Portland cement clinker, granulated blast-furnace slag, steelmaking slag and flyash were used for mixed cement production of different mixture ratio.

All components of mixed cements were grinded in VIS mills not only for fineness needed but also for their mechanoactivation. The real size of BET surface area is 10000…10400 sq. m/kg, which is an indication of high defect structure and related concentration of surface specific sites.

3   Results and Discussion

During hardening of mixed binding systems several chemical reactions of hydration are connected with various hydraulic activity of their components. Each chemical reaction of hydration, which occurs in a mixed binding system is interconnected with one process of structure formation and affects its development. The character of structure formation of mixed cements with mineral additives of different nature enables to divide them into two types: with a single-stage or multi-stage structure formation (Garkavy 2005). The nature of the mineral admixture shows that mixed cements belong to one or another type.

The multi-stage structure formation means the cyclicity in occurrence of certain structural conditions of mixed binding system. The number of stages of structure formation is determined by the composition of mixed cement such as coagulation or coagulation-condensation contacts.

Prerequisite for the development of multi-stage structure formation is a significant distinction of the apparent activation energies of hydrate-formation of mixed binding components (Garkavy 2005).

Values of apparent activation energy for components of mixed binding is shown in Table 1.

Table 1. Activation energy of hydration of the components of the mixed cements

Component of mixed cement	Activation energy of hydration for the period of, J/mol		
	Induction	Acceleration	Damping
Portland cement clinker	15703	14610	23736
Flyash	47837	17288	31563
Granulated blast-furnace slag	15668	14703	23898
Steelmaking slag	43675	19953	35595

The contribution of components of the mixed cement to the formation of inter-particle contacts can be estimated by the quantity of instantaneous power of structure formation $W\eta$. This value is the product of the rate constant of structure formation process $k\eta$ and the value of the apparent activation energy of this process $E\eta$, i.e. $W\eta = k\eta * E\eta$ (Garkavy 2005).

By the synchronous development of the processes of hydrate and structure formation, in other words, by the one-stage structure formation the correlation is present:

$$W_{\eta mix} < W_{\eta cl} + W_{\eta min.ad} \tag{1}$$

where $W_{\eta mix}$, $W_{\eta cl}$ and $W_{\eta min.ad}$ – instantaneous power of structure formation of the mixed cement, clinker component and mineral admixture, respectively.

Consequently, by one-stage structure formation both components of the mixed cement are involved in the formation of structure, but the predominant influence has a component, which possesses a high instantaneous power structure formation. In case of breach of synchronicity of these processes, that is, by the multi-stage structure formation, the relation is implemented:

$$W_{\eta mix} = W_{\eta cl} + W_{\eta min.ad} \tag{2}$$

It means that in the structure formation components of the mixed cement are consistently involved, the component with the greater instantaneous power of structure formation contributes the most.

The correlation (2) which takes into account the mass fraction φ of the mineral admixture in the mixed cement has the form:

$$W_{\eta mix} = (1 - \varphi) \cdot W_{\eta cl} + \varphi \cdot W_{\eta min.ad} \tag{3}$$

where φ - the mass fraction of mineral admixture in the mixed cement.

From the formula (3) follows, that ceteris paribus, the nature of structure formation is determined by the ratio of the components of the mixed cement. According to the result of the experimental research, the multi-stage structure formation is typical of the mixed cements based on steelmaking slags ($\varphi > 0.3$) and fly ash ($\varphi > 0.5$) (Garkavi et al. 2010).

Thus, the Eq. (3), along with the ratio of the quantity of activation energy of hydrate formation components of the mixed cement can be an indicator of multi-stage structure formation in the binder systems. It is obvious that in the process of hardening of mixed cements with the diverse character of structure formation, intermediate and final structural states with diverse thermodynamic stability are formed. It predetermines their various physical-mechanical and operational characteristics and creates the prerequisites for the creation of rational compositions of mixed cements which have desired properties.

As a result of numerous experimental data, it has been detected (Garkavi et al. 2010; Garkavi et al. 2013), that the strength of stone which is formed in the process of hardening of mixed binders, has an extreme character which depends on the mass fraction of the mineral admixture in the mixed cement.

The extreme dependences, which are shown in Fig. 1, are approximated by the empirical equation:

$$R_{mix} = R_{cl} \cdot exp(b \cdot \phi + c \cdot \phi^2), \tag{4}$$

where R_{mix} – activity of the mixed cement R_{cl} - activity of Portland cement clinker; b, c - constants depending on the nature and dispersion of the mineral admixture; φ - the mass fraction of mineral admixture of mixed cement.

From the correlation (4) follows that the value b/c characterizes the mass fraction of mineral admixture of mixed cement φ cr, in which its activity equal to the activity of the Portland cement with no additives. According to the Eq. (3), in the mixed cement the conditions for the development of multi-stage formation are created when the content of mineral admixture is more than φ cr.

4   Conclusion

In agreement with the nature of hardening of mixed cements, they can be divided into three groups with the following rational fields of application:

Group 1 – mixed cements which contain the mineral admixtures $\varphi \leq \varphi cr$ are characterized by single-stage structure formation, they comply with Eq. (1). The mixed cements of this group are substitutes of cements with no additives.

Group 2 - mixed cements with multi-stage structure formation, which satisfy the relation (3). These cements should be used for the manufacture of concretes, which hardened by the heat treatment, as well as for slow-hardening concretes for massive structures.

Group 3 – cements content a high mass fraction of mineral admixture (80%). These cements form the weak strength structure of hardening with low thermodynamic stability. The cements of this group should be used for the production of concrete and low-grade concretes and solutions.

The offered classification of mixed cements allows to identify their rational compositions, on the basis of the properties of mineral admixture and applications.

References

Garkavy MS (2005) Thermodynamic analysis of structural transformations in the binder systems. MSTU, Magnitogorsk, 243 p

Garkavi MS, Hripacheva IS (2010) Optimization of mixed binders with the use of dump steelmaking slag. Build Mater 2:56

Garkavi MS, Hripacheva IS, Artamonov AV (2013) Centrifugal impact grinding cements. Cement Appl 4:106–109

Use of Slags in the Production of Portland Cement Clinker

V. Konovalov[✉], A. Fedorov, and A. Goncharov

Belgorod State Technological University named after V.G. Shukhov,
Belgorod, Russia
konovalov52@mail.ru

Abstract. The use of technogenic raw materials as input products in the production of portland cement provides for considerable reduction of energy consumption during clinker burning. The study reveals the mineral formation features caused by the change of the liquid phase composition and crystallization of silicate phases. The use of slags in raw mixes increased the mechanical strength of cements by over 50%.

Keywords: Slag · Cement · Clinker phases · Microstructure

1 Introduction

Any waste can be considered as secondary material resources, which may be fully or partially (as additives) used in production (Klassen et al. 2003). In terms of their physical and chemical properties, slags are similar to igneous rocks used in the production of construction materials. Being exposed to high-temperature treatment and containing basic calcium silicates, they considerably reduce fuel consumption during clinker burning (Vvedensky 1978; Kopeliovich et al. 1998).

2 Methods and Approaches

The study included the methods of chemical analysis of input products and clinkers carried out according to GOST 5382-93. The X-ray phase analysis was performed via the powder diffraction technique using DRON-3 M. The polished sections were studied in reflected light using a universal polarizing microscope NU-2 by Karl Zeiss Jena. The etching of polished sections was carried out via the universal etching agent, i.e. M.I. Strelkov's reagent. The thermal test was carried out using the scanning calorimeter STA 449 F1 Jupiter® by NETZSCH in inert media. Thin slag structures were studied on a scanning electron microscope MIRA3 TESCAN.

© The Author(s) 2019
S. Glagolev (Ed.): ICAM 2019, SPEES, pp. 327–330, 2019.
https://doi.org/10.1007/978-3-030-22974-0_79

3   Results and Discussion

Besides others, the LLC South-Ural Mining and Processing Company utilizes smelter slags as input products. The chemical composition of slags and raw materials given in Table 1 indicates a possibility of replacing the natural constituent, i.e. clay, and slightly reducing the consumption of a carbonate component.

Table 1. Chemical composition of input products, %.

Raw mix components	CaO	Al_2O_3	Fe_2O_3	SiO_2	MgO	PPP
Blast furnace slag	39.70	8.80	2.09	39.08	4.08	0.17
Open-hearth slag	32.36	3.96	21.20	19.76	11.18	3.61
Limestone	52.12	0.24	0.18	1.53	0.48	43.11
Clay	7.57	12.8	6.32	50.45	3.97	12.80

According to X-ray phase analysis, the phase composition of dump blast furnace non-granulated and granulated slag differs only in the intensity of new crystal growths. The main identifiable minerals include gehlenite, akermanit, quartz and melilite minerals.

The DTA method of a dump blast furnace slag in the range of 600–700 °C demonstrates a slight vague exo-effect caused by ferrous iron oxidation. The same temperature range indicates the decomposition of a secondary calcium carbonate followed by the insignificant loss of a sample mass. At 840–900 °C the glass phase is crystallized with further heat release.

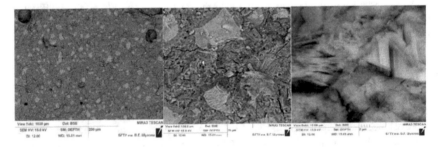

Fig. 1. Microstructure of dump blast furnace slag

The process of devitrification for the blast furnace granulated slag has more expressed exothermic maximum, which is caused by high concentration of a glass phase.

The open-hearth slag acts as a correcting ferrous additive, besides hematite and magnetite is rich in calcium ferrite, monticellite, diopside and magnesium oxide in the form of a periclase.

Figure 1 shows the structure of dump slags representing the conglomerate of crystalline phases and melting particles. The phase formation analysis was carried out in raw mixes having similar chemical composition with various ratio of clay and slag components (Table 2).

Table 2. Raw mixing ratio

Raw mix, No.	Limestone	Clay	Blast furnace slag	Open-hearth slag
1	75%	22%	0%	3%
2	61%	0%	33%	6%

The clear exo-effect is observed in a clay-based raw mix (No. 1) at 1227 and 1256 °C caused by the formation of belite phase, as well as endothermal melting effects at 1288 and 1300 °C. According to (Kougiya, Ugolkov 1981), the exotherm of belite mass crystallization at higher temperatures improves the synthesis of alite and its formation in a fine-crystalline state.

Within sintered materials cooled at 1250 °C the crystallization of phase C3A and C4AF is recorded, the reflection intensity of aluminate phase considerably increases with temperature rise. When clay is replaced with slag in a mix the exothermal processes characterizing the formation of silicate phases are weakly expressed within the range of 1000 and 1200 °C. The melting in these mixes is recorded at 1259 and 1308 °C.

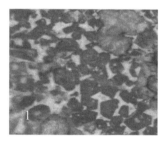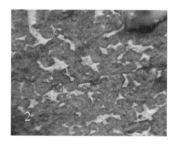

Fig. 2. Micrographs of polished section of sample clinkers: identifications are given in Table 1

Within sintered materials №2 cooled at 1250 °C the aluminate phase prevails thus leading to the appearance of ferrous phases – C6A2F and C4AF with temperature rise. The basic clinker fusion is formed at 1327 °C. These differences in A-F formation are clear with the increase of slag concentration in raw mixes. The gehlenite and mayenite is observed at 1200 °C in clay-containing mixes along with belite phase, which is not observed in slag-containing samples. The introduction of slag intensifies the formation of belite phase at early burning stages. Some changes in the composition of a 'liquid' phase, which increase its temperature and molten viscosity with the increase of slag composition in a raw mix, also affect the features of crystallization of clinker phases.

Figure 2 shows micrographs of polished sections of sample clinkers, which demonstrates the crystallization difference of alite phase.

Clinkers from clay-based raw mixes have clear monadoblastic texture. Clinkers from slag-based raw mixes are different in terms of the number of alite growths.

The strength of cements was defined in small samples from cement paste (1:0) with water-cement ratio of 0.28. The results given in Table 3 show that the use of slags in raw materials instead of clay positively affects the activity of clinkers. The strength improvement within a 28-day interval at full replacement of a clay component with slag made 62%. The heat burning input of limestone-slag mixes may be reduced by over 0.85 mJ/t.

Table 3. Cement stone strength, W/C = 0.28

Mix No., Identifications are given in Table 1	Tensile strength, MPa, days		
	2	7	28
1	7	14	32
2	30	37	52

4 Conclusions

The integral analysis of raw materials and calcined products indicates the possibility of full replacement of clay in a raw mix for the production of portland cement clinker. This contributes to the improvement of qualitative parameters of a calcined product and to the reduction of its production cost.

Acknowledgements. The study is implemented in the framework of the Flagship University Development Program at Belgorod State Technological University named after V.G. Shukhov, using equipment of High Technology Center at BSTU named after V.G. Shukhov.

References

Klassen VK, Borisov IN, Klassen AN, Manuylov VE (2003) Features of mineral formation in slag-containing raw mixes of various basicity. Bull High Educ Inst Constr Ser 7:56–58
Kopeliovich VM, Zdorov AI, Zlatkovsky AB (1998) Utilization of industrial wastes in cement production. Cement 3:174
Kougiya MV, Ugolkov VL (1981) Differential thermal analysis of portland cement raw mixes. Cement 11:19–21
Vvedensky VG (1978) Environmental and economic efficiency of waste utilization. Complex Use Mineral Raw Mater 3:59

Geopolymerization and Structure Formation in Alkali Activated Aluminosilicates with Different Crystallinity Degree

N. Kozhukhova[1(✉)], V. Strokova[1], I. Zhernovsky[1], and K. Sobolev[2]

[1] Belgorod State Technological University named after V.G. Shukhov,
Belgorod, Russia
kozhuhovanata@yandex.ru
[2] University of Wisconsin-Milwaukee, Milwaukee, USA

Abstract. The work presents the results of grain-size analysis of alkali-activated aluminosilicate suspensions with different crystallinity degree. It is found that the crystallinity degree of aluminosilicate is inversely proportional to its solvability in strong alkaline substance. The mechanism of geopolymeric system formation during the geopolymerization process has been suggested.

Keywords: Aluminosilicates · Crystallinity degree · Structure formation · Geopolymerization

1 Introduction

The application of colloid and nano-sized silicate and aluminosilicate components for the synthesis of effective binding systems is one of the most attractive directions in the science of construction materials (Vivian et al. 2017; Dmitrieva et al. 2018; Sobolev 2016).

The earlier studies (Shekhovtsova et al. 2018; Galindo Izquierdo et al. 2009) discussed various factors influencing the ability of alkali activated cements to form the aluminosilicate structures from the anthropogenic aluminosilicates, in particular fly-ash (Kozhukhova et al. 2018; Wang et al. 2018).

2 Materials and Methods

To estimate the viability of the research hypothesis three types of natural aluminosilicates with a different crystallinity degree were used:

– Obsidian - effusive rock of an acidic composition and amorphous structure;
– Pearlite - effusive rock of an acidic composition and crypto-crystalline structure;
– Crouan – intrusive compound acidic rock with a hollow crystalline structure.

To prepare alkaline silicate suspensions the equal volumes (50 g) corresponding to each sample of milled aluminosilicate material were placed into glass bottles and mixed with 50% NaOH water solution.

S. Glagolev (Ed.): ICAM 2019, SPEES, pp. 331–334, 2019.
https://doi.org/10.1007/978-3-030-22974-0_80

These suspensions were mixed for three days (72 h) using a LS-110 mixing device. The specific surface and the average size of grains of aluminosilicate powders was performed using a laser analyzer ANALYSETTE 22 NanoTec plus.

3 Results and Discussions

This report is based on the hypothesis that during the alkaline activation of aluminosilicate particles the dissolving process is gradual, starting with the dissolution of surface layers. As the result, the alkali aluminosilicate gel is formed which acts as a binding base for further geopolymerization. At the same time, the aluminosilicate component crystallinity degree influences its solubility in the alkaline medium.

To test the research hypothesis the average size of the aluminosilicate particles average size was determined as well as the specific area in the initial condition and also after the alkaline activation (Table 1).

Table 1. The particle size and specific area of alumosilicate powders after the alkaline activation

№	Aluminosilicate type	Average particle size, µm		Change, %	Specific area, m^2/kg		Change, %
		Before activation	After activation		Before activation	After activation	
1	Crouan	11	9	−16	910	1003	10
2	Perlite	11	14	21	838	774	−7
3	Obsidian	14	17	23	785	642	−18

The results of grain size analysis for crouan after the alkaline activation prove that the average particle size is reduced relative to the particle size before activation. It may be caused by crouan grains subsolution resulting in the reduction of the particle size.

At the same time in the alkaline activated suspensions of perlite and obsidian there is a tendency of particle size increase in comparison with those before activation.

Hence, for crouan there is an increase of the specific area of the solid phase after the activation and a decrease of average particle size.

In case of perlite and mostly obsidian, the alkaline activation has a reverse effect: the specific area decreases and the average particle size increases (Fig. 1).

The received data of the grain size analysis suggested a scheme of aluminosilicate structure formation that occurs during the geopolimerization. This scheme includes two simultaneous processes: the dissolution of the aluminosilicate component and the formation of the alkali aluminosilicate gel, which is the chemical interaction of alkali aluminosilicate gel with unreacted grains. The chemical interaction causes the formation of a «gel layer – unreacted grain» in investigated system.

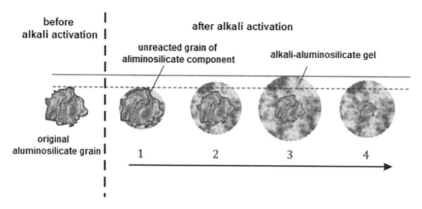

Fig. 1. The structure formation mechanism in the system "gel layer – unreacted grain of aluminosilicate component" during the alkali activation.

The lower aluminosilicate crystallinity degree results in a high intensity of the dissolution of aluminosilicate particles and a thicker surface gel layer based on the newly formed compound.

4   Conclusion

The crystallinity degree of aluminosilicates is inversely proportional to the reactivity in alkali systems, which is controlled by the solubility degree in highly alkali medium. Based on this observation, it was suggested that the geopolimerization scheme in the system «gel layer – unreacted grain of aluminosilicate component» occurs during the alkaline activation.

Acknowledgements. The work has been fulfilled within the project Federal Target Program of Research and Development on "Priority Development Fields of science and technology sector in Russia for 2014–2020", unique project number is RFMEFI58317X0063.

References

Dmitrieva TV, Strokova VV, Bezrodnykh AA (2018) Influence of the genetic features of soils on the properties of soil-concretes on their basis. Constr Mater Products 1:69–77

Galindo Izquierdo M, Querol X, Davidovits J, Antenucci D, Nugteren HW, Fernández-Pereira C (2009) Coal fly ash-slag-based geopolymers: microstructure and metal leaching. J Hazard Mater 166(1):561–566

Kozhukhova NI, Zhernovsky IV, Sobolev KG (2018) Effect of variations in vitreous phase of low-calcium aluminosilicates on strength properties of geopolymer systems. Bull BSTU named after V.G. Shukhov 3:5–12

Shekhovtsova J, Zhernovsky I, Kovtun M, Kozhukhova N, Zhernovskaya I, Kearsley PE (2018) Estimation of fly ash reactivity for use in alkali-activated cements - a step towards sustainable building material and waste utilization. J Cleaner Prod 178:22–33

Sobolev K (2016) Modern developments related to nanotechnology and nanoengineering of concrete. Front Struct Civil Eng 10(2):131–141

Vivian F-I, Pradoto R GK, Moini M, Kozhukhova M, Potapov V, Sobolev K (2017) The effect of SiO_2 nanoparticles derived from hydrothermal solutions on the performance of portland cement based materials. Front Struct Civil Eng 11(4):436–445

Wang YS, Provis JL, Dai JG (2018) Role of soluble aluminum species in the activating solution for synthesis of silico-aluminophosphate geopolymers. Cement Concrete Compos 93:186–195

Matrix Instruments for Calculating Costs of Concrete with Multicomponent Binders

T. Kuladzhi[1]([✉]), S.-A. Murtazaev[2,3,4], S. Aliev[2], and M. Hubaev[2]

[1] Lomonosov North (Arctic) Federal University, Arkhangelsk, Russia
kuladzhit@list.ru
[2] Millionshchikov Grozny State Oil Technical University, Grozny, Russia
[3] Ibragimov Complex Research Institute, RAS, Grozny, Russia
[4] Academy of Sciences of the Chechen Republic, Grozny, Russia

Abstract. The paper describes methodology for determining the costs of construction products - concrete with multicomponent binders, containing the matrix formula by Professor M.D. Kargopolov, which is recommended a modern micro-prediction tool for the production efficiency of building materials and products, allowing simultaneous calculations of their costs taking into account all the volumes of material costs: cement, fittling, components of binders, etc., as well as wage costs, depreciation etc.

Keywords: Concretes with multicomponent binders ·
Solar thermal processing · Micro-prediction tools for production efficiency ·
Composite building materials · Matrix formula by professor M.D. Kargopolov

1 Introduction

Measures are taken in the building complex of Russia at the state level to improve the efficiency and the system for determining the costs of building products. The estimated cost of construction work includes: direct costs, overhead costs and estimated profit (The Civil Code… 1996; On approval… 2004; Taymaskhanov et al. 2012).

To ensure the effectiveness of the production of innovative products the matrix formula by Professor Kargopolov is recommended, providing transparency and accuracy of calculations of the costs of production, taking into account all the territorial conditions affecting the estimated cost of production (Kuladzhi 2014; Kargopolov 2001).

2 Methods and Materials

D.Sc.(Economics) Kargopolov in (Kargopolov 2001) showed that any economic system of economic entities can be represented as a scheme of interacting objects producing a specific product (as part of X), from which part of the output - W in the studied economic system is used inside the system, and the other - (Y) as the final product is taken outside this system. Therefore, to assess the magnitude of costs and

S. Glagolev (Ed.): ICAM 2019, SPEES, pp. 335–338, 2019.
https://doi.org/10.1007/978-3-030-22974-0_81

results of the effectiveness of each such product, the entire production process Π of an organization can be represented by a general structure.

Matrix formula by Professor M.D. Kargopolov is given by:

$$P = (E - AT) - 1 * DT * C \qquad (1)$$

where: $P = \|p_j\|$; $j = \overline{1,n}$ – desired column vector of production (full) cost per unit of production (works, services);

E – single matrix n × n;

$A = \|a_{ij}\|$, $i = \overline{1,n}$, $j = \overline{1,n}$ – matrix n × n of consumption rates of own production resources;

$D = \|d_{ij}\|$, $i \in L \ U \ R$, $j = \overline{1,n}$ – matrix of consumption rates of the primary resources (L – variable, R – constant),

T – transposition mark for matrices A and D.

$C = \|c_i\|$, $i \in L \ U \ R$ – column vector of the wholesale and procurement prices of primary resources, and if resources are represented by value indicators in matrix D, then in matrix C these resources, respectively, should be denoted by number – 1 (one).

3   Results and Discussion

To calculate the total cost of production of reinforced concrete floor slabs using helioforms, we used actual data on labor costs and costs of materials for the manufacture of reinforced concrete floor slabs of type 10-60.12 under the conditions of Argun reinforced concrete products and structures (Table 1).

Thus, Table 1 presents matrix D in such a way that it consistently (from simple to complex products – a reinforced concrete slab) reveals indicators of the consumption of materials for:

- production of process water: 1 column;
- steam production: 2 column;
- production of complex binders (dry mix: cement, filler and additives "Bio-NM), made in the scientific laboratory of building department of the Grozny State Oil Technical University named after Academician M.D. Millionshchikov (Kargopolov 2001): 3, 4 columns
- production of components of concrete mixes (dry mix: complex binders - CB and crushing residue): 5–6 columns;
- production of 1 m^3 of concrete products under solar thermal treatment: 7–8 columns.

Table 1. Matrix D of consumption rates of primary resources, incl. purchased, for the production of reinforced concrete products and matrix C of wholesale procurement prices of primary resources

Costs	Water	Steam	Binder composition		Dry concrete mix		Reinforced concrete products with solar thermal treatment		№	Matrix C
			KB 100	KB3 50	KB 100	KB3 50	KB 100	KB3 50		Price
Capital investments in helioforms (per 1 m^3 of reinforced concrete products), thousand rubles/m^3	0	0	0	0	0	0	0,012	0,012	1	0,012
Cement, thousand rubles/t	0	0	0,5	0,254	0	0	0	0	2	5
Crushing residue, thousand rubles/t	0	0	0	0	1,5	1,524	0	0	–	0,25
Filler, thousand rubles/t	0	0	0	0,254	0	0	0	0	2	1,5
BIO-NM additive, thousand rubles/t	0	0	0,01	0,00508	0	0	0	0	4	22
Fitting, thousand rubles/t	0	0	0	0	0	0	0,065	0,065	5	5
Electricity and fuel	0,0124	0,0414	0	0	0	0	0,2794	0,2794	6	1
Water, thousand rubles/t	0,02365	0	0	0	0	0	0	0	7	1
Wages, thousand rubles	0,01	0,025	0	0	0	0	0,3439	0,3439	8	1
Equipment maintenance costs (127.8% of salary), thousand rubles	0,0128	0,032	0	0	0	0	0,4395	0,4395	9	1
Shop expenses (25% of salary), thousand rubles	0,0025	0,0064	0	0	0	0	0,086	0,086	10	1
* Deduction for social insurance (34% for 2011 *), thousand rubles	0,0034	0,0085	0	0	0	0	0,1169	0,1169	11	1
Plant costs (20%), thousand rubles	0,002	0,005	0	0	0	0	0,06878	0,06878	12	1
Other, thousand rubles	0,0137	0,05	0	0	0	0	0	0	13	1

4   Conclusions

Considering that the balance equation of the Nobel Prize winner in economics V.V. Leontiev is a macro-prediction tool for output of products at the national and world levels, the matrix formula by Professor M.D. Kargopolov should be considered as an instrument for micro-prediction of cost indicators of products of any economic entities - companies, households and other subjects, including products of cluster subjects.

References

Kargopolov MD (2001) Interoperable balances of costs and results of production: theory and practice. Monograph. Arkhangelsk: Publishing House of AGSTU, 182 p

Kuladzhi TV (2014) Methodology for evaluating the effectiveness of design solutions in the construction complex. North (Arctic) Federal University - Arkhangelsk: NAFU Publishing House, 296 p

On approval and implementation of the Methodology for determining the cost of construction products in the territory of the Russian Federation (together with "MDS 81-35.2004 ..."). Resolution of the State Construction Committee of Russia of 05.03.2004 No. 15/1 (as amended on 06.16.2014)

Taymaskhanov KE, Bataev DK-S, Murtazaev S-AY, Saidumov MS (2012) Justification of the economic efficiency of the production of concrete composites based on technogenic raw materials. Questions Econ Law 2:124–128

The Civil Code of the Russian Federation (part two) dated January 26, 1996 No. 14-FZ (as amended on July 29, 2017) (as amended and added, entered into force on December 30, 2018)

Characterisation of Perovskites in a Calcium Sulfo Aluminate Cement

G. Le Saout[1(⌧)], R. Idir[2], and J.-C. Roux[1]

[1] C2MA, IMT Mines Ales, Univ Montpellier, Ales, France
gwenn.le-saout@mines-ales.fr
[2] CEREMA, DIM Project Team, Provins, France

Abstract. Calcium sulfo aluminate cement ($C\bar{S}A$) is a promising low CO_2 footprint alternative to Portland cement. The phase assemblage of a commercial $C\bar{S}A$ cement was investigated by a combination of XRD, SEM-EDX and selective extraction techniques. This study focused on the composition of perovsite phases present in the cement.

Keywords: Calcium sulfo aluminate cement · Perovskite · X-ray diffraction · Scanning electron microscope

1 Introduction

Calcium sulfo aluminate cements $(C\bar{S}A)$[1] were developed by the China Building Material Academy in the seventies. $C\bar{S}A$ have many specific properties compared to Portland cement as fast setting, rapid hardening, shrinkage reduction. This special cement used alone or in combination with calcium sulphates and Portland cement has found applications such as airport runways and roads patching, selfleveling mortars, tile adhesives grouts… (Zhang et al. 1999). This is also a promising low CO2 footprint alternative to Portland cement due to the difference in the amount of energy used to produce $C\bar{S}A$ cements (lower kiln temperatures and energy at the mill to grind). The main raw materials used for making $C\bar{S}A$ cements are bauxite, limestone, clay, and gypsum and this leads to a mineralogical composition very different than Portland cement. While many studies have been carried out on the characterization of Portland cement, few are available concerning $C\bar{S}A$. In this study, we report a characterization of a commercial $C\bar{S}A$ using Rietveld quantitative analysis and scanning electron microscopy.

---

[1] Standard cement chemistry notation is used.

As per this simplified notation: C = CaO, A = Al_2O_3, F = Fe_2O_3, S = SiO_2, $\bar{S}$ = SO_3 and T = TiO_2.

© The Author(s) 2019
S. Glagolev (Ed.): ICAM 2019, SPEES, pp. 339–343, 2019.
https://doi.org/10.1007/978-3-030-22974-0_82

2 Methods and Approaches

The C$\overline{S}$A cement was from a commercial supplier and the chemical characteristics of the cement are given in Table 1.

Table 1. Mineralogical and chemical compositions of the C$\overline{S}$A. Mineralogical composition determined by XRD/Rietveld analysis. Chemical analysis by X-ray fluorescence (DIN 51001)

Minerals	Mass %	Oxides	Mass %
Anhydrite	18.3	SiO_2	8.42
Gypse	2.9	Al_2O_3	19.1
Ye'elimite	31.4	Fe_2O_3	6.94
Belite	21.2	TiO_2	0.76
Perovskite	11.3	K_2O	0.08
Ferrite	5.4	Na_2O	0.02
Merwinite	1.9	CaO	44.9
Calcite	3.5	MgO	1.27
Magnesite	1.5	SO_3	15.2

[a]Loss on ignition measured by calcination until 1025 °C according to ISO 12677

In order to analyze mineralogical composition, X-ray diffraction was performed on cement with a diffractometer BRUKER D8 Advance. Powder samples were analyzed using an incident beam angle (Cu Kα, $\lambda = 1.54$ Å) varying between 5 and 70°. Software X'Pert High Score was used to process diffraction patterns and crystals were identified using the Powder Diffraction File database. Rietveld analysis allowed obtaining mass fractions of crystalline phases in the cement.

For the microscopical investigations, powder samples were impregnated using a low viscosity epoxy and polished down to 0.25 μm using diamond pastes. The samples were further coated with carbon (~ 15 nm) and examined using a Quanta 200 FEG scanning electron microscope (SEM) from FEI coupled to an Oxford Xmax N 80 mm^2 energy dispersive X-ray spectroscopy (EDX) analyser.

In order to improve the characterization of the cement, two different selective dissolution methods were used. In the first method, the silicate phases were removed in a solution of acid salicylic in methanol (Hjorth and Lauren 1971). In addition, a second selective dissolution method was used to get mainly perovskite phases in C$\overline{S}$A by removing ye'elimite, anhydrite, gypsum with 5% Na_2CO_3 solution (Wang 2010). The method was modified to prevent precipitation of $CaCO_3$ by washing the filtered suspension with 6% acetic acid. The filter paper and contents were placed in an oven at 105 °C until a constant weight was reached.

3 Results and Discussion

The main phases observed in the experimental diffraction pattern (Fig. 1a) are the orthorhombic ye'elimite $C_4A_3\bar{S}$ with small amount of the pseudo cubic form, belite β and α'_{H-} C_2S and perovskites from the $C\bar{S}A$ clinker. Anhydrite II $C\bar{S}$ is also present as mineral addition. Perovskite family has crystal structures related to the mineral perovskite CT. Ferrite phase $Ca_2(Al_xFe_{1-x})_2O_5$ is usually present in $C\bar{S}A$ and its structure is derived from that of perovskite by the substitution of Al and Fe for Ti, together with ordered omission of oxygen atoms, which causes onehalf of the sheets of octahedral in perovskite to be replaced by chains of tetrahedral (Taylor 1997). To obtain a good Rietveld refinement, it is also necessary to add a perovskite phase CT (Alvarez- Pinazo et al. 2012). The titanium dioxide is present by the use of bauxite, which usually contains some TiO2, as raw materials in the manufacturing process of $C\bar{S}A$ clinker. The peaks associated with this cubic phase are confirmed in the XRD pattern of the $C\bar{S}A$ after the extraction of the main phases (Fig. 1b).

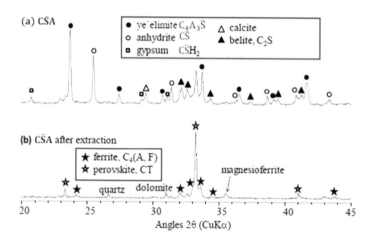

Fig. 1. Diffraction pattern of the cement as received (a), after extraction (b)

However, with the assumed stoichiometry of perovskite CT, the elemental oxide composition TiO_2 calculated from phase content deduced by Rietveld analysis is strongly overestimated in comparison with XRF analysis.

EDX analysis on polished section of $C\bar{S}A$ (Fig. 2a) and $C\bar{S}A$ after extraction revealed an average composition of ferrite $Ca_{1.99}Al_{0.39}Si_{0.10}Fe_{1.35}Ti_{0.08}Mg_{0.07}O_{5.00}$ not far from the brownmillerite series $Ca_2(Fe_{2-x}Al_x)O_5$. However, the average composition of perovskite $Ca_{2.00}Al_{0.30}Si_{0.22}Fe_{1.06}Ti_{0.40}Mg_{0.05}O_{5.31}$ is very different to the $Ca_2Ti_2O_4$ composition. These compositions were similar to those observed in the ferrite/perovskite phases of a calcium aluminate cement (Gloter et al. 2000). We also observed in some grains perovskite lamellae with high amount of Ti on the scale of few micrometers (Fig. 2b).

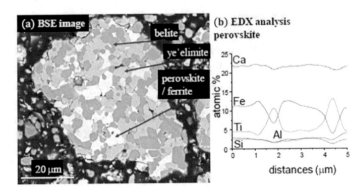

Fig. 2. Back scattering image of the cement (a), example of EDX analysis on perovskite (b)

4 Conclusions

The phase assemblage of a commercial calcium sulfo aluminate cement has been investigated with a special attention to the ferrite- perovskite phases. The ferrite composition is closed to the brownmillerite $Ca_2(Fe_{2-x}Al_x)O_5$ whereas the perovskite shows heterogeneity with important substitution of Ti by Fe, Al and Si.

Acknowledgements. The authors acknowledge A. Diaz (C2MA, IMT Mines Ales) for sample preparation for SEM experiments

References

Alvarez- Pinazo G, Cuesta A, Garcia-Maté M, Santacruz I, Losilla ER, De la Torre AG, Leon-Reina L, Aranda MAG (2012) Rietveld quantitative phase analysis of Yeelimitecontaining cements. Cem Concr Res 42:960–971

Gloter A, Ingrin J, Bouchet D, Scrivener K, Colliex C (2000) TEM evidence of perovskite-brownmillerite coexistence in the $Ca(Al_xFe_{1-x})O_{2.5}$ system with minor amounts of titanium and silicon. Phys. Chem. Miner 27:504–513

Hjorth L, Lauren K-G (1971) Belite in Portland cement. Cem Concr Res 1:27–40

Taylor HFW (1997) Cement Chemistry. Thomas Telford, London

Wang J (2010) Hydration mechanism of cements based on low-CO_2 clinkers containing belite, ye'elimite and calcium alumino ferrite. PhD dissertation, University of Lille I, France

Zhang L, Su M, Wang Y (1999) Development of the use of sulfo- and ferroaluminate cements in China. Adv Cem Res 11:15–21

Geonics (Geomimetics) as a Theoretical Basis for New Generation Compositing

V. Lesovik[1]([✉]), A. Volodchenko[1], E. Glagolev[1], I. Lashina[1], and H.-B. Fischer[2]

[1] Belgorod State Technological University Named After V.G. Shukhov, Belgorod, Russia
naukavs@mail.ru
[2] Bauhaus-Universität Weimar, Weimar, Germany

Abstract. The article introduces basic principles of a new transdisciplinary research area geonics (geomimetics) in the construction material science. This research area differs from bionics, which uses knowledge about nature to solve engineering problems. The purpose of geonics (geomimetics) is to solve engineering problems with the account of knowledge about geologic and cosmochemical processes.

Keywords: Geonics · Anthropogenic methasomatism in the construction material science · The law of consanguinity · Construction composites

1 Introduction

Nowadays many scientists are developing construction composites of new generations (Delgado et al. 2015; Ahadi 2011). Basing on the theoretical background of geonics (geomimetics) it is possible to reduce power consuming in construction materials production by application of energy of geological and cosmochemical processes, energy efficient raw material, specially treated with geological and cosmochemical processes.

So, production technologies for a wide range of composite binding materials (including water resistant and freeze proof gypsum binders) with the application of new raw material (Lesovik et al. 2014; Zagorodnyuk et al. 2018) with high free internal power have been suggested. Clay rocks with an incomplete stage of mineral formation and a sediment genesis zone, effusive rocks with an amorphous and cryptocrystalline structure, quartz rocks of greenschist coal ranging with crystalline defects and inclusions of mineral formation, gas and air inclusions, some kinds of anthropogenic materials and others refer to these materials (Elistratkin and Kozhukhova 2018).

2 Methods and Approaches

The law of consanguinity has been formulated within the theoretical backgrounds of geonics (geomimetics). It implies designing layered composites and maintenance systems at nano-, micro-, and macro- levels similar to basic matrix that enhances

© The Author(s) 2019
S. Glagolev (Ed.): ICAM 2019, SPEES, pp. 344–347, 2019.
https://doi.org/10.1007/978-3-030-22974-0_83

significantly material adhesion and durability. The implementation of this law allows creating a composite which components have close deformation and temperature characteristics.

The theory of anthropogenic methasomatosis in the construction material science has been suggested. This stage of composite materials evolution is characterized by composite ability to adjust to changing conditions during building and construction service. Designing construction composites with the account of anthropogenic methasomatosis theory in the construction material science provides possibility of self-healing defects, which appear during construction and building service and design, they are so-called "smart" composites.

3 Results and Discussion

Principles for productivity enhancement of wall materials manufacturing using sandy and clay rocks with an incomplete stage of clay formation and industrial wastes in hydrothermal processing at atmospheric pressure have been suggested. It has been found that aeolian-sedentary-diluvia clay rocks of the Quaternary period are the products of the initial clay formation stage including metastable atelene minerals of nano-level and non-rounded finely dispersed quartz and are suitable as a raw material for autoclaved silicate materials production.

Application of sandy and clay rocks instead of traditional quartz sand in silicate materials production enhances raw mixture formation, hardens raw bricks by 4–11 times, that allows developing the production technique of highly-hollowed construction products.

Application of these rocks allows widening the range of autoclaved raw material base, decrease energy-output ratio of their production, improve the ecological state of the environment and create comfortable conditions for human life.

Within the law of consanguinity it has been found that rocks independent of their genesis (magmatic, metamorphic and aqueous) with banded texture, whose layers are introduced by minerals whose stress related characteristics and thermal-expansion coefficient differ significantly, are non-durable and have anisotropy coefficient 7–9 higher comparing with prototypes with anisotropy coefficient 2–3. Application of the suggested mixtures increases brickwork breaking limit by 3–5 times. It is explained by the contact zone microstructure, for example, ceramic brick and binding matrix. The designed binding matrix and wall material are nearly sole block sample and constructions with a traditional binding matrix have a distinct contact zone – the weakest place of the samples. This law allows creating restoration mixtures, plastering materials, brickwork and restoration compositions of new generations for every walling.

With the account of geonics theoretical background a wide range of acoustic, insulating, and construction insulating composites based on foam glass have been suggested.

Designing construction composites with the account of anthropogenic methasomatosis theory in the construction material science provides possibility of defects self-healing which appear during construction and building service and design so-called "smart" composites.

These materials are designed with the account of the system interacting with the environment. This system allows reacting to the external actions by defects "self-healing" and having positive impact on the triad "a man-material-environment".

This approach has been tested on the composite bindings based on calc-sinter, which create favorable conditions for at the early stages of structure formation and system hardening. It decreases stress in the hardening composite and as a consequence decreases the amount and the size of micro-cracks that predetermines technical and economic efficiency of the composite binding based on tuff, especially in hot climate. Volcanic tuff is known to be a heteroporous rock. Pore space of this rock is rather complex in form and combines pores of different sizes.

Water in this rock is in complex interaction with its mineral grid, whose boundaries and ratio are relative and change constantly: vapour; chemically and physically bounded water; free or gravitational water.

In a hot climate, with a deficit of liquid phase in the concrete itself, tuff particles in the binding mixture composition, during the hardening process will release the capillary water, it will activate structure formation processes and synthesize more massive structure of the materials during concrete hardening and utilization.

The micro-cracks appearing during service at different conditions will self-heal by interaction of water in tuff particles with unreacted cement minerals. During service tuff particles release saved capillary water and that will result in structure formation activation and synthesizing more massive materials structure during concrete hardening and operation. These are so-called smart composites.

4   Conclusions

Hence, monodisciplinary and interdisciplinary approaches in the construction material science promoted developing a wide range of construction composites which main task is construction of solid and durable structures. Design and creation of materials for environment optimization "a man-material-environment" is a complex problem, which requires united work of scientists of different fields. A single way to solve this problem is transdisciplinary approach as a way of widening the scientific world view, which requires considering phenomena beyond a single science.

Acknowledgements. The study is carried out in the framework of the State Task of the RF Ministry of Education and Science No. 7.872.2017/4.6. Development of principles for the design of ecologically positive composite materials with prolonged bioresistance 2017–2019.

References

Ahadi P (2011) Applications of nanomaterials in construction with an approach to energy issue. Adv Mater Res 261–263:509–514

Delgado JMPQ, Cerný R, de Lima AGB, Guimarães AS (2015) Advances in building technologies and construction materials. Adv Mater Sci Eng 2015:1–3 no. 312613

Elistratkin MY, Kozhukhova MI (2018) Analysis of the factors of increasing the strength of the non-autoclave aerated concrete. Constr Mater Prod 1(1):59–68

Lesovik VS, Chulkova IL, Zagorodnjuk LH, Volodchenko AA, Popov DY (2014) The role of the law of affinity structures in the construction material science by performance of the restoration works. Res J Appl Sci 9(12):1100–1105

Zagorodnyuk L, Lesovik VS, Sumskoy DA (2018) Thermal insulation solutions of the reduced density. Constr Mater Prod 1(1):40–50

Regularities in the Formation of the Structure and Properties of Coatings Based on Silicate Paint Sol

V. Loganina$^{(\boxtimes)}$, E. Mazhitov, and V. Demyanova

Department "Quality Management and Technology of Construction Production", Penza State University of Architecture and Construction, Penza, Russia
loganin@mail.ru

Abstract. The authors established a higher quality of the appearance of coatings based on silicate paint sol in comparison with coatings based on silicate paints. Information is provided on the regularities in the formation of the quality of the appearance of coatings on the basis of silicate paint sol is that the silicate paint sol has a higher value coefficient of the wetting and spreading on the cement substrate in comparison with silicate paint. It was established that when using a paint brush on the cement substrate, there has been some slowing of the rise time of the structure of silicate paint compositions based on polysilicate solutions. A higher value of the adhesion work based on polysilicate solutions were indentified, indicating the strong force of the paint and the cement substrate.

Keywords: Sol silicate paint · Polysilicate binder · Coatings · Wetting · Coating quality

1 Introduction

The problem of reliability and durability of protective decorative coatings of exterior walls of buildings is one of the topical scientific and technical objectives in materials science. It is known that the longevity of coatings depend on the type of binder, the technology of applying the paint composition, operating conditions, etc. (Ailer 1982). In the practice of finishing works, silicate paints, which are a suspension of pigments and fillers in liquid glass of potassium, proved to be very useful. To improve the performance of coatings based on silicate paints, it is important to develop methods for modifying liquid glass. Analysis of patent and scientific-technical literature shows that one way of modification is the introduction of organosilicon and other polymeric compounds into the binder. In works (Figovsky et al. 2012) there is an increased durability of silicate coatings when imposing of polymeric compounds. It is of interest to use polysilicates in silicate paints as film-forming substances that provide higher performance properties of coatings. However, at present time the questions of the formation of the structure and properties of coatings based on sol silicate paints have not been studied, the questions of mechanism for improving the operational properties of coatings based on them have not been considered. Polysilicates are characterized by

S. Glagolev (Ed.): ICAM 2019, SPEES, pp. 348–351, 2019.
https://doi.org/10.1007/978-3-030-22974-0_84

a broad range of the degree of polymerization of anions and are dispersions of colloidal silica in an aqueous solution of alkali metal silicates. We have established a paint composition based on a polysilicate binder obtained by mixing liquid glass with a silica sol (Loganina et al. 2018a, b). It was found, that coatings based on polysilicate solutions are characterized by faster curing. Films based on polysilicate solutions have higher cohesive strength. The tensile strength of a film based on potassium liquid glass is Rp = 0.392 MPa, and the tensile strength of a film based on a polysilicate solution (15% Nanosil 20) is 1.1345 MPa. The paint forms a coating characterized by a uniform homogeneous surface.

To study the regularities in the formation of the quality of the appearance of coatings on the based on sol silicate paint, the character of filling on a porous cement substrate was considered.

2 Methods and Approaches

The character of the filling of the sol of the silicate paint was evaluated. The method of determining filling consisted in applying five parallel strips of paint and determining the degree of spreadability according to the number of stuck bands. Paint with an operating viscosity was applied to a glass plate measuring $200 \times 100 \times 1.2$ mm. The spreading of five parallel strips was evaluated on a ten-point scale of filling.

The surface tension of the paint was determined by the drop method (stalagmometric method). Work of adhesion of the paint to a cement substrate was calculated using the Dupre - Young thermodynamic equation. The wetting operation and spreading coefficient were determined.

The quality of the appearance of the coatings was estimated from the surface roughness Ra by the method of scanning probe microscopy (SPM) (Chizhik and Syroezhkin 2010).

3 Results and Discussion

Previously, the rheological type of the solutions was determined. The rheological properties were evaluated by the indicators of conventional viscosity according to B3-4, critical shear stress with instrument Reotest-2. It is found, that all systems are typical pseudoplastic bodies. In the region of slow flow, the viscosity of polysilicate solutions gradually declined with increasing shear stress.

Analysis of data (Table 1) shows, that silicate paints based on polysilicate solutions have a long filling time. Thus, the time for bottling for a paint based on liquid glass is 6 min, and for a paint based on a polysilicate solution - 8 min 40 s. The degree of filling is satisfactory (no more than 10 min). For paints based on a polysilicate solution, a large work of adhesion to the substrate is characteristic. So, work of the adhesion of paint based on the potassium polysilicate solution to the substrate is 108.17 mN/m, while work of the adhesion of the paint based on potassium liquid glass is 96.82 mN/m. A higher value of work of the adhesion of paint based on polysilicate solutions indicates a stronger interaction of the paint and the cement substrate.

Table 1. Test results

The name of indicators	Name of the paint composition	
	Based on potassium liquid glass	Based on potassium polysilicate solution
Surface roughness, Ra, [µm]	16,208	10,880
The contact angle of wetting	50,9	51,6
Surface tension of the paint composition, [mN/m]	59,38	66,73
Filling the colorful composition *	7 min 40 s 9	8 min 40 s 9
Adhesion work, [mN/m]	96,82	108,17
Wetting operation, [mN/m]	37,44	41,44
Cohesion work, [mN/m]	118,76	133.46
Coefficient of spreading [mN/m]	−21,94	−25,29
Coefficient of wetting	0,815	0,81

Note: * Above the line are the values of time of restoration of the paint structure, below the line - the value of filling

The work of wetting paints based on polysilicate solution is higher, which indicates better wetting of the paint surface of the cement substrate. Thus, the work of wetting the sol of silicate paint on the basis of potassium polysilicate solution is 41, 44 mN/m, and on the basis of potassium liquid glass - 37.44 mN/m. When the sol of the silicate paint is applied to the substrate, the wetting and spreading coefficient increases, which indicates more favorable conditions for the formation of the quality of the appearance. The surface roughness of the coating based on silicate paint is Ra = 16.208 µm, and based on the potassium polysilicate solution, Ra = 10.880 µm. The quality of the appearance of the surface of the coatings formed by the sol with silicate paint, in accordance with GOST 9.032-74 ** "Unified system for protection against corrosion and aging. Coatings for paint and varnish, Groups, technical requirements and designations" is graded IV class, and on the basis of liquid glass - V class.

Testing of solution samples, colored with sol by silicate paint, was carried out for frost resistance by alternating thawing and freezing. Appearance of coatings was assessed according to GOST 6992-68. "Coatings for paint and varnish. Test method for resistance to atmospheric conditions". It was found, that coatings on the basis of the developed composition had withstood 40 test cycles, while the coating condition after 40 test cycles was estimated at I.1 points, which corresponds to the coating condition with no color change, chalking, mud retention.

To assess the waterproof properties of coatings, tests were carried out of solution samples stained with silicate and sol silicate paints. After curing of the coatings, water absorption was determined upon capillary suction. It was found, that water absorption by capillary suction of samples stained with silicate paint is 4.4%, and stained with silicate paint - 4.6%. The lower value of water saturation of samples colored with sol by silicate paint indicate a change in the pore size in the coating structure as compared to the coating based on silicate paint.

Higher waterproof properties of coatings based on sol silicate paint are caused, in our opinion, by the structure of the coating. Scanning probe microscopy (SPM) methods were used to estimate the local structure of the coating surface. It is established, that the surface of coatings based on potassium liquid glass contains a certain number of pores of the nanometric range, differing in size and shape. The maximum pore size is 19.8 μm. Pores with a diameter of 18.85 to 19.6 μm are mainly present, whereas in the coating based on the potassium polysilicate solution there are two groups of pores: from 19.25 to 19.8 μm and from 20.0 to 20.6 μm. The value of the maximum pore size is 21.2 μm. In the coating based on the polysilicate solution is observed a more uniform pore size distribution.

Coatings based on the developed paint are characterized by high adhesion (1.1–1.3 MPa), coefficient of vapor permeability - 0,00878 mg/m * hPa.

4   Conclusions

The properties of the paint and coating based on it meet the requirements for coatings for exterior decoration of buildings, have higher adhesion, sufficient vapor permeability.

References

Ailer P (1982) The chemistry of silica (Transl. from English), Part 1. Mir, Moscow, p. 416

Chizhik SA, Syroezhkin SV (2010) Methods of scanning probe microscopy in micro- and nanomechanics. Instr Meas Methods 1:85–94

Figovsky O, Borisov Yu, Beilin D (2012) Nanostructured binder for acid-resisting builder materials. J Sci Israel-Technol Advant 14(1):7–12

Loganina VI, Kislitsyna SN, Mazhitov YB (2018a) Development of sol-silicate composition for decoration of building walls. Case Stud Constr Mater 9:e00173

Loganina VI, Kislitsyna SN, Mazhitov YB (2018b) Properties of polysiylate binders for sol-silicate pains. Adv Mater Res 1147:1–4

Influence of Sodium Oxide on Brightness Coefficient of Portland Cement Clinker

D. Mishin[✉] and S. Kovalyov

Department of Technology of Cement and Composite Materials,
Chemical Technology Institute, Belgorod State Technological University
named after V.G. Shukhov, Belgorod, Russia
mishinda.xtsm@yandex.ru

Abstract. The paper is devoted to the possibility of adjusting the reflection factor of portland cement clinker. For this purpose, Na_2CO_3 is introduced into the slurry of CJSC Belgorod Cement Plant. The influence of Na_2O on brightness coefficient of the crushed clinker is established at a burning temperature of 1250–1300 °C. With the increase of Na_2O concentration up to 1–2% the brightness coefficient is reduced and the increase of Na_2O in the range of 3.5–7% leads to sharp increase of the reflection factor and CaO_{free} content.

Keywords: Sodium oxide · Brightness coefficient · White cement · Free calcium oxide

1 Introduction

Modern rotary furnaces of cement industry are characterized by accumulation and circulation of alkali salts in the furnace system (Luginina 2002). As a result, the R_2O content of furnace charge may reach 3.5% before the sintering zone, and in some plants this value may even reach 10%. The formation of calcium aluminate ferrite is not observed in furnace charges of alumina industry characterized by high Na_2O content due to the formation of aluminates and sodium ferrites (Lisiyenko 2004). Hence, as may be expected, the accumulation and circulation of alkali salts will lead to the situation that occurs in furnaces of alumina industry.

Raw mixes from analytical reagents used to obtain C_4AF are characterized by the formation of sodium compounds instead of calcium compounds (Kovalyov 2015). It will allow obtaining a clinker of lighter shade.

Thus, the purpose of this study is to determine the possibility of adjusting the brightness coefficient of portland cement by introducing sodium oxide into the raw mix.

2 Methods and Approaches

The dried slurry of CJSC Belgorod Cement Plant with the following modular characteristics: KH = 0.91; n = 2.23; p = 1.29 was used as a raw mix (Table 1).

© The Author(s) 2019
S. Glagolev (Ed.): ICAM 2019, SPEES, pp. 352–355, 2019.
https://doi.org/10.1007/978-3-030-22974-0_85

Table 1. Chemical composition of slurry of CJSC Belgorod Cement Plant, %

Losses on ignition	SiO_2	Al_2O_3	Fe_2O_3	CaO	MgO	K_2O	Na_2O	SO_3	Other
34.8	14.23	3.59	2.78	43.12	0.6	0.4	0.11	0.09	0.37

Sodium carbonate of CH classification was introduced into the raw mix in the amount of 0.5; 1; 2; 3.5; 5; 7% of Na_2O in equivalent of ignited basis. Tablets weighing 2 g were formed from obtained mix under manual pressure. In order to avoid Na_2O volatilization during roasting, the samples were covered with a platinum cup. Roasting of samples was carried out in a laboratory furnace with isothermal time of 20 min. The heating rate of the furnace made 10 °C/min.

The influence of alkali salts on the formation of aluminate-ferrite phase was estimated according to the content of free calcium oxide and the brightness coefficient of samples. The CaO_{free} content in a clinker was defined through ethyl-glycerate method (Butt and Timashev 1973). The whiteness (brightness coefficient) of a clinker was defined via FB-2 reflection meter on a reference polished barium sulfate plate.

3 Results and Discussion

The compositions burned at 1250 and 1300 °C were analyzed. The roasting temperature was chosen based on the following: at given temperatures the formation of belite was complete, and the synthesis of alite had not started yet. As a result, the formation of silicate phases would not change the content of free calcium oxide.

At a temperature of 1250 °C (Fig. 1) at the initial stage (up to 1% Na_2O) the content of free calcium oxide is reduced alongside with the brightness coefficient (BC). The brightness coefficient almost does not change with Na_2O concentration in the range of 1–3.5% and the content of free calcium oxide. With the introduction of over 3.5% of Na_2O the CaO_{free} content and the brightness coefficient sharply increase.

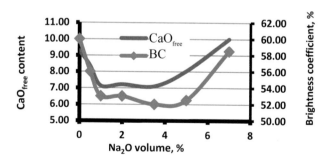

Fig. 1. Influences of Na_2O on brightness coefficient and content of free calcium oxide at 1250 °C roasting temperature.

At a roasting temperature of 1300 °C (Fig. 2) a similar situation is observed. With the amount of introduced Na_2O up to 1–2% the considered characteristics are reduced. Sharp increase of reflection factor and CaO_{free} content was also observed in the range of 3.5–7% of Na_2O.

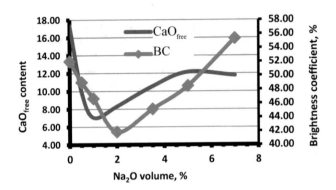

Fig. 2. Influence of Na_2O on reflection factor and content of free calcium oxide at 1300 °C roasting temperature.

The relatively low content of free calcium oxide without alite formation may be caused by the formation of intermediate phases, for example carbonate spurrite. The presence of Na_2CO_3 in the system fosters its formation. Hence, the increase of Na_2O content at the initial stage leads to CaO_{free} decrease.

The sharp increase of free calcium oxide content starting from Na_2O concentration of approximately 3.5% confirms the formation of ferrites or sodium aluminates instead of calcium aluminate ferrites in clinker systems, since such interaction results in the emission of additional amounts of calcium oxide and may be presented by the following reactions:

$$3CaO \cdot Al_2O_3 + Na_2O = 2NaAlO_2 + 3CaO \tag{1}$$

$$4CaO \cdot Al_2O_3 \cdot Fe_2O_3 + 2Na_2O = 2NaFeO_2 + 2NaAlO_2 + 4CaO \tag{2}$$

The change dependence of brightness coefficient within samples on Na_2O content in a clinker correlates with the change dependence of free calcium oxide content. This may be explained by the fact that CaO_{free} crystals are white. Therefore, the increase or reduction of its content leads to the corresponding change of the brightness coefficient of a clinker. Based on above conclusions, the sharp increase of the brightness coefficient with the increase in Na_2O concentration of over 3.5% may also be caused by the complexity of aluminate-ferrite phase. This phase is the most colored part of a clinker (Zubekhin et al. 2004). Hence, the change of its composition and quantity resulting from the formation of aluminates and sodium ferrites will be characterized by the corresponding change of the brightness coefficient of a clinker.

4   Conclusions

The introduction of up to 2–3.5% of Na_2O leads to the decrease in the brightness coefficient of samples due to the reduction of free calcium oxide content. Apparently, this is caused by the formation of intermediate phases – carbonate spurrite.

The Na_2O content of over 3.5% increases the brightness coefficient of samples. This is caused by the formation of aluminates and sodium ferrites instead of aluminates and calcium aluminate ferrites.

Acknowledgements.   The study is implemented in the framework of the Flagship University Development Program at Belgorod State Technological University named after V.G. Shukhov, using the equipment of High Technology Center at BSTU named after V.G. Shukhov.

References

Butt YM, Timashev VV (1973) Practical Guide on Chemical Technology of Binding Materials: Study Manual. Higher School, p 504

Zubekhin AP, Golovanova SP, Kirsanov PV (2004) White portland cement. Publishing House of Izvestiya Vuzov Journal, Rostov-On-Don, North Caucasian Region, p 264 (Zubekhin AP (ed))

Kovalyov SV (2015) Essentially new method of clinker bleaching with high iron content. In: Kovalyov SV, Mishin DA (eds) International seminar-competition of young scientists and graduate students working in the field of binding materials, concrete and dry mixes: collection of articles: 180 pages. Alitinform Publishing House, Saint Petersburg, pp 29–37

Lisiyenko VG (ed) (2004) Rotary furnaces: thermal engineering, management and ecology. Teplotekhnik, p 688 (Reference edition. Lisiyenko VG, Shchelokov YM, Ladygichev MG)

Luginina IG (2002) Selected articles. Publishing House of BSTU named after V.G. Shukhov, Belgorod, p 302

Production of Bleached Cement

D. Mishin$^{(\boxtimes)}$ and S. Kovalev

Department of Technology of Cement and Composite Materials,
Institute of Chemical Technology, Belgorod State Technological University
named after V.G. Shukhov, Belgorod, Russia
mishinda.xtsm@yandex.ru

Abstract. The present paper studies a new method of bleaching that allows extending the range of raw materials for white cement. This paper investigates the separate introduction of mineralizers into raw slurry of CJSC Belgorod Cement plant containing 2.78% of Fe_2O_3. The separate introduction of mineralizers provides bleaching effect and complete consumption of free calcium oxide at the temperature of 1250 °C. The bleaching is caused by the reduced content and changed composition of ferroaluminate phase.

Keywords: Clinker bleaching · White cement · Separate introduction · Na_2CO_3 · CaF_2 · Mineralizers

1 Introduction

Currently, white cement is one of the most demanded building materials due to its wide building and technological properties. The demand for this cement grows with the rate of city development. However, the development of white cement industry is limited by strict requirements to the quality of raw materials. To produce cement of graded whiteness, the content of iron oxide in the clinker should not exceed 0.5% (Zubekhin et al. 2004). The number of stock deposits that can provide such content of iron oxide in clinker is relatively low and will gradually decrease with their depletion.

Thus, the development of new technologies extending the range of raw material sources for white cement by involving components with higher iron content becomes very urgent.

2 Methods and Approaches

Materials and Methods. The raw mixture with high content of Fe_2O_3 was represented by dried slurry of CJSC Belgorod Cement plant (Table 1) with the following module characteristics: LSF = 0.91; n = 2.23; p = 1.29.

The raw mixture was doped by mineralizers Na_2CO_3, CaF_2 and $2C_2S \cdot CaF_2$ calculated as ignition basis over 100%: The final mixture contained 3.5% of sodium carbonate as R_2O, 1.5% of calcium fluoride and 8.11% of synthesized $2C_2S \cdot CaF_2$ (1.5% if expressed as CaF_2). The mineralizers were introduced separately (Mishin et al. 2016).

© The Author(s) 2019
S. Glagolev (Ed.): ICAM 2019, SPEES, pp. 356–359, 2019.
https://doi.org/10.1007/978-3-030-22974-0_86

Table 1. Chemical composition of slurry of CJSC Belgorod Cement plant [%]

Loss on ignition	SiO₂	Al₂O₃	Fe₂O₃	CaO	MgO	K₂O	Na₂O	SO₃	other
34.8	14.23	3.59	2.78	43.12	0.6	0.4	0.11	0.09	0.37

The effect of mineralizers on the phase composition of the clinker was established using XRD analysis using ARL™ X'TRA Powder Diffractometer (Switzerland).

The completion of clinker formation process was assessed by the content of free calcium oxide. The content of CaO_{free} was determined by ethyl-glycerate method (Boutt and Timashev 1973).

The whiteness grade (brightness coefficient) of the clinker was determined by FB-2 reflection meter using reference polished plate of barium sulphate.

3   Results and Discussion

The implementation of separate introduction of mineralizers leads to bleaching of the samples at the temperature of 1250–1300 °C (Fig. 1(1)). The increase of the burning temperature from 1250 to 1300 °C causes gradual return of clinker color to typical black (Fig. 1(1), (2) and (3)).

Fig. 1. Appearance of clinker samples produced by separate introduction at burning temperature of: (1) 1250 °C; (2) 1275 °C; (3) 1300 °C.

In this temperature interval, complete consumption of free calcium oxide is observed, which indicates the completion of clinker formation processes (Table 2).

Table 2. Characteristics of clinker samples produced by separate introduction of $2C_2S·CaF_2$ mineralizer

Burning temperature [°C]	1250	1250	1275	1300	1400*
Cooling method	(water)	(air)	(air)	(air)	(air)
Content of CaO_{free} [%]	0.55	0.5	0.3	0.31	0.10
Brightness coefficient [%]	46	41	37.5	30	31

* common plant clinker

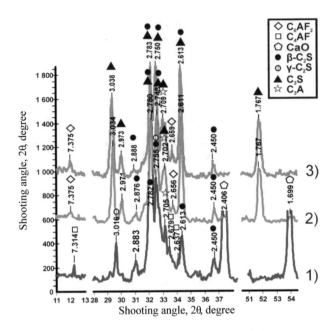

Fig. 2. Phase composition of clinker samples: (1) no additives at 1250 °C; produced by separate introduction at: (2) 1250 °C; (3) 1300 °C.

According to the phase composition analysis, at the burning temperature of 1250 °C (Figs. 2(1), (2)), in the samples produced by separate introduction, the composition of ferroaluminate phase C_4AF (d, Å: 7.314; 2.679; 2.637) shifts into the region of compositions with a higher iron content and approaches that of C_6AF_2 (d, Å: 7.375; 2.656). The intensity of reflections of the ferroaluminate phase decreases, which tells about the decrease in its content in the clinker. Ferroaluminate phase is the most coloring among the clinker components (Zubekhin et al. 2004). In this connection, the decreased amount of ferroaluminate phase leads to increased brightness coefficient of the clinker.

The increased burning temperature up to 1300 °C (Fig. 2(3)) causes increased intensity of ferroaluminate phase reflections (d, Å: 7.375; 2.659), which witnesses its increased content in the clinker. As a result, the brightness coefficient of the clinker reduces from 41% down to 30%.

4   Conclusions

The separate introduction of mineralizer allows bleaching the clinker with the iron oxide content of more than 0.5%. With the iron oxide content in raw mixture of 2.78%, the brightness coefficient increases from 31% up to 46%.

The increase in the brightness coefficient is connected with decreased content of calcium ferroaluminate phases in the clinker: C_4AF, C_6AF_2.

The implementation of separate introduction of mineralizer will allow extending the range of sources for white cement production involving raw components with increased iron content.

The separate introduction of mineralizer in the presence of increased iron oxide content allows achieving complete consumption of free calcium oxide at 1250 °C.

Acknowledgements. The work is realized in the framework of the Program of flagship university development on the base of Belgorod State Technological University named after V.G. Shoukhov, using equipment of High Technology Center at BSTU named after V.G. Shoukhov.

References

Boutt YM, Timashev VV (1973) Practical guide on chemical technologies of binders: guide. Higher school, Moscow, 504 p. (in Russian)

Mishin DA, Kovalev SV, Chekulaev VG (2016) Reason for reduced efficacy of mineralizers for burning Portland cement clinker. Bulletin of BSTU n.a. V.G. Shoukhov, no 5, p 161–166 (in Russian)

Zubekhin AP, Golovanova SP, Kirsanov PV (2004) White Portland cement. Zubekhin AP (ed) J. "Bulletin of HEIs North-Caucasus region", Rostov on Don, 264 p (in Russian)

Multicomponent Binders
with Off-Grade Fillers

S.-A. Murtazaev[1,2,3], M. Salamanova[1,2(✉)], M. Saydumov[1],
A. Alaskhanov[1], and M. Khubaev[1]

[1] Millionshchikov Grozny State Oil Technical University, Grozny, Russia
Madina_salamanova@mail.ru
[2] Ibragimov Complex Research Institute, RAS, Grozny, Russia
[3] Academy of Sciences of the Chechen Republic, Grozny, Russia

Abstract. The paper deals with issues related to development of multicomponent binders (MCB) and high-quality concretes based on them. The production of such binders is based on the use of finely divided mineral additives of natural and technogenic origin. Particular attention is paid to the aggregate, the strength of coarse aggregate should be at least 20% higher than the strength of concrete, and the maximum particle size should not exceed 8–20 mm. At present, considerable experience was accumulated for production of multicomponent binders, and the results of studies conducted in this direction showed that the raw material potential of the Republic allowed obtaining high-quality class B30-40 concrete, and if we expanded the geography of the use of natural resources by regions of the North Caucasus, we could produce concretes with higher strength.

Keywords: High-quality concretes · Composite binders ·
Reactive mineral components · Volcanic ash · Thermal power plant (TPP) ash ·
Fractionated filler

1 Introduction

Concrete is one of the oldest materials, but its potential and possibilities seem inexhaustible (Murtazaev et al. 2016; Nesvetaev et al. 2018; Stelmakh et al. 2018), since at all times of its existence and in the future this material will occupy a leading place among a huge variety of building compositions.

The active component of concrete is cement. It is known that varying finely dispersed mineral additives in its composition results in modern composite materials, which properties will vary in wide ranges (Udodov 2015; Salamanova et al. 2017).

In accordance with GOST 31108-2003, granulated slag, fuel ashes, including acidic or basic fly ash, microsilica, burnt clay, burnt shale, marl, quartz sand, etc. are used as mineral components—main components of cement (Udodov 2015; Murtazaev et al. 2016). Various mineral additives can be used as auxiliary components of cement, which will not significantly increase the water demand of cement and reduce durability of concrete.

S. Glagolev (Ed.): ICAM 2019, SPEES, pp. 360–364, 2019.
https://doi.org/10.1007/978-3-030-22974-0_87

2   Methods and Materials

As part of the work carried out in this direction, we developed formulations of multicomponent binders, which include mineral additives of natural and technogenic origin.

The North Caucasus has large reserves of natural raw materials for these developments, the chemical analysis of the mineral components used in the studies is shown in Table 1.

Table 1. The chemical composition of mineral components, wt.%

Type	MgO	Al_2O_3	SiO_2	K_2O	CaO	Fe_2O_3	TiO_2	SO_3	LOI
TPP ash	2,49	23,89	42,88	0,48	4,6	7,95	0,11	0,66	16,9
Volcanic ash	0,20	13,57	73,67	6,00	1,79	1,52	2,85	-	0,40
Limestone flour	0,72	1,55	5,05	0,6	90,14	1,4	-	0,49	-
Quartz powder	6,32	14,99	73,83	1,83	0,6	0,97	1,32	0,14	-

3   Results and Discussion

To produce multicomponent binders, the additives under study were ground in VM-20 laboratory ball vibratory mill for 30 and 40 min. Figure 1 shows dependence of specific surface of mineral additives on the grinding time.

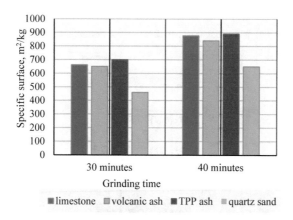

Fig. 1. Specific surface of mineral components

To determine the optimal degree of saturation of Portland cement (PC) – mineral powder (MP) system (PC:MP), samples were prepared from the proposed multicomponent binder formulations and properties (Table 2).

Table 2. Properties of multicomponent binders (MCB)

No.	Mineral Powder	PC:MP	Normal density, %	Setting time, hour-min		Activity, MPa
				start	end	
1	Limestone flour	70/30	25,5	2-05	3-00	35,8
2		60/40	26,8	2-15	3-20	30,4
3	Quartz powder	70/30	24,6	1-30	2-10	41,8
4		60/40	27,0	1-55	2-50	39,7
5	TPP Ash	70/30	26,4	2-10	3-15	34,1
6		60/40	28,1	2-25	3-35	28,2
7	Volcanic ash	70/30	25,2	1-35	2-15	42,6
8		60/40	26,5	2-05	3-00	40,3
9	–	100	25,0	2-20	3-40	48,0

The results of the studies showed that the most rational are the compositions of binders using mineral powders of volcanic ash and quartz powder with a ratio of 70:30%, with a specific surface of 876 m^2/kg and 650 m^2/kg, respectively, with a typical increase in the activity of the binder and a slight increase in normal thickness, and 30% of portland cement are saved.

Next, a concrete mixture with P2 mobility mark was produced, the samples were subjected to heat and humidity treatment (HHT) in a steam chamber at $2 + 3 + 7 + 2$ h at an isothermal holding temperature of 80 °C. Table 3 shows the experimental compositions and properties of the studied concretes.

Table 3. The compositions of the studied concretes

No composition	Mineral powder	Consumption of materials, kg/m^3				Average density, kg/m^3	Compressive strength, MPa	
		MCB-70	ACS	DS	B		After HHT	Age 28 days
1	Limestone flour	450	1100	680	220	2430	43,3	38,4
2	Quartz powder	450	1100	680	210	2410	50,2	45,9
3	Volcanic ash	450	1100	680	215	2415	52,1	46,5
4	TPP ash	450	1100	680	230	2420	37,7	35,9
5	PC	450	1100	680	200	2420	51,5	48,6

Note: PC – Portland cement; ACS – Alagir crushed stone fraction 5–20 mm; FS – fractionated fine filled based on the sands of the Alagir and Chervlensk deposits.

We established that the strength of concrete after HHT is 12% higher than the indicators of the strength of concrete after 28 days of natural hardening. The use of MCB-70 with volcanic ash showed the best results on the compressive strength of concrete in comparison with other additives and slightly inferior to similar indicators of control samples (Murtazaev et al. 2016; Stelmakh et al. 2018). The study of operational characteristics (Table 4) showed that the indicators of these properties depend on the composition of the MCB-70 and its activity, as well as on the type and value of the porosity of the material.

Table 4. Operational properties of concrete using MCB-70

Indicators	Mineral powder			
	Limestone flour	Volcanic ash	TPP ash	Quartz powder
MCB-70 activity, MPa	35,8	42,6	34,1	41,8
Compressive strength, MPa	38,4	46,5	35,9	45,9
Flexural strength, MPa	4,1	4,9	3,8	4,4
Porosity, %	9,7	7,6	12,4	6,9
Frost resistance, cycle	F300	F350	F200	F350
Pressure, MPa	1,4	1,8	1,2	1,8
Water absorption, %	4,2	3,5	5,2	3,6
Water resistance, Kr - softening coefficient	0,79	0,89	0,63	0,90

4 Conclusions

Multicomponent binders based on mineral powders of natural and man-made origin allow to obtain high-quality concrete of class of strength B30-40, including for high monolithic construction.

References

Udodov SA (2015) Re-introduction of plasticizer as a tool for controlling the mobility of concrete mix. In: Proceedings of the Kuban State Technological University, no 9, pp 175–185

Murtazaev S-Y, Salamanova MS, Bisultanov RG, Murtazaeva TS-A (2016) High-quality modified concretes using a binder based on a reactive active mineral component. Stroitelnyematerialy, no 8, pp 74–80

Nesvetaev G, Koryanova Y, Zhilnikova T (2018) On effect of superplasticizers and mineral additives on shrinkage of hardened cement paste and concrete. In: MATEC Web of Conferences 27, Cep, 27th R-S-P Seminar, Theoretical Foundation of Civil Engineering (27RSP), TFoCE 2018, p 04018

Stelmakh SA, Nazhuev MP, Shcherban EM, Yanovskaya AV, Cherpakov AV (2018) Selection of the composition for centrifuged concrete, types of centrifuges and compaction modes of concrete mixtures. In: Kim Y-H, Parinov IA, Chang S-H (ed) Physics and Mechanics of New Materials and Their Applications (PHENMA 2018) Abstracts & Schedule, p 337

Salamanova M, Khubaev M, Saidumov M, Murtazaeva T (2017) Self-consolidating concretes with materials of the Chechen Republic and neighboring regions. Int J Environ Sci Educ 11 (18):12719–12724

High-Quality Concretes for Foundations of the Multifunctional High-Rise Complex (MHC) «Akhmat Tower»

S.-A. Murtazaev[1,2,3], M. Saydumov[1(✉)], A. Alaskhanov[1], and M. Nakhaev[4]

[1] Millionshchikov Grozny State Oil Technical University, Grozny, Russia
saidumov_m@mail.ru
[2] Ibragimov Complex Research Institute, RAS, Grozny, Russia
[3] Academy of Sciences of the Chechen Republic, Grozny, Russia
[4] Chechen State University, Grozny, Russia

Abstract. The paper presents results of studies of monolithic concrete mixes and concretes produced with the integrated use of local natural and technogenic raw materials, including waste scrap and crushed bricks. We developed optimal compositions of monolithic concretes and studied their technological and physical-mechanical properties.

Keywords: High-strength concrete · High-quality concrete mix · Filled binder · Technogenic wastes · Mineral technogenic filler · Monolithic concrete

1 Introduction

The modern materials science and construction now deal with an important national economic and engineering problem: development of efficient technologies for producing high-strength monolithic concretes through the integrated use of technogenic raw materials to obtain secondary raw materials for concrete, while eliminating the enormous environmental damage caused by waste "cemeteries" (Bazhenov et al. 2006; Salamanova et al. 2017; Lesovik et al. 2012; Murtazaev et al. 2009; Kaprielov et al. 2018) in Russia and the world and, in particular, for the Chechen Republic, considering construction of 435-m high-rise complex «Akhmat-Tower» in Grozny City (Udodov 2015; Koryanova et al. 2018; Salamanova and Murtazaev 2018).

2 Methods and Materials

Our experimental studies used local additive-free Portland cement of PC 500 D0 grade as a binder. Natural sand from the Chervlensky deposit of the Chechen Republic was used as a fine filler. Local gravel of 5–20 mm fractions from the Argunsky and Sernovodsky deposits of the Chechen Republic and imported crushed stone of 5–20 mm

S. Glagolev (Ed.): ICAM 2019, SPEES, pp. 365–368, 2019.
https://doi.org/10.1007/978-3-030-22974-0_88

fraction from granite-diabase rocks of the Alagirsky deposit of the Republic of North Ossetia-Alania were used as a coarse filler.

As plastifying agents in accordance with GOST 24211-2008 "Additives for concrete. General technical requirements" modern additives of various manufacturers of building chemicals (POLYPLAST, TOKAR, etc.) were used.

The raw materials for the production of dispersed technogenic mineral fillers (DTMF) were local materials, mainly technogenic, namely concrete scrap, crushed ceramic bricks (CCB), ash and slag mixture from the Grozny heat and power plant (HPP) and very small non-conditioned quartz sands ones were used in comparative tests.

All the DTMFs were ground for 5 min in MV-20-EX laboratory vibratory ball mill with a loading volume of 5–6 L to obtain a specific surface of 450–600 m^2/kg.

3   Results and Discussion

The Filled Binder (FB) formulation was developed and investigated with activity 60–71 MPa with concrete scrap and CCB fillers with ratio 70:30%. The proportion of the mixture of filler in FB was 25 and 40% by weight of the binder.

Due to the fact that for designing the underground part of the Akhmat Tower multifunctional complex, concrete of different strength classes (B40, B75-B80) was laid, the task was to develop high-quality concrete mixes (HCM), starting from the middle B40-B50 classes and ending with high-strength concrete of B80-B90 classes, with the integrated use of local raw materials, including with technogenic nature.

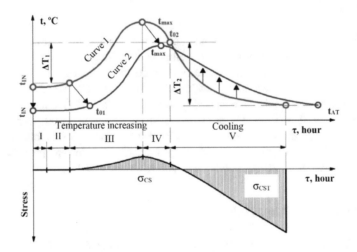

Fig. 1. Temperature change and stress characteristic in fresh concrete with limited deformation: σ_{CS} - internal compressive stresses; σ_{CST} - the same, tensile; t_{IN} - initial temperature of concrete mixture; tmax - the same, maximum; t_{AT} - ambient temperature (air); I, II, ... V - stages (periods) of process of heat dissipation of the concrete mix in time; curve 1 - kinetics of heat dissipation of concrete with PC; curve 2 - the same with FB

In massive building constructions, such as the Akhmat Tower MFC base plate, because of their large dimensions, as a rule, the heat from cement hydration is slowly released into the air or into adjacent structural elements, as a result of which the core of the monolithic element heats up much faster and stronger than the shell. Therefore, we investigated thermophysical processes of the developed compositions.

Figure 1 schematically shows the dependence of temperature and voltage due to external pressure generated by this concreting technology.

The dependence of heat release curves in time is conventionally divided into 5 stages (Table 1).

Table 1. The main stages (periods) of the process of heat dissipation in time of concrete mixes with various binders

Stage №	Duration, h		Description
	With PC	With FB	
1	0–2	0–8	The initial stage without raising the temperature of the concrete mix (dormant period). This period is significantly increased due to the use of surface-active substances (surfactants) in the composition of FB, hardening retarders, etc.
2	2–6	8–15	Temperature increase due to hydration of the binder, no measurable stress, because in the plastic concrete, thermal expansions are converted to relative compression. At the end of this stage, the temperature is referred to as "the first temperature at zero stress" t_{01}
3	6–13	15–24	Further heating of the concrete, the strength of the concrete increases and a compressive stress is formed, partially decreasing because of relaxation. Stage III ends when the maximum temperature tmax is reached
4	13–24	24–72	Heat transfer prevails: the temperature of concrete and compressive stress in concrete decrease, a part of compressive stress decreases because of relaxation. The "second temperature at zero stress" t_{02} is reached, which significantly exceeds t_{01} in cooling rate and age of concrete
5	24–72	72–144	Further cooling and increasing tensile stress, which are partially reduced due to relaxation. If the tensile stress reaches the tensile strength of the concrete under tension (at $\Delta T_{КРИТ}$) through cracks are formed

According to calculations (Kaprielov et al. 2018), the temperature difference between the upper surface layers of concrete slab and outside air $\Delta T1$ should be no more than 20 °C, and the difference between the side layers $\Delta T2$ - no more than 30 °C.

4 Conclusions

Thus, analysis of data confirms the effectiveness of the use of FB with DTMF in high-quality concretes used for concreting massive structures. We established that the peak of the maximum heat release tmax from the exotherm of cement in (massive) concrete on HB was reduced by 30–35% in comparison to the concrete with PC (from 70–75 °C to 50–55 °C).

References

Bazhenov YM, Demyanova BC, Kalashnikov VI (2006) Modified high-quality concretes. Publishing House of the Association of Construction Universities, Moscow, 368 p

Kaprielov SS, Sheinfeld AV, Al-Omais D (2018) Experience in the production and quality control of high-strength concrete on the construction of the high-altitude complex "OKO" in Moscow-City International Business Center. Industrial and Civil Construction, no 1, pp 18–24

Koryanova YI, Rezantsev NE, Shumilova AS (2018) Materials and structures used in the construction of high-rise buildings - from tradition to innovation. Alley Sci 6(4):95–99

Lesovik BC, Murtazaev SAY, Saydumov MS (2012) Construction composites based on screenings of crushing of concrete scrap and rocks. MUP "Typography", Groznyy, 192 p

Murtazaev S-AY, Bataev DK-S, Ismailova ZK (2009) Fine-grained concretes based on fillers from secondary raw materials. Comtechprint, Moscow, 142 p

Salamanova M, Khubaev M, Saidumov M, Murtazayeva T (2017) Int J Environ Sci Educ 11 (18):12719–12724

Salamanova MS, Murtazaev SAY (2018) Clinker-free binders based on finely dispersed mineral components. In: Collection: Ibausil Conference Proceedings, pp 707–714

Udodov SA (2015) Re-introduction of plasticizer as a tool for controlling the mobility of concrete mix. In: Proceedings of the Kuban State Technological University, no 9, pp 175–185

Designing High-Strength Concrete Using Products of Dismantling of Buildings and Structures

T. Murtazaeva[1(✉)], A. Alaskhanov[1], M. Saidumov[1],
and V. Hadisov[1,2]

[1] Millionshchikov Grozny State Oil Technical University, Grozny, Russia
tomamurtazaeva@mail.ru
[2] Ibragimov Complex Research Institute, RAS, Grozny, Russia

Abstract. The paper presents an analysis of experience of using the products of dismantling of buildings and structures, the technology of recycling of secondary raw materials to produce secondary raw materials for concrete. We presented results of tests of heavy concrete based on filled binders using the products of processing of concrete and brick scrap.

Keywords: Building demolition products · Concrete scrap · Brick scrap · Ecology · Recycling · Secondary aggregate · Fine ground aggregate filled with binder

1 Introduction

A great interest for application of the secondary product of crushing concrete scrap of dismantling buildings and structures, according to the authors of the paper, is related to the possibility of its application as a fine-milled mineral component in filled binders, based on the use of which it is possible to produce high-strength concrete, including monolithic high-rise construction (Bazhenov et al. 2011; Batayev et al. 2017; Murtazaev et al. 2009).

2 Methods and Materials

The following materials were used as raw materials for concrete: natural sand from the Chervlenskoye deposit, crushed stone from gravel of 5–20 mm fractions from the Argunsky and Sernovodsky deposits, imported crushed stone of a 5–20 mm fraction from granite-diabase rocks of the Alagirsky deposit of the Republic of North Ossetia-Alania, local non-additive portland cement of brand PC 500 D0, plastification additives Polyplast and D-5.

© The Author(s) 2019
S. Glagolev (Ed.): ICAM 2019, SPEES, pp. 369–371, 2019.
https://doi.org/10.1007/978-3-030-22974-0_89

3   Results and Discussion

To obtain the optimal formulations of high-strength concrete (HSC) with the integrated use of local raw materials, including of technogenic nature, compositions of filled binders (FB) with fine-milled mineral filler of technogenic nature (MFTN) of HB-75:25 and HB-60:40grades, allowing to obtain high-strength cement stone with noticeably smaller pores and less shrinkage (Murtazaev et al. 2009; Udodov 2015; Murtazaev and Salamanova 2018).

From the test results of HSC based on FB, it can be seen that the dynamics of the strength of concrete on FB are noticeably different from the dynamics of the growth of concrete strength on Portland cement.

It was established that the process of durability of concrete on FB at an early age (1–3 days) is accelerated by 1.5–2 times. So, concrete on FB at the age of 1 day has a strength of about 33–36% of the designed, and 3 days old - this indicator reaches 70%. 7-day strength of concrete, produced using FB, is about 85–90% of the designed, which is significantly higher than traditional compositions on ordinary Portland cement. These indicators for concrete on Portland cement at the age of 1, 3 and 7 days are about 24, 35 and 70% of the designed strength, respectively.

4   Conclusions

The optimal formulations of highly mobile concrete mixtures were designed using local natural and technogenic raw materials with a grade of P5 cone sediment and persistence for more than 8 h to obtain HSC classes of compressive strength up to B60-B80 with unique performance properties.

References

Bazhenov YM, Bataev DK-S, Mazhiev KN (2011) Fine-grained concretes from recycled materials for the repair and restoration of damaged buildings and structures. FE "Sultanbegova Kh.S.", Grozny, 342 p

Batayev DK-C, Saidumov MS, Murtazaeva TS-A (2017) Recipes of high-strength concretes on technogenic and natural raw materials. In: Materials of the All-Russian Scientific and Practical Conference Dedicated to the 60th Anniversary of the Building Department of Millionshchikov GSTU, 12–13 October 2017, Bisultanova P.Sh., Grozny, pp 109–117

Murtazaev S-AY, Bataev DK-S, Ismailova ZK (2009) Fine-grained concretes based on fillers from secondary raw materials. Comtechprint, Moscow, 142 p

Udodov SA (2015) Re-introduction of plasticizer as a tool for controlling the mobility of concrete mix. In: Proceedings of the Kuban State Technological University, no 9, pp 175–185

Murtazaev SAY, Salamanova MS (2018) Clinker-free binders based on finely dispersed mineral components. In: Collection: Conference Proceedings, pp 707–714

Estimation of Rheo-Technological Effectiveness of Polycarboxylate Superplasticizer in Filled Cement Systems in the Development of Self-compacting Concrete for High-Density Reinforced Building Constructions

T. Nizina[✉], A. Balykov, V. Volodin, and D. Korovkin

Department of Building Structures, Ogarev Mordovia State University,
Saransk, Russia
nizinata@yandex.ru

Abstract. Analysis of rheo-technological effectiveness of the polycarboxylate superplasticizer Melflux 5581 F in the microcalcite-filled cement systems was performed. Optimal quantities of polycarboxylate superplasticizer and carbonate filler were determined, which allow obtaining highly mobile cement-mineral suspensions at reduced water content, which is an important step in the development of self-compacting concrete mixtures.

Keywords: Cement-mineral suspension · Rheology · Efficiency · Superplasticizer · Microcalcite · Self-compacting concrete

1 Introduction

Currently, there is an active growth in the area of concrete development, which the world technological community clearly classifies as cement composites of the new generation with high strength, workability, volume stability and durability (Collepardi 2006; Nawy 2001; Nizina and Balykov 2016; Nizina et al. 2017; Sivakumar et al. 2014; Tran and Kim 2017). A special place among the concretes of the new generation is occupied by Self-Compacting Concrete (SCC) – Selbstverdichtender Concrete (SVB, German), Betonautoplacant (BAP, French), which are currently fairly widespread abroad. This term, proposed in 1986 by the Japanese professor H. Okamura (Okamura and Ouchi 2003), combines concrete mixtures with high workability characteristics (standard cone flow over 55–60 cm at water-to-cement ratio reduced to 0.35–0.4 or less), which are due to the high deformability of the suspension matrix, along with its high resistance to segregation or separation during movement.

A number of papers experimentally proved that one of the basic principles for producing self-compacting concrete mixtures is the presence of a significant amount of dispersed micro-particles of cement or mineral fillers (mainly 1–100 microns in size) in their formulation, which, together with Portland cement, increase the volume of a dispersion-water suspension. At the same time, not all dispersed fillers are capable of providing a higher flowability in suspension with a superplasticizer (SP) as compared with cement suspensions. The research (Kalashnikov et al. 2014) shows that carbonate rocks (limestone, marble, dolomite), which include particles with a significant proportion of positively charged active centers, are the most compatible with anionic superplasticizers.

Thus, the generate of self-compacting concrete mixtures must begin with the development of rheological active formulation of the filled cement binders, which allow, when mixing together with the superplasticizer, to form aggregate resistant suspensions, which have a high concentration of the solid phase, low values of the shear stress limit and plastic viscosity at high gravitational fluidity under its own weight. At the same time, the efficiency of plasticizers in such systems will depend on many factors – addition procedure and optimal quantity of the plasticizer, rheological activity of the fillers used, etc.

The purpose of the study is to determine the optimal quantity of polycarboxylate superplasticizer and carbonate filler in cement-mineral suspensions when developing self-compacting concrete mixtures.

2 Methods and Approaches

To prepare suspensions, Portland cement CEM I 32.5R (C) by Mordovtsement PJSC (GOST 31108) was used. The mineral part included a carbonate filler from KM100 microcalcite (MKM) by Polipark LLC with the dosage of $0 \div 300\%$ of Portland cement weight ($0 \div 75\%$ of solid phase weight) with the variability pitch of 100%. Melflux 5581 F (SP) polycarboxylate superplasticizer by BASF Construction Solutions (Trostberg, Germany) was used as a plasticizer.

The study was carried out with the fixed water-solid ratio W/S = 0.15 with the varying factors being:

- ratio MKM/C, $x_1 = 0 \div 3.0$ relative units;
- ratio SP/(C + MKM), $x_2 = 0 \div 1.5\%$.

3 Results and Discussion

The study results were used to develop an experimentally statistical (ES) model describing the changes in the flow diameter of cement-carbonate suspensions (Portland cement + microcalcite) [D_f^{HC}, mm] from the Hegermann cone from the content of the varying factors x_1 and x_2:

$$
\begin{aligned}
D_f^{HC} = {} & 288.9 + 52.9 \cdot x_1 + 64.8 \cdot x_2 + 19.8 \cdot x_1 \cdot x_2 - 51.33 \cdot x_1^2 \\
& - 89.3 \cdot x_2^2 + 37.97 \cdot x_1^3 + 32.34 \cdot x_2^3 - 59.06 \cdot x_1 \cdot x_2^2 \\
& - 15.19 \cdot x_1^2 \cdot x_2 + 31.64 \cdot x_1^2 \cdot x_2^2.
\end{aligned}
\tag{1}
$$

Using the polynomial (1), isolines have been built, which reflect the changes in the flow diameter of cement-carbonate suspensions from the Hegermann cone depending on the content of microcalcite and Melflux 5581 F superplasticizer (Fig. 1). It has been found that for the constant water/solid ratio of W/S = 0.15 relative units, an increase in the dosage of the superplasticizer and the mineral filler (microcalcite) causes a significant increase in the flow diameter of the cement-mineral suspension.

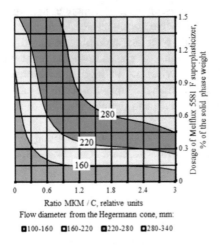

Fig. 1. Changes in the flow diameter of cement-carbonate suspensions from the Hegermann cone depending on the superplasticizer dosage and microcalcite filling degree

Analysis on Fig. 1 shows that cement suspensions without microcalciteat the specified water content W/S = 0.15 begin to flow under the action of gravity and the value of the Hegermann cone flow diameter D_f^{HC} = 100÷135 mm with superplasticizer quantities exceeding 0.1÷0.5% the mass of the solid phase. However, even an increase in the superplasticizer content to 1.0÷1.5% does not allow achieving self-compacting of suspensions (D_f^{HC} = 160 ÷ 210 < 280 mm).

It was determined (Fig. 1) that the level of rheo-technological indices specified for self-compacting suspensions (D_f^{HC} ≥ 280 mm) is achieved at the microcalcite content x_1 = 0.82÷3.0 relative units (82÷300% of the Portland cement mass or 45÷75% of the solid phase mass (C + MKM)) and the quantity of Melflux 5581 F superplasticizer x_2 = 0.45÷1.5% of the solid phase mass, wherein when decreasing the indicator x_1, an increase in the indicator x_2 in the specified ranges ((x_1 = 3.0 relative units; x_2 = 0.45%) → (x_1 = 0.82 relative units; x_2 = 1.5%)) is required.

4   Conclusions

The analysis of research results found the optimal levels of the varying factors that allow reaching the self-compacting of suspensions for the flow diameter from the Hegermann cone above 280 mm and the water/solid ratio of 0.15 relative units: the dosage of Melflux 5581 F superplasticizer is 0.5÷1.0% of the solid phase weight; the filling degree of suspension with microcalcite is at least 105% of the Portland cement weight.

Acknowledgements. The reported study was funded by RFBR according to the research project № 18-29-12036.

References

Collepardi M (2006) The New Concrete. Grafiche Tintoretto, Villorba

Kalashnikov VI, Moskvin RN, Belyakova EA, Belyakova VS, Petukhov AV (2014) High-dispersity fillers for powder-activated concretes of new generation. Syst Methods Technol 2 (22):113–118

Nawy EG (2001) Fundamentals of High-Performance Concrete. Wiley, New York

Nizina TA, Balykov AS (2016) Experimental-statistical models of properties of modified fiber-reinforced fine-grained concretes. Mag Civil Eng 2:13–25. https://doi.org/10.5862/MCE.62.2

Nizina TA, Ponomarev AN, Balykov AS, Pankin NA (2017) Fine-grained fibre concretes modified by complexed nanoadditives. Int J Nanotechnol 14:665–679. https://doi.org/10.1504/IJNT.2017.083441

Okamura H, Ouchi M (2003) Self-compacting concrete. J Adv Concr Technol 1:5–15

Sivakumar N, Muthukumar S, Sivakumar V, Gowtham D, Muthuraj D (2014) Experimental studies on high strength concrete by using recycled coarse aggregate. Res Inventy: Int J Eng Sci 4:27–36

Tran NT, Kim DJ (2017) Synergistic response of blending fibers in ultra-high-performance concrete under high rate tensile loads. Cement Concr Compos 78:132–145

Parameters of Siliciferous Substrate
of Photocatalytic Composition Material
as a Factor of Its Efficiency

Y. Ogurtsova[✉], E. Gubareva, M. Labuzova, and V. Strokova

Department of Materials Science and Technology,
Belgorod State Technological University named after V.G. Shoukhov,
Belgorod, Russia
ogurtsova.y@yandex.ru

Abstract. The article presents the results of the determination pf the properties of the photocatalytic composite material (PCM) of the "TiO_2 – SiO_2" system synthesized by the sol-gel method. The characteristics of siliciferous raw material - diatomite and silica clay, as substrates in the composition of PCM - mineral composition, microstructural features, composition and concentration of active centers on its surface are determined. The dependences of the elemental composition of the surface, the features of the microstructure and photocatalytic activity of PCM on the properties of siliciferous raw material are found. The research shows that the use of diatomite makes it possible to obtain PCM with better characteristics, which is caused by a higher content of the amorphous phase, a more developed and chemically active surface of the particles.

Keywords: Siliciferous raw material · Titanium dioxide ·
Sol-gel · Photocatalysis · Microstructure · Activity

1 Introduction

The production of photocatalytic composite materials (PCM) of the "TiO_2 – SiO_2" system is aimed at increasing the efficiency of photocatalytic decomposition of pollutants (Arai et al. 2006; Guo et al. 2016).The peculiarities of physical and chemical interaction of siliciferous and titanium-containing components in the synthesis and use of PCM directly affect its photocatalytic activity. In this regard, it is important to study the influence of the properties of siliciferous raw materials on the final characteristics of PCM.

2 Methods and Approaches

As a siliciferous raw material and as a substrate in the composition of the photocatalytic composite material, the Diasil diatomaceous fine dispersed powder (specific surface S_s = 1.39 m^2/g) was used (Diamix, Ulyanovsk region, Russia); fine-ground silica clay (S_s = 1.08 m^2/g) (Alekseevskii deposit, Mordovia, Russia) were used. The determination of the mineral composition of siliciferous raw materials was carried out using an

S. Glagolev (Ed.): ICAM 2019, SPEES, pp. 376–380, 2019.
https://doi.org/10.1007/978-3-030-22974-0_91

ARL 9900 WorkStation X-ray fluorescence spectrometer. The peculiarities of microstructure and elemental composition of the surface were studied with the help of TESCAN MIRA 3 LMU high resolution scanning electron microscope. The acid-base characteristics of the surface of siliciferous raw materials were studied using the indicator method (Nechiporenko 2017; Nelyubova et al. 2018).

The production of composite material of the TiO_2 – SiO_2 system was obtained by the sol-gel method using a titanium-containing organic precursor—titanium butoxide $Ti (OC_4H_9)_4$ (TBT) (TU 6-09-2738-89, "PROMHIMPERM", Russia). It was dissolved in ethanol, and then the siliciferous substrate (SS) material was introduced into the resulting solution, in a ratio of "TBT/SS" - 4/1. After stirring it on a magnetic mixer, the material was dried and burned at 550 °C.

Then, the tablets were prepared from the obtained materials of the "TiO_2 – SiO_2" system. White cements CEM I 52, 5 R (Adana, Turkey) was used as a binder. The ratio TiO_2 – SiO_2/cement is 1.3/1. The photocatalytic activity was determined using the photocatalytic decomposition method of the organic pigment Rhodamine B (Rhodamine B, $C_{28}H_{31}ClN_2O_3$). The pigment was applied to the tablets at a concentration of $4 \cdot 10^{-4}$ mol/l. The samples were kept for 4 and 26 h under ultraviolet radiation (UV-A, 1.1 ± 0.1) W/m^2). The evaluation of color change, as an indicator of the effectiveness of self-cleaning of the surface, was carried out according to the Lab color space (coordinate a) using software (Guo et al. 2016).

3 Results and Discussion

Diatomite is a sedimentary biogenic rock consisting of microscopic siliciferous shells of algae (diatoms) with a valve size of 5–200 mcm. The presence of nanoscale pores and elements is shown on the Fig. 1a. Silica clay is of sedimentary biogenic and chemogenic origin, composed mainly of opal-cristobalite silica particles with a size of less than 5 mcm. It is a microporous rock and the content of organic fragments is insignificant (Fig. 1b). The mineral composition of the raw material is similar; a higher content of the amorphous phase is found in the composition of diatomite (Table 1).

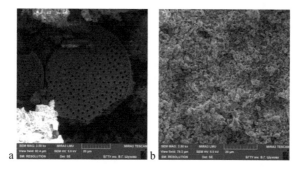

Fig. 1. Microstructure of siliciferous raw materials: a – diatomite, b – silica clay

Table 1. Mineral composition of siliciferous raw material

Siliciferous raw material	α-Quartz	Cristobalite low	Tridimite low	Illite 2M1	Albite	Amorphous
Diatomite	9.35 ± 0.72	40.57 ± 4.22	4.21 ± 0.90	2.44 ± 0.89	0.53 ± 0.20	42.90
Silica clay	17.36 ± 0.72	31.63 ± 1.42	14.63 ± 1.31	3.42 ± 0.65	0.54 ± 0.12	32.42

The presence of a high concentration of acid sites characterized by proton acidity (Brønsted) is noted on the surface of diatomite (Table 2).

Table 2. The composition and concentration of active centers on the surface siliciferous raw materials (q · 10^{-3}, mEq/g)

Siliciferous raw material	pK_a				
	−0.29 Lewis bases	+0.80 Bronsted acids	+7.15	+12.00 Bronsted bases	+12.80
Diatomite	53.11	490.43	54.09	47.95	55.80
Silica clay	115.90	No data	20.51	98.68	13.41

The photocatalytic activity of PCM based on diatomite (Table 3) is high and close to the control specimen – the industrial nano-sized Aeroxide TiO_2 P25 photocatalyst.

Table 3. Elimination of Rhodamine B, %

Time	Aeroxide TiO_2 P25	Diatomite	TiO_2–SiO_2 (diatomite)	Silica clay	TiO_2–SiO_2 (Silica clay)
4 hours	29	9	29	1	5
26 hours	91	61	86	43	68

The analysis of the peculiarities of the microstructure and elemental composition of the surface of synthesized PCM based on diatomite (Fig. 2a) and silica clay (Fig. 2b) shows that the silica particles are partially covered with titanium-containing new formations. The surface of PCM particles based on diatomite is more developed; the distribution of the titanium-containing phase is more even.

Fig. 2. Microstructure of PCM on the basis of: a – diatomite, b – silica clay (with mapping by elements)

4   Conclusions

The siliciferous raw material differs in morphology, concentration of acid-base centers and content of the amorphous phase: the surface of the diatomite is more developed, characterized by a high concentration of proton acid centers; it has a higher content of the amorphous phase in its composition. As a result, the photocatalytic activity of PCM synthesized on the basis of diatomite is higher by 20% in comparison with PCM on the basis of silica clay. In order to improve the efficiency, it is advisable to consider the possibility of pre-activation of silica clay, which will allow using this waste (by-product) rock to produce modern self-cleaning materials.

Acknowledgements. The research was carried out with financial support from Russian Science Foundation grant (project № 19-19-00263).

References

Arai Y, Tanaka K, Khlaifat AL (2006) Photocatalysis of SiO_2-loaded TiO_2. J Mol Catal A: Chem 243:85–88

Guo M-Z, Maury-Ramirez A, Poon CS (2016) Self-cleaning ability of titanium dioxide clear paint coated architectural mortar and its potential in field application. J Clean Prod 112:3583–3588

Nechiporenko AP (2017) Donor-acceptor properties of surface of solid-phase systems. Indicator method, 1st edn. Lan, St. Petersburg

Nelyubova VV, Sivalneva MN, Bondarenko DO, Baskakov PS (2018) Study of activity of polydisperse mineral modifiers via unstandardized techniques. J Phys: Conf Ser 118:012029

Properties Improvement
of Metakaolin-Zeolite-Diatomite-Red Mud
Based Geopolymers

F. Rocha[1(✉)], C. Costa[1], W. Hajjaji[1], S. Andrejkovičová[1],
S. Moutinho[2], and A. Cerqueira[1]

[1] GeoBioTec, Geosciences Department, University of Aveiro,
3810-193 Aveiro, Portugal
tavares.rocha@ua.pt
[2] RISCO, Civil Engineering Department, University of Aveiro,
3810-193 Aveiro, Portugal

Abstract. Addition of pozzolanic materials increases the mechanical characteristics of construction materials and contributes towards a higher durability. Metakaolin is an artifical pozzolan obtained by calcination of kaolinitic clays at an adequate temperature. Geopolymers are inorganic materials from mineral origin, composed of a precursor, an alkaline activator and a solvent. New geopolymer formulations were designed by sodium silicate/NaOH/KOH activation of metakaolin, zeolites, diatomites and red mud mixtures. The effects of source materials on the microstructure and mechanical properties were studied. Mineralogical and chemical compositions were assessed as well as microstructure, specific surface area, compressive strength and adsorption. In general, incorporation of red mud, zeolite filler and diatomites to metakaolin in medium of alkali activators of low concentration provided formation of more eco-friendly materials with high mechanical resistances and water treatment capabilities.

Keywords: Geopolymers · Metakaolin · Alkaline activation ·
Additives formulations

1 Introduction

Addition of pozzolanic materials increases the mechanical characteristics of construction materials and contributes towards a higher durability. Metakaolin is an artifical pozzolan obtained by calcination of kaolinitic clays at an adequate temperature. Metakaolin is the structurally disordered product obtained following the dehydroxylation of kaolin, more precisely of its essential component kaolinite - $Al_2Si_2O_5(OH)_4$.

Geopolymers are inorganic materials from mineral origin, composed of a precursor, an alkaline activator and a solvent. The process of geopolymerization involves a chemical reaction that takes place in an alkaline medium, resulting in the formation of inorganic polymers that have silicon and aluminum as main constituents (Si+Al), connected by oxygen ions. The solution becomes alkaline using activators such as sodium and potassium hydroxides and/or sodium and potassium silicates.

S. Glagolev (Ed.): ICAM 2019, SPEES, pp. 381–384, 2019.
https://doi.org/10.1007/978-3-030-22974-0_92

Depending on the composition, geopolymer characteristics can be attained in terms of physical, chemical and mechanical performance (Mackenzie and Welter 2014). Moreover, geopolymers have the advantage to be possibly formulated from a wide range of aluminosilicate minerals.

2 Methods and Approaches

New geopolymer formulations were designed by sodium silicate/NaOH/KOH activation of metakaolin, zeolites, diatomites and red mud mixtures. The effects of source materials on the microstructure and mechanical properties were studied.

Geopolymers were prepared using commercial metakaolin (1200S, AGS Mineraux, France, D50 = 1.1 μm, bulk density = 296 g/dm^{-3}), diatomite from Rio Maior and Amieira deposits (Portugal), zeolite (ZeoBau micro 50, from Nižný Hrabovec, Zeocem, Slovakia, CEC = 83 meq/100 g, SSA = 1663 m^2/kg, particle size 0–0.05 mm, bulk density = 500–600 g/dm^{-3}, more information about Nižný Hrabovec deposit is available on http://www.iza-online.org/natural/), red mud from aluminium metallurgies wastes, hydrated sodium silicate (Merck, Germany Merck, Germany; 8.5 wt% Na$_2$O, 28.5 wt% SiO$_2$, 63 wt% H$_2$O) and sodium and potassium hydroxide (ACS AR Analytical Reagent Grade Pellets).

The role of the above ingredients in the preparation of geopolymers is as follows: metakaolin was used as a precursor of aluminium; diatomite was used as a precursor of silica; zeolite was used as filler with high specific surface area and cation exchange capacity; red mud as precursor of aluminium and iron; sodium silicate was used as a source of silicon and sodium and potassium hydroxide as alkaline activators for dissolution of aluminosilicate. Water was the reaction medium.

The X-ray diffraction was conducted on Philips X'Pert diffractometer using CuKα radiation at a speed of 0.02°/s. The X'Pert HighScore (PW3209) program was used to analyze XRD peaks. The chemical composition (major elements) of materials was analysed using PANalytical Axios X-ray fluorescence spectrometer. Loss on ignition was determined by heating the samples in an electrical furnace at 1000 °C during 3 h. The microstructural characterization was carried out by scanning electron microscopy (SEM – Hitachi, SU 70) and energy dispersive X-ray spectrometry (EDS – EDAX with detector Bruker AXS, software: Quantax) operated at 3–30 kV.

The specific surface area of geopolymers (heated at 200 °C during 6 h) was determined by BET method (Brunauer, Emmett and Teller, nitrogen gas adsorption at 77 K) (Brunauer et al. 1938). Compressive strength tests were carried out on 3 probes of individual geopolymer on (SHIMADZU: AG-IC 100 kN) equipment, by applying a maximum force of 5 kN at the speed of 50 N/s according to the Standard EN 1015-11. Adsorption experiments were carried out in batch using nitrate solutions of Pb^{2+}, Zn^{2+}, Cu^{2+}, Cd^{2+} and Cr^{3+}.

3 Results and Conclusions

Performed analyses confirm that zeolite particles are responsible for higher amount of crystalline phases producing more compact and firm microstructure of blended geopolymers relative to that of reference MK100. SEM analysis reveals that incorporation of 50 wt% of zeolite to metakaolin geopolymer (MK50) leads to the denser geopolymer matrix related to microstructure when compared to 100 wt% metakaolin in reference geopolymer (MK100). Dense microstructure of MK50 manifests in its considerably lower water adsorption and specific surface area values.

All geopolymers containing zeolite show increase in compressive strength compared to pure metakaolin one, with optimal ratio metakaolin precursor/zeolite filler 50:50 providing the highest compressive strength (8.8 MPa at 28 days). The adsorption of heavy metals increases as the amount of metakaolin in the structure increases.

Diatomite enriched formulations showed different evolutions according to the Na or K activator; those K activated showed higher compressive strength.

The well-known geopolymer composition (amorphous phase predominant over residual quartz, illite and anatase) is slightly affected by the red mud, despite it provides aluminum and iron oxides and oxy-hydroxide (hematite, goethite, gibbsite and boehmite). These phases, in fact, are to a limited extent involved in the geopolymerization process and the alkaline aluminosilicate amorphous phase is about 90 wt% in all samples. The structural features of the amorphous geopolymer, as resumed by the broad hump of XRD patterns, are not modified by addition of RM.

The physical properties of geopolymers are not significantly affected by RM, as all samples exhibit high values of water absorption and low apparent density. The chemical stability is good: sodium leaching test gave leachate concentrations close to 100 ppm without evidences of deterioration of mechanical performance. Red mud influences the mechanical strength during curing (especially at the higher amounts of RM but the lower additions, cured for 28 days showed good compressive strength).

In general, incorporation of red mud, zeolite filler and diatomites tometakaolin inmedium of alkali activators of low concentration provided formation of more eco-friendly materials with high mechanical resistances and water treatment capabilities.

References

Brunauer S, Emmett PH, Teller EJ (1938) Adsorption of gases on multimolecular layers. J Am Chem Soc 60:309–319

Mackenzie K, Welter M (2014) Geopolymer (aluminosilicate) composites: synthesis, properties and applications. Adv Ceram Matrix Compos. 445–470

Features of Production of Fine Concretes Based on Clinkerless Binders of Alkaline Mixing

M. Salamanova[1,2(✉)], S.-A. Murtazaev[1,2,3], A. Alashanov[1], and Z. Ismailova[1]

[1] Grozny State Oil Technical University named after Academician M.D. Millionshchikov, Grozny, Russia
Madina_salamanova@mail.ru
[2] Complex Research Institute named after H.I. Ibragimov, Russian Academy of Sciences, Grozny, Russia
[3] Academy of Sciences of the Chechen Republic, Grozny, Russia

Abstract. The paper showed relevance and potential for the development of clinkerless alkaline activation, since at the present the production of currently leading "constructional" binder, Portland cement, has been increasing, and carbon dioxide released during cement production has a negative effect on the ecological situation of separate countries and the whole world. The world community has long been concerned about the problem of switching to clinkerless binders and building composites to replace resource-intensive cement, at least in those areas of construction that do not need its high technical and functional properties. We gave formulations and properties of clinkerless alkaline activation based on highly dispersed mineral components, effective compositions of fine concretes based on the use of the proposed clinkerless alkaline activation cements were obtained. It was theoretically substantiated and practically proved that Brønsted acid sites on the surface of highly active powders accelerated synthesizing silica gel, supported polymerization of silicon-oxygen anions, enhanced ion exchange reactions and stabilized intergranular contact formation.

Keywords: Portland cement · Clinkerless binders · Mineral powders · Alkaline mixer · Liquid glass · High dispersion

1 Introduction

The environmental problem, created by the cement industry, is associated with consumption of large volumes of raw materials and release of huge amounts of carbon dioxide and dust during production of mineral binders into the atmosphere. A promising direction for solving this problem is the use of alkaline activation binders, which can be produced both from wastes of the fuel and energy industry, if present in the region, and using highly dispersed aluminosilicate additives, which chemical composition is characterized by an increased content of aluminum and silicon (Nikiforov et al. 2011; Salamanova et al. 2015; Strokova et al. 2013).

© The Author(s) 2019
S. Glagolev (Ed.): ICAM 2019, SPEES, pp. 385–388, 2019.
https://doi.org/10.1007/978-3-030-22974-0_93

2   Methods and Materials

This paper presents results of researches on the formulation and study of the properties of alkaline mixing binders using of sedimentary and magmatic rocks: quartz sand, limestone, volcanic tuff and silicified marl.

3   Results and Discussion

To prepare highly dispersed powders from the studied rocks, they were preliminarily crushed in a jaw crusher, and then subjected to fine grinding for 1 h in a vibratory ball mill.

At the next stage, a number of Bronsted active crystallization centers on the surface of the mineral powder was studied by the method (Strokova et al. 2013) of determining the exchange capacity with respect to calcium ions (Table 1).

Table 1. Surface activity of fine powders

No.	Mineral powder	Coefficient of activity Ka,%	Coefficient of hydraulic activity G_{HD}	Number of active centers of crystallization, mg eq/g	Specific surface of powders, m^2/kg
1	Quartz sand (QS)	22	1,76	21	810
2	Volcanic Tuff (VT)	37	1,90	34	905
3	Limestone (L)	8	1,24	12	1060
4	Thermo-activated marl (TM) at 700 °C	62	2,03	42	1150
5	Sandstone (S)	10	1,44	16	1020

The coefficient of activity Ka,% The coefficient of hydraulic activity GAMD The number of active centers of crystallization, mg eq/g Specific surface of powders, m^2/kg.

Thus, analyzing the results obtained we can state that the activity coefficients, including the number of active crystallization centers, largely depend on the degree of disclosure of defects formed during their grinding, in combination, all this contributes to an increase in the reactivity of the powders used in concrete as highly dispersed additives.

After confirming the reactivity of the proposed powders, samples of 10 cm in size were prepared using the mixture: a highly dispersed component, fractionated sand, obtained by mixing in a ratio of 55:45% of screening crushing of rocks of the Argun field and fine sand of the Chervlensky field. The mixing was carried out with liquid glass, sodium hydroxide and accelerator of the precipitation of silica gel with sodium fluoride in specified proportions. The prepared samples were hard in normal conditions at a temperature of 20 ± 2 °C, but after 2 days the samples were placed in an oven at a temperature of 40–50 °C for several days. The test results of the investigated fine-grained concretes based on alkaline activation binders are shown in Table 2.

Table 2. Properties of fine concrete based on clinkerless binders of alkaline activation

No compound	Materialconsumption per 1 m³					Density of concrete, kg/m³	Strength at compaction, MPa at the age, days	
	Fine powders	Fractionated sand	Na₂SiO₃	NaOH	Na₂SiF₆		7	28
1	QP - 480	1700	72,0	6,0	42,0	2240	11,8	24,7
2	TM - 480	1700	72,0	6,6	41,4	2250	31,4	40,5
3	VT - 480	1700	84,0	6,0	30,0	2246	25,3	34,2
4	L - 480	1700	60,0	3,6	56,4	2235	5,9	14,1
5	S - 480	1700	60,0	3,8	56,2	2239	6,6	15,9

Note: QP - quartz powder; TM - thermo-activated marl at 7000 °C; VT - volcanic tuff; L - limestone flour; S - sandstone; Na₂SiO₃ - sodium liquid glass; Na₂SiF₆ - sodium silicofluoride; NaOH - sodium hydroxide.

The alkaline activation binder with the use of fine powders of thermally activated marl showed the best results, so these particular samples of concrete were examined by Quanta 3D scanning electron microscope (Fig. 1).

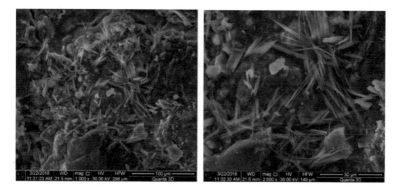

Fig. 1. Micrographs of concrete on thermo-activated alkali activated marl

In the contact area we determined a fairly strong accretion of particles of binder and quartz sand, no defects on the surface in the form of growths or cracks, irregularities of various shapes and sizes, and that individual particles have a needle-fibrous, vitreous structure, which indicated an increased binder activity.

High strength results of fine concrete using clinkerless alkaline mixing on the basis of thermally activated marl are explained by formation of a durable geopolymer stone, represented by a frame aluminosilicate shutter alkaline medium with the formation of a 3D aluminosilicate hydrogel (Murtazaev and Salamanova 2018; Soldatov et al. 2016).

4   Conclusions

Thus, we theoretically justified and practically proved that Brønsted acid sites on the surface of highly active powders accelerated process of synthesizing silica gel, promoted the polymerization of silicon-oxygen anions, enhanced ion exchange reactions and stabilized intergranular contact formation.

The results of the researches significantly expanded the field of application of clinkerless binders on a liquid-glass binder and might enable partial replacement of expensive and energy-intensive portland cement in the construction industry.

References

Murtazaev S-A, Salamanova M (2018) Prospects for the use of thermoactivated raw materials of aluminosilicate nature. Volga Sci J 46(2):65–70

Nikiforov EA, Loganina VI, Simonov EE (2011) The influence of alkaline activation on the structure and properties of diatomite. Bull BSTU Named After V.G. Shukhov (2):30–32

Salamanova MSh, Saidumov MS, Murtazaeva TS-A, Khubaev MS-M (2015) High-quality modified concretes based on mineral additives and superplasticizers of various nature. Sci Anal Mag "Innov Invest" (8):159–163

Soldatov AA, Sariev IV, Zharov MA, Abduraimova MA (2016) Building materials based on liquid glass. In: The collection: actual problems of construction, transport, engineering and technosphere safety materials of the IV-th annual scientific-practical conference of the North-Caucasian Federal University. N.I. Stoyanov (executive editor), pp 192–195

Strokova VV, Zhernovskiy IV, Maksakov AV (2013) Express-method for determining the activity of silica raw material for production, granulated nanostructuring aggregate. Constr Mater (1):38–39

Impact of Thermal Modification on Properties of Basalt Fiber for Concrete Reinforcement

V. Strokova, V. Nelyubova[✉], I. Zhernovsky, O. Masanin,
S. Usikov, and V. Babaev

Department of Materials Science and Materials Technology,
Belgorod State Technological University named after V.G. Shukhov,
Belgorod, Russia
vvnelubova@gmail.com

Abstract. The paper shows feasibility and efficacy of thermal modification of basalt fiber to increase its corrosion resistance and durability in a cement matrix. The authors justify the mechanism of phase and structure transformation of the fiber subsurface layer providing its increased physicochemical properties.

Keywords: Basalt fiber · Fiber · Thermal modification · Oxidation

1 Introduction

Under the conditions of real operation, the elements of concrete structures suffer cyclic, fatigue, impact, stretching and twisting loads, which causes uncontrolled cracking and consequent destruction of the cement matrix (Lesovik et al. 2018).

This problem was solved by using fibers to reinforce concrete matrix, which prevents brittle fracture of concrete and enables control of cracking (Klyuyev et al. 2013).

The analysis of approaches to enhancing the corrosion resistance of glass fiber in alkaline medium demonstrates that the thermal treatment of basalt fiber without its crystallization with initiation of a number of processes that increase alkali resistance and strength of the fibers should be considered as the most advanced and economically feasible method for increasing the stability of basalt fiber for its consequent application for concrete production on the basis of cement binders. (Knot'ko et al. 2007; Knotko et al. 2011)

According to the accepted working hypothesis, the increase of alkali resistance of fiber surface should be achieved by thermal treatment. This method is considered to be the simplest, affordable and economically reasonable. Its technical performance is predetermined by a range of physicochemical processes taking place in glass fibers during thermal treatment: oxidation, structure and diffusive rearrangement of ions in the material, annealing and densification of the structure, pre-crystallization and crystallization, etc.

S. Glagolev (Ed.): ICAM 2019, SPEES, pp. 389–392, 2019.
https://doi.org/10.1007/978-3-030-22974-0_94

2   Methods and Approaches

The main component in the work was basalt fiber produced at Novgorod glass fiber plant using machine factory BASK by blowing the melt by vertical air stream.

The method for investigation of thermal treatment impact on the fiber properties included step-wise heating of the fiber from 300 to 700 °C (700 °C is the working temperature of the basalt fiber, T_{work}) with the step of 100°. The isothermal holding time was 30 min, which considered the microscopic dimensions of fibers and striving to provide high productivity of the thermal treatment process. The fiber was cooled down in air by convective heat exchange. The fiber specimens then were tested for alkali resistance in model solutions represented by cement water extract.

3   Results and Discussion

According to preliminary data on the resistance of the fiber from different producers in alkaline and acidic medium, the chosen basalt fiber stands out for its high alkali and acid resistance. This is explained by the wash-out of cations persisting in vitrified phase sue to alkaline hydrolysis into the silicon dioxide solution with formation of aluminates and zincates with consequent liberation of anions SiO_4^{4-}, $Si_2O_5^{2-}$ and SiO_3^{2-}. Insoluble complex aluminate and zincate salts accumulate on the chemically active fiber surface, which inhibits further leaching of silicon. In the absence of further introduction of alkaline component for supporting necessary pH, the decomposition retards.

The corrosion of basalt fiber and its consequent disruption is caused by both interaction with hydroxyl groups of silicon-oxygen radicals and capability of cations dissolved in water alkaline medium to exchange with cations comprising the basalt fiber.

Noteworthy, during the utilization of basalt fiber in real conditions in concrete, the fiber dissolution degree will be not that significant, since the solution processes will die down along with curing and solidification of the cement. Nevertheless, the chemical processes involving the fiber that take place in concrete during its life should not be underestimated. This necessitates the development of the fiber modification method to improve its resistance to aggressive alkali medium.

The analysis of obtained data (Fig. 1) confirmed that thermal treatment of fiber at 500 °C was the most effective, as it promoted its alkali resistance as compared to non-treated fiber by 35.3% (after 28 days). Further increase of the treatment temperature was useless because, firstly, the fiber alkali resistance reduced; secondly, it was not economically reasonable. The effect of increased alkali resistance at 300 and 400 °C amounted to 26.7 and 30%, respectively.

Heating of basalt fibers leads to negligible reduction of their mass due to loss of adsorbed and chemically bound water, while further heating increases the mass by binding of air oxygen during divalent iron FeO oxidation into trivalent Fe_2O_3. The increased oxidation degree of iron cations leads to their decreased ion radius and reduction of the coordination number with formation of groups of iron-oxygen tetrahedrons $[FeO_4]^-Na^+$ that are embedded into the glass structure forming complex iron-aluminum-silicon-oxygen framework and increasing the degree of association, stability

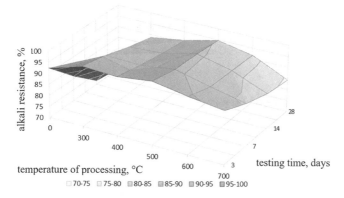

Fig. 1. Alkali resistance of fiber versus preliminary treatment temperature and testing time

and density of the structure. Similar structural restructuring triggers the diffusion redistribution of modifying cations (Na^+, Ca^{2+}) in the fiber.

The thermal treatment is accompanied by microdefect healing, structure densification and its approaching to stable equilibrium state.

The processes described above are most active at 500 °C—the temperature comparable with the glass transition temperature T_g (corresponds to the viscosity of $10^{12.3}$ Pa s and transition of glass from solid into plastic state)—which conditions the highest alkali resistance of fibers and stability of their work in cement concretes.

At higher temperatures (600–700 °C) and decreased viscosity (10^{10}–10^7 Pa s), the surface of basalt fibers suffers the manifestation of structural pre-crystallization and crystallization processes accompanied by the formation of diverse defects, which increase the free energy of the material and makes it more chemically active. This results in accelerated interaction with alkali and increased fiber mass losses. Besides, the thermal treatment at high temperature leads to the strength loss of basalt fibers: at 600 °C to negligible strength decrease; at 700 °C to the loss of 50% of initial strength.

Thermal treatment of basalt fibers at 500 °C initiated a number of important physicochemical and structural processes leading to the modification of glass fiber surface with acquisition of higher chemical resistance.

The holding close to the glass transition temperature T_g leads to gradual densification of the structure and its transition into more stable equilibrium state with higher chemical resistance. During the structure stabilization, numerous microdefects heal inside and on the surface of the fiber, which results in its decreased free energy and chemical activity. The structure densification also decreases the rate of numerous diffusion processes, which are the basis of any chemical reactions with solid phases.

Activation of iron ion oxidation during thermal treatment leads to substantial restructuring of the surface layer of basalt fiber: modifying ions of Fe^{2+} after oxidation to Fe^{3+} embed as iron-oxygen tetrahedrons $[FeO_4]^{5-}$ into aluminum-silicon-oxygen lattice of the glass, thus increasing the degree of its association and chemical resistance.

Preliminary 30-min processing of the basalt fiber at 500 °C by 35% increases its resistance in alkaline medium of curing cement and can be recommended as effective, simple and economically reasonable method for modifying basalt fiber for its effective implementation as a micro-reinforcing component of fibrous concretes.

4 Conclusions

Thus, the processes initiated by thermal treatment promote the corrosion resistance of fibers in alkaline medium, which prolongs the corrosion resistance and increases the efficacy of fiber application as a micro-reinforcing component for cement composites.

Acknowledgements. The work has been fulfilled within the project Federal Target Program of Research and Development on "Priority Development Fields of science and technology sector in Russia for 2014–2020", the unique project number is RFMEFI58317X0063.

References

Klyuyev SV, Klyuyev AV, Sopin DM, Netrebenko AV, Kazlitin SA (2013) Heavy loaded floors based on fine-grained fiber concrete. Mag Civil Eng 38(3):7–14

Knot'ko AV, Garshev AV, Davydova IV, Putlyaev VI, Ivanov VK, Tret'yakov YuD (2007) Chemical processes during the heat treatment of basalt fibers. Prot Metals 43(7):694–700

Knotko AV, Pustovgar EA, Garshev AV, Putlyaev VI, Tret'yakov YD (2011) A protective diffusion layer formed on surface of basaltic fiberglass during oxidizing. Prot Metals Phys Chem Surf 47(5):658–661

Lesovik VS, Glagolev ES, Popov DY, Lesovik GA, Ageeva MS (2018) Textile-reinforced concrete using composite binder based on new types of mineral raw materials. In: IOP conference series: materials science and engineering, vol 327, no 3

Activation of Cement in a Jet Mill

S. Titov[✉] and A. Kazakov

Federal State Budgetary Educational Institution, Russian University of Transport
(MIIT), Moscow, Russia
titovs3094@yandex.ru

Abstract. Mechanical treatment of cement in a jet mill leads to an increase in
the strength of cement-sand stone in compression and during bending. The effect
of increased activity is achieved by changing the shape of cement particles from
angular to rounded.

Keywords: Cement · Jet mill · Cement activity · Compressive strength ·
Bending strength · Particle shape factor

1 Introduction

Activation is a set of measures aimed at increasing the activity of cement by exposing it
to various mechanical methods. As a result of such exposure, physical and chemical
processes of different nature occur in the raw materials, leading to a change in the
characteristics of the final product (Kuznetsova and Sulimenko 1985). As a rule, during
mechanical activation, cement is crushed with an increase in its specific surface.

In a jet mill, activation occurs due to processing of cement in turbulent swirling
flows (Korchakov 1986), leading not only to grinding, but also to transformation of the
shape of particles from angular to rounded.

2 Methods and Approaches

The analysis of the shape of cement particles modified in a jet mill was performed using
the proposed shape criterion:

$$k_i = (S/P^2)^i \tag{1}$$

where S – area of the particle in the plane of its cross section, P – length of the
perimeter in the cross section of the particle; i – order of the criterion ($i = 1/n\ldots,1,\ldots,n$,
where n – counting numerals).

To determine the degree of influence of the particle shape on the properties of
cement systems, standard samples were tested and the results obtained were processed
using mathematical statistics methods.

S. Glagolev (Ed.): ICAM 2019, SPEES, pp. 393–394, 2019.
https://doi.org/10.1007/978-3-030-22974-0_95

3   Results and Discussion

It was found that processing in a jet type mill leads to an increase in dispersity of cement from 5% to 0.8–1.2% on sieve No. 008 with a non-significant change in normal density of the cement paste.

Increasing the fineness of cement grinding causes, as a rule, an increase in mixing water consumption to obtain a mixture of equal workability. But the experiment showed the opposite – the water-cement ratio of the solution prepared on cement, processed in a jet mill, not only did not increase, but also decreased. This is due to the change in the shape of cement particles from angular to rounded, which was statistically justified using the proposed shape criterion k_i (1).

The results of the tests showed an increase in strength of the samples based on portland cement processed in a jet mill for compression by 38–68% and by bending by 12–25% compared to non-activated cement.

4   Conclusions

The increase in compressive strength and bending of the samples based on cement processed in a jet mill is explained by the change in the shape of its particles from angular to rounded, which is associated with the characteristics of the vortex activation method.

Acknowledgements. The authors express their gratitude to the research supervisor, D.Sc. (Eng), prof. Kondrashchenko V.I.

References

Korchakov VG (1986) Aerodynamics of flows in jet mills when grinding silicate materials, Ph.D. (Eng) thesis, Kharkov, 168 p

Kuznetsova TV, Sulimenko LM (1985) Mechanical activation of portland cement raw mixtures. Cement, no 4, pp 20–21

The Law of Similarity and Designing High-Performance Composites

A. Tolstoy$^{(\boxtimes)}$, V. Lesovik, E. Glagolev, and L. Zagorodniuk

Belgorod State Technological University named after V.G. Shukhov,
Belgorod, Russia
tad56@mail.ru

Abstract. The increasing requirements for the quality of building products and constructions condition the importance of developing new ways of controlling the processes of shaping the structure of composite materials. This paper considers the aspects of dealing with the tasks of designing high-performance powder composites factoring in the law of similarity.

Keywords: Law of similarity · High-performance materials · Powder concrete · Technogenic raw materials

1 Introduction

In our days, erecting special-purpose buildings and structures with complex design, as well as the so-called unique buildings, calls for using high-performance powder concrete, which composition differs from the composition of traditional normal concrete in the increased proportion of cement, higher fineness of grain, complex composition and increased dispersity of aggregate (Bazhenov et al. 2007; Tolstoy et al. 2018a, b; Lesovik et al. 2015). The role of each of the source materials, as well as that of the mechanism of interaction between them, increases manifold, with fundamental concepts of the law of similarity can explain the nature of the processes that occur.

The theoretical foundation for designing high-quality composites is a new cross-disciplinary scientific school: geonics (geomimetics), which employs the results of studies of natural processes and rocks to create building materials of the future (Lesovik 2014; Elistratkin and Kozhukhova 2018; Dmitrieva et al. 2018). This allowed to develop a system for designing powder concrete using raw materials that were specially prepared through geological processes, i.e. they are genetically activated (Tolstoy et al. 2014).

A separate school is further distinguished: crystal energy science, the science about the modern approaches to estimating the quantitative indicators of the properties of materials and explaining the processes of hydrate formation (Lesovik and Evtushenko 2002).

The criteria of applicability of mineral components should be: abundance, accessibility, cost, constant composition (Fig. 1).

S. Glagolev (Ed.): ICAM 2019, SPEES, pp. 395–398, 2019.
https://doi.org/10.1007/978-3-030-22974-0_96

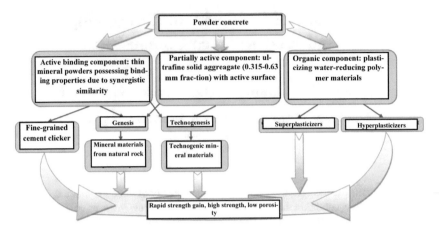

Fig. 1. Types of raw materials for making powder concrete

Here it is necessary to use the "experience" of geological processes, for example, high-performance siltstones that are similar to powder concrete in their strength and other properties. To create a strong and lasting composite of powder concrete, it is necessary to ensure the reliable physical, mechanical and performance characteristics of the material structure taking into account all energy parameters of all the substances involved, as well as the compliance of other properties. This structure should be similar to natural materials in terms of the main properties and genesis.

2   Methods and Approaches

The experiments involved mineral admixtures containing the aluminate and carbonate components, as well as off-the-shelf polymer ones: Melflux 2651, Melment, and finely ground quartzite sandstone, a by-product rock from the Kursk Magnetic Anomaly (KMA).

3   Results and Discussion

At the moment there is no unified approach to the technique of producing high-performance compositions. Some researchers address the issue from the perspective of technological mechanics, other approaches consider the rheological properties of the system, optimization of grain size composition, and the relation between the optimal structure and extreme properties of components (the "law of gauge" by I.A. Rybiev (Rybev 1999)).

The proposed methods of controlling the formation of structure in curing systems of powder mixes with technogenic components allow to achieve the composite strength of 80–100 MPa (Table 1).

Table 1. Comparative figures of normal concrete and powder concrete

Parameter	Values	
	Normal concrete	Powder concrete
Average density, kg/m^3	2200–2500	2300
Compressive strength, MPa	10–50	97.5
Water retention capacity, %	78–80	90
Strength-density ratio	0.17	0.36
Water resistance grade. W	2–4	4
Freeze-thaw resistance grade, F	50–150	300
Wear capacity, kg/m^2	0.7–0.8	0.36
Shrinkage	No fractures	
Thermal conductivity ratio, W/(m K)	0.8–1.2	1.29

A feasible foundation can be represented by the method of estimating the composition of high-density concrete taking into account the law of similarity of properties (Tolstoy et al. 2018a, b).

Dense structure of powder concrete is characterized by a virtually complete absence of pores and microfractures. It was possible to implement this by using the adjusted composition of the curing mass, introducing the necessary number of fine-grained technogenic components, their densest packing, and the self-compacting effect of curing.

Study of the micrographs of the curing compositions revealed the following:

– the microstructure of the cement rock obtained through intergrinding possesses higher homogeneity than that obtained through separate grinding of components;
– growth of needle-shaped crystals that permeate the volume of the material's structure was detected;
– dense neogeneses are present near the aggregate grains;

4   Conclusions

This way, introducing the theory of geonics (geomimetics), particularly the law of similarity, allows to obtain high-performance powder concretes, all components of which possess a high adhesion and similar deformation and thermal characteristics. The first attempt to test this law allowed to obtain composites with the maximum compressive strength of up to 100 MPa.

Acknowledgements. The work has been fulfilled within the project Federal Target Program of Research and Development on "Priority Development Fields of science and technology sector in Russia for 2014–2020", unique project number is RFMEFI58317X0063.

References

Bazhenov IM, Demianova VS, Kalashnikov VI (2007) Modifitcirovannye vysokoprochnye betony [Modified high-performance concretes], ASV, Moscow, 368 p

Dmitrieva TV, Strokova VV, Bezrodnykh AA (2018) Influence of the genetic features of soils on the properties of soil-concretes on their basis. Constr Mater Prod 1(1):69–77

Elistratkin MYu, Kozhukhova MI (2018) Analysis of the factors of increasing the strength of the non-autoclave aerated concrete. Constr Mater Prod 1(1):59–68

Lesovik VS (2014) Geonika (Geomimetika) [Geonics (Geomimetics)]. Primery realizatcii v stroitelnom materialovedenii [Examples of application in building materials science]. Bulletin of BSTU named after V.G. Shoukhov, Belgorod, 206 p

Lesovik VS, Zagorodniuk LK, Chulkova IL, Tolstoy AD, Volodchenko AA (2015) Srodstvo struktur, kak teoreticheskaia osnova proektirovaniia kompozitov budushchego [Similarity of structures as a theoretical foundation of designing the composites of the future]. Stroitelnye materialy [Building materials], no 9, pp 18–22

Lesovik BC, Evtushenko EI (2002) Stabilizatciia svoistv stroitelnykh materialov na osnove tekhnogennogo syria [Stabilizing the properties of building materials on the basis of technogenic raw materials]. Izvestiia VUZov [Bul. HEIs]. 12:40–44

Rybev IA (1999) Otkrytie zakona stvora i vzaimosviaz ego s zakonom kongruentcii v stroitelnom materialovedenii [Discovering the gauging law and its relation to the law of congruence in building materials science]. Stroitelnye materialy [Build Mater] 12:30–31

Tolstoy AD, Lesovik VS, Milkina AS (2018a) Osobennosti struktury betonov novogo pokoleniia s primeneniem tekhnogennykh materialov [Specifics of the structure of new-generation concretes using technogenic materials]. Russ Automob Highw Ind J Sect III Constr Arch. 15(4, 62):588–595

Tolstoy AD, Lesovik VS, Kovaleva IA (2014) Organomineralnye vysokoprochnye dekorativnye kompozitcii [Organic and mineral high-performance decorative compositions]. Bull BSTU Named After V.G. Shoukhov 5:67–69

Tolstoy AD, Lesovik VS, Glagolev ES, Krymova AI (2018b) Synergetics of hardening construction systems. In: IOP conference series: materials science and engineering, vol 327, p 032056. https://doi.org/10.1088/1757-899x/327/3/032056

Genesis of Clay Rock of the Incomplete Stage of Mineral Formation as a Raw Material Base for Autoclaved Materials

A. Volodchenko[(✉)] and V. Strokova

Belgorod State Technological University named after V.G. Shukhov,
Belgorod, Russia
volodchenko@intbel.ru

Abstract. We illustrated possible application of clay rock of incomplete stage of mineral formation to produce autoclaved silica materials. A distinguishing feature of this rock was the presence of thermodynamically unstable compounds. Weathering processes resulted in partial disintegration of the rock, formation of defects in the crystalline lattice of the rock-forming minerals, and partial amorphization of both silicates and aluminosilicates, which reduced the energy potential of the raw materials. Thus, it was possible to use hydrothermal conditions to boost the formation of neogeneses of cementing compounds and reduce power consumption of the production of autoclaved materials.

Keywords: Lime · Clay rock · Autoclaved silica

1 Introduction

The traditional technology of manufacturing of silica materials is based on using lime and silica sand as raw materials, and the reserves of the latter are getting depleted. Furthermore, this technology is marked with a high power consumption. The consumption of power for grinding the binder and hydrothermal treatment can be reduced by using highly reactive polymineral raw materials, which would allow not only to replace silica sand but also intensify the technological processes. A prospective raw material for producing autoclaved silica materials is the clay rock of the incomplete stage of mineral formation, which is currently unconventional in the building industry (Lesovik 2012; Lesovik et al. 2014; Alfimova and Shapovalov 2013; Volodchenko et al. 2015; Volodchenko et al. 2016; Alfimov et al. 2006; Alfimova et al. 2013). As compared with the traditional silica sand, this rock ensures the synthesis of neogeneses in a more complex system "CaO–[SiO$_2$–Al$_2$O$_3$–(MgO)]–H$_2$O", which the reduction of power consumption of manufacturing and improving physical, mathematical, and performance properties of both solid and cellular autoclaved materials (Volodchenko and Lesovik 2008a, 2008b; Volodchenko 2011).

The purpose of this paper is to study the genesis of clay rock of the incomplete stage of mineral formation as raw material for producing autoclaved silica materials.

© The Author(s) 2019
S. Glagolev (Ed.): ICAM 2019, SPEES, pp. 399–402, 2019.
https://doi.org/10.1007/978-3-030-22974-0_97

2 Results and Discussion

Some of the most proliferated rocks are deposits of clay. They are formed through weathering of igneous and metamorphic rocks.

The process of weathering of felsic rock passes the following stages: physical weathering (clastic stage), start of the chemical weathering with illite formation (siallite stage), chemical weathering with kaolinite formation (acidic siallite stage), and complete decomposition of silicates with the formation of aluminum, ferrous, and silicon oxides (allite stage). Weathering of rock results in the formation of minerals predominantly of the montmorillonite group.

The processes of weathering can be represented as the destruction of the crystalline structure of the source minerals followed by a transfer through the pseudo-amorphous state and then the formation of the crystalline structure of new rock-forming minerals. However, the ideal amorphous (non-crystalline) state, as well as crystalline state, is not achieved and the transition through it is considered conditional. This results in the change of the source framework structure of the feldspar into the layered structure of clay minerals.

This transition results in rocks that can be characterized as products of the intermediary stage of weathering that occupy a significant stretch in the line of sub-stance transformation and predominate in nature. The rock-forming minerals of these formations include the X-ray amorphous phase, illites, mixed-layer minerals, and imperfectly structured kaolinite and montmorillonite. The formation of these compounds is accompanied by disordering of the crystalline structure of the source minerals, which increases the thermodynamic instability of the rock.

The rocks in the intermediary stage of weathering that were mechanically and chemically activated by exogenous geologic processes, i.e. those that belong to the starting stage of chemical weathering that results in the formation of defects in the structure of silicates and aluminosilicates and the formation of illites, are virtually never used. However, due to their thermodynamic instability, these rocks are the most efficient raw material as components of the systems that must be highly reactive. In particular, such rocks can be used as components of hydro-thermal curing binders, which allows to manage the processes of the formation of the structure of new-generation autoclave materials.

Based on the data about the genesis of the weathering rind rocks, the processes of phase formation in artificial systems of autoclaved curing and how they compare with their natural counterparts of mineral formation, a diagram of the exogenous processes of the genesis of clay rocks as a raw material base for autoclaved materials is proposed (Fig. 1).

The rocks that formed after the second (siallite stage) and partially the third (acidic siallite) stage of weathering belong to the rocks of the incomplete stage of mineral formation that contain compounds of the intermediary stage of weathering: the source minerals have decomposed while the final ones have not formed yet. Such minerals are characterized by the presence of crystalline structure defects and, consequently, thermodynamic instability. This kind of rock can either remain at the place of its formation

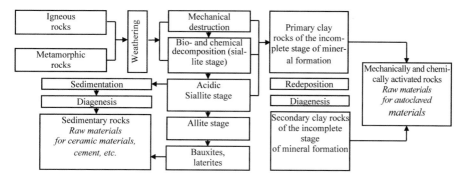

Fig. 1. Diagram of the exogenous processes of the genesis of clay rocks as raw materials base for autoclaved materials

or redeposit in new places as the result of transfer and diagenesis. These rocks are useful as raw materials for autoclaved curing materials.

Thus, weathering of source rocks of different compositions leads to the formation of mineral resources that are different in their composition and industrial significance. If one considers the weathering rind profile, it can be seen that at the moment the building materials industry is using the lower horizon (source rocks) and products of the final stage of weathering: kaolinite and montmorillonite clays as well as bauxites.

Clay rocks of the incomplete stage of mineral formation are some of the most inconsistent in terms of composition and the structure of the sedimentation mass deposits. In terms of the crystallochemical features of the rock-forming minerals, it is necessary to note the high content of the crystalline structure defects, the dis-order of the interlayer aluminosilicate stacks in the layered structure of minerals, etc. The crystalline lattice of the typical clay minerals that belong to layered silicates is characterized by the alteration of silicon-oxygen tetrahedra and hydroxyl octahedra. To some degree, clay minerals can be considered objects, where the nanoparticles (elementary layer packets) are packed into the microscopic structures, where the properties of the individual constituent particles are concealed due to a strong interparticle interaction.

3 Conclusions

Thus, lay rocks of the incomplete stage of mineral formation are energy-saturated deposits. Due to the composition and crystallochemical features of the rock-forming minerals, these rocks can be recommended as energy-efficient raw materials for the production of autoclaved curing materials. A distinguishing feature of this rock is the presence of thermodynamically unstable compounds. At the same time, aluminosilicates are characterized by variable chemical composition and the imperfection of their crystalline lattice. Weathering processes result in partial disintegration of rock, formation of defects in the crystalline lattice of the rock-forming minerals, and partial amorphization of both silicates and alumino-silicates, which reduces the energy potential of the rock-forming minerals. Thus, it is possible to use hydrothermal conditions to boost the formation of neogeneses of cementing compounds and reduce the power consumption of the production of autoclaved materials.

Acknowledgements. The work is realized in the framework of the Program of flagship university development on the base of the Belgorod State Technological University named after V.G. Shukhov, using equipment of High Technology Center at BSTU named after V.G. Shukhov.

References

Alfimov SI, Zhukov RV, Volodchenko AN, Yurchuk DV (2006) Technogenic raw materials for silicate hydration hardening. Mod High Technol 2:59–60
Alfimova NI, Shapovalov NN (2013) Materials autoclaved using man-made aluminosilicate materials. Fundam Res 6(3):525–529
Alfimova NI, Shapovalov NN, Abrosimova OS (2013) Operational characteristics of silica brick, manufactured using man-made aluminosilicate materials. Bulletin of BSTU named after V.G. Shukhov, vol 3, pp 11–14
Lesovik VS (2012) Geonik. Subject and tasks. Publisher Belgorod State Technological University. VG Shukhov, p 219
Lesovik VS, Volodchenko AA, Svinarev AA, Kalashnikov NV, Rjapuhin NV (2014) Reducing. World Appl Sci J 31(9):1601–1606
Volodchenko AA, Lesovik VS, Volodchenko AN, Glagolev ES, Zagorodnjuk LH, Pukharenko YV (2016) Int J Pharm Technol 8(3):18856–18867
Volodchenko AA, Lesovik VS, Volodchenko AN, Zagorodnjuk LH (2015) Int J Appl Eng Res 10(24):45142–421149
Volodchenko AN, Lesovik VS (2008a) Increasing production efficiency of autoclave materials. In: Proceedings of the higher educational institutions. Building, vol 9, pp 10–16
Volodchenko AN, Lesovik VS (2008b) Autoclave silicate materials with nanometer-sized materials. Build Mater 11:42–44
Volodchenko AN (2011) Features magnesia clay interaction with the calcium hydroxide in the synthesis and the formation of tumors microstructure. Bulletin of Belgorod State Technological University. VG Shukhov, vol 2, pp 51–55

Abnormal Mineral Formation in Aluminate Cement Stone

I. Zhernovsky[✉], V. Strokova, V. Nelyubova, Yu. Ogurtsova, and M. Rykunova

Belgorod State Technological University named after V.G. Shukhov, Belgorod, Russia
zhernovsky.igor@mail.ru, zhernovskiy.iv@bstu.ru

Abstract. The paper describes the results of the study on the correlation between the strength properties and the concentration of calcium aluminate hydrates during long hardening of aluminate cement in aqueous medium. It shows the abnormal nature of hydrate mineral formation in the system of hardening aluminate cement.

Keywords: Aluminate cement · Calcium aluminate hydrates

1 Introduction

The properties of any composite materials directly depend on the characteristics of their base matrix. This also fully applies to composites based on binding agents with various composition, which quality indicators are set as a result of system hardening (Strokova et al. 2015; Kozhukhova et al. 2016; Nelyubova et al. 2017; Dmitrieva et al. 2018; Shulpekov et al. 2018). In this regard the study of phase formation in such systems seems quite relevant.

Construction materials based on aluminate cements have different properties than those based on portland cement. In particular, they are characterized by a high thermal stability and resistance to acid corrosion (Kuznetsova 1986; Kuznetsova et al. 1989).

The calcium aluminate hydrates, which are formed through hardening of aluminate cement, represent the following phases: CAH_{10}, C_2AH_8, C_3AH_6, AH_3 (gibbsite). From this point onward the 'cement' notation in formulas of chemical compounds will be used ($C - CaO$, $A - Al_2O_3$, $H - OH$).

The hydrate mineral formation in aluminate cements is characterized by the dependence of C-A-H phases on temperature. At a temperature below 15 °C the main reaction-active CA phase is hydrated according to the following scheme: $CA + 10H \rightarrow CAH_{10}$. At room temperature the C-A-H is formed according to the following scheme: $2CA + 11H \rightarrow C_2AH_8 + AH_3$. At higher temperatures (above 28 °C) the metastable hydrates CAH_{10} and C_2AH_8 will transform into stable hydrate C_3AH_6 according to the following equations: $3CAH_{10} \rightarrow C_3AH_6 + 2AH_3 + 18H$ and $3C_2AH_8 \rightarrow 2C_3AH_6 + AH_3 + 9H$ (Rashid et al. 1994).

Thus, according to insights into hydration of aluminate cements, the hydrate phases during long hydration shall be presented as follows: $C_3AH_6 + AH_3$.

© The Author(s) 2019
S. Glagolev (Ed.): ICAM 2019, SPEES, pp. 403–406, 2019.
https://doi.org/10.1007/978-3-030-22974-0_98

The paper provides the results of hydration hardening of aluminate cement with long-term exposure to aqueous medium, which contradict the above ideas.

2 Materials and Methods

The aluminate cement GTS-50 produced by JSC Pashiysky Cement and Metallurgical Plant (Perm Region, Russian Federation) was used in the study as a binding agent. Cubes with a 2 cm side are formed from cement paste (water-cement ratio equal 0.3). After hardening on the 1st day the samples were taken from a mold and placed into the desiccator with 100% air humidity. After 28 days the samples were dried in a drying cabinet at 80 °C and placed in water, from which they were taken for further study in 1, 2, 3 and 4 months.

The study methods included the compression strength test on PGM-100 MG4 press (average against three measurements) and the quantitative full-scale XRF to define potential changes in mineral composition.

The diffraction spectra of samples were obtained via the ARL X'TRA diffractometer (λ_{Cu}) in the range of diffraction angles $2\theta° = 4\text{-}64$, step angle – 0.02°. The diffraction spectra were smoothened prior to treatment.

The quantitative full-scale XRF was carried out via the DDM v.1.95e software for a difference curve derivative (Solovyov 2004).

3 Results and Discussion

According to XRF, the mineral composition of GTS-50 cement is presented by the following crystal formations (wt.%): CA (28.3), CA_2 (9.2), $C_{12}A_7$ (1.4), akermanite-gelenite (15.6), $\beta\text{-}C_2S$ (10.0), $\alpha'_H\text{-}C_2S$ (10.2), wollastonite 2M (7.6), dolomite (10.6), perovskite (7.2).

X-ray diagnostics of mineral phases in hydrated samples of aluminate cement indicated the presence of CAH_{10}, C_2AH_8, C_3AH_6, AH_3, akermanite-gelenite, $\beta\text{-}C_2S$, $\alpha'_H\text{-}C_2S$, wollastonite 2M, dolomite and perovskite.

Due to lack of C_2AH_8 structural model, the quantitative XRF was combined with the approach suggested by Cuberos et al. (2009).

As data show (Table 1), for nearly 5 months (28 days in damp atmosphere and 4 months in water) CAH_{10} remains the main hydrate phase. The reduction of CAH_{10} concentration on the 2nd month in water, which correlates well with the increase of crystal gibbsite (AH_3), does not lead to the similar increase of C_3AH_6. Most likely it is caused by the formation of C_3AH_6 in its cryptocrystalline state.

The abnormal nature of hydrate mineral formation includes the increase of CAH_{10} metastable hydrate concentration starting from the 3rd month. At the same time the temporal change of C_3AH_6 concentration coincides well with the change of compression strength of the studied samples (Fig. 1).

Table 1. Concentration of hydrate phases (wt. %)

Time in aqueous medium, months	CAH_{10}	C_2AH_8	C_3AH_6	AH_3
0	18.8	3.5	3.1	2.0
1	26.5	4.4	8.2	5.9
2	9.0	2.1	6.8	25.1
3	33.5	3.5	4.7	6.3
4	37.2	5.8	5.6	3.0

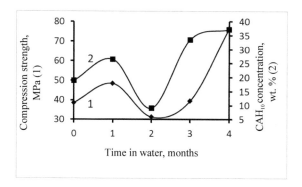

Fig. 1. Compression strength of aluminate cement stone and CAH_{10} concentration depending on time

4 Conclusions

The paper describes the results of the study on the correlation between the strength properties and the concentration of calcium aluminate hydrates during long hardening of aluminate cement in aqueous medium. At present it is impossible to give unambiguous interpretation of the observed abnormal compositional changes of hydrate phases of hydrated aluminate cement during long hardening.

Acknowledgements. The study is carried out in the framework of the State Task of the RF Ministry of Education and Science No. 7.872.2017/4.6. Development of principles for the design of ecologically positive composite materials with prolonged bioresistance 2017–2019.

References

Cuberos AJM, De la Torre ÁG, Martín-Sedeño MC, Moreno-Real L, Merlini M, Ornez LM, Aranda MAG (2009) Phase development in conventional and active belite cement pastes by Rietveld analysis and chemical constraints. Cem Concr Res 39:833–842

Dmitrieva TV, Strokova VV, Bezrodnykh AA (2018) Influence of the genetic features of soils on the properties of soil-concretes on their basis. Constr Mater Prod 1(1):69–77

Kozhukhova NI, Chizhov RV, Zhernovsky IV, Strokova VV (2016) Structure formation of geopolymer perlite binder vs. type of alkali activating agent. ARPN J Eng Appl Sci 11 (20):12275–12281

Kuznetsova TV (1986) Aluminate and sulfo-aluminate cements, Moscow

Kuznetsova TV, Kudryashov IV, Timashev VV (1989) Physical chemistry of binding materials, Moscow

Nelyubova V, Pavlenko N, Netsvet D (2015) Cellular composites with ambient and autoclaved type of hardening with application of nanostructured binder. In: IOP Conference series: materials science and engineering, vol 96, no 1, p 012010

Nelyubova VV, Strokova VV, Sumin AV, Jernovskiy IV (2017) The structure formation of the cellular concrete with nanostructured modifier. Key Eng Mater 729:99–103

Rashid S, Barnes P, Bensted J, Turrillas X (1994) Conversion of calcium aluminate cement hydrates re-examined with synchrotron energy-dispersive diffraction. J Mater Sci Lett 13:1232–1234

Shulpekov AM, Lepakova OK, Radishevskaya NI (2018) Phase – and structural formation in the TIO2-AL-C system in the SHS process. Chem Bull 1(1):4–11

Solovyov LA (2004) Full-profile refinement by derivative difference minimization. J Appl Crystallogr 37:743–749

Strokova VV, Botsman LN, Ogurtsova YN (2015) Impact of epicrystallization modifying on characteristics of cement rock and concrete. Int J Appl Eng Res 10(24):45169–45175

Structural Transformations
of Low-Temperature Quartz During
Mechanoactivation

I. Zhernovsky[(✉)] and V. Strokova

Belgorod State Technological University named after V.G. Shukhov,
Belgorod, Russia
zhernovsky.igor@mail.ru

Abstract. The paper presents the study of mechanoactivation impact on crystal structure of α-quartz. The volume of silicon-oxygen tetrahedron SiO_4^{4-} is accepted as the structural parameter depending on the mechanoactivation degree. The paper compares the dependence of this parameter on temperature, pressure and time of mechanoactivation of α-quartz in a planetary mill.

Keywords: Quartz · Crystal structure · Silicon-oxygen tetrahedron · Mechanoactivation

1 Introduction

The mechanoactivation dispergation of quartz materials is a widespread method of technological processing of this mineral raw material in various fields of technological mineralogy.

In practice of technical petrogenesis of cast stone, forming the basis of synthesis of inorganic silica-containing binding agents (Dmitrieva et al. 2018), the mechanoactivation dispergation of quartz raw material holds a special place.

The result of almost a century-long study of phase and structural transformations of low-temperature quartz during mechanoactivation is the amorphicity of a surface layer of quartz particles and the formation of nanosized β-quartz crystals in α-quartz matrix (Zhernovsky et al. 2018). At the same time there was no study on structural transformations of α-quartz within a quartz particle matrix during mechanoactivation. The only exception is the study by Archipenko et al. (1987, 1990) concerning phase transformations in mechanoactivated α-quartz. The task of this study is to fill the gap in this matter partially.

2 Materials and Methods

The grinding of hydrothermal quartz (Ural) was carried out in the PULVERISETTE 6 classic line planetary mill (Fritsch, Germany) with lining and grinding bodies of tungsten carbide. The milling time made 3, 12, 30, 60, 120, 180, 240, 300 and 360 min.

S. Glagolev (Ed.): ICAM 2019, SPEES, pp. 407–410, 2019.
https://doi.org/10.1007/978-3-030-22974-0_99

The diffraction spectra of samples are obtained using ARL 9900 Workstation (λCo). Shooting interval – $2\theta°$:8–80, step angle – 0.02. The specification of structural parameters was carried out in DDM v.1.95e software for the difference curve derivative (Solovyov 2004). 174-ICSD data (P$3_2$21) were used as a structural model. Profile function – pseudo-Voigt. The specification of the profile parameters was carried out in the anisotropic approximation. Thermal corrections were specified in anisotropic option. Table 1 shows the experimental results.

Table 1. Coordinates of α-quartz atoms after mechanoactivation

	Milling time, min								
	3	12	30	60	120	180	240	300	360
Si^* x/a	0.4499	0.463	0.460	0.466	0.465	0.466	0.468	0.469	0.470
O x/a	0.403	0.416	0.413	0.419	0.418	0.419	0.422	0.422	0.423
O y/b	0.2660	0.2702	0.2697	0.2715	0.2668	0.272	0.2697	0.2721	0.2707
O z/c	0.7950	0.7921	0.7919	0.7905	0.7886	0.7868	0.7866	0.7887	0.7876

Note: Si $y/a = 0$, Si $z/c = 2/3$.

3   Results and Discussion

The volume of silicon-oxygen tetrahedrons was chosen as a structural and sensitive quartz parameter on mechanoactivation influence since it depended on several elementary structural parameters: bond length Si-O, bond angle O-Si-O and twist angle of tetrahedrons (Goryaynov and Ovsyuk 1999).

It is known that mechanoactivation is the result of two processes influencing the material – local thermal influence and impact pressure. Hence, it is advisable to consider structural changes of quartz during mechanoactivation in comparison with changes during thermal and baric impacts. The following were used as references on changes of quartz structural parameters: at a thermal influence – the work of Kihara (1990), at a baric impact – works of Levien et al. (1980), Hazen et al. (1989) and Glinnemann et al. (1992).

The given results show that mechanoactivation alongside with thermal and baric influences transforms the structure of α-quartz by reducing the volume of SiO_4-tetrahedrons. At the same time the dependences of volumes of α-quartz silicon-oxygen tetrahedrons on severity of exposure differ a lot (Figs. 1 and 2).

The energy accumulation by quartz during mechanoactivation, as well as under thermal and baric influence most likely happens due to the reduction of Si-O bond length.

The energy accumulated by quartz during mechanoactivation can be estimated by comparing the dependences (Figs. 1a and 2). This comparison confirms that material energy saturation after three-hour mechanoactivation is equivalent to its heating up to 500 °C.

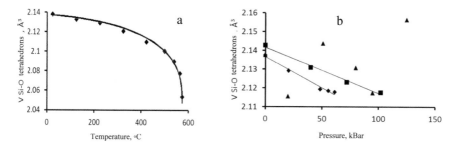

Fig. 1. Calculated dependences of volumes of α-quartz silicon-oxygen tetrahedrons on thermal (*a*) and baric (*b*) influences. (*b*) indicates the following data: ◆– according to Levien et al. (1980), ▲ – according to Hazen et al. (1989), ■ – according to Glinnemann et al. (1992).

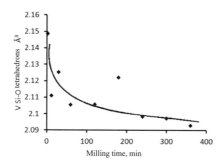

Fig. 2. Dependence of volumes of α-quartz silicon-oxygen tetrahedrons on mechanoactivation time

4 Conclusions

The change of volume of silicon-oxygen tetrahedrons may be considered as an indicator of the α-quartz crystal structure response to the thermal and baric influence, as well as to the mechanoactivation. The difference of these mechanisms of influence is demonstrated by different dependences of this parameter of α-quartz crystal structure on the severity of exposure.

Acknowledgements. The work was performed within the Federal Target Program of Research and Development on Priority Development Fields of Science and Technology Sector in Russia for 2014–2020, unique project number: RFMEFI58317X0063. The authors would like to express gratitude to the Doctor of Geology and Mineralogy, Prof. N.V. Zubkova (MSU) for meaningful consultations.

References

Archipenko DK, Bokyi GB, Grigorieva TN, Koroleva SM, Yusupov TS (1987) On a new quartz phase stable at room temperature and found during tribo-processing (via x-ray diffraction). Reports of the USSR Academy of Sciences, vol 296, pp 1370–1374

Archipenko DK, Bokyi GB, Grigorieva TN, Koroleva SM, Yusupov TS, Shebanin AP (1990) Deformed quartz structures obtained after mechanoactivation. Reports of the USSR Academy of Sciences, vol 310, pp 874–877

Dmitrieva TV, Strokova VV, Bezrodnykh AA (2018) Influence of the genetic features of soils on the properties of soil-concretes on their basis. Constr Mater Prod 1:69–77

Glinnemann J, King HE, Schulz H, Hahn T, La Placa SJ, Dacol F (1992) Crystal structures of the low-temperature quartz-type phases of SiO2 and GeO2 at elevated pressure. Zeitschrift fur Kristallographie 198:177–212

Goryaynov SV, Ovsyuk NN (1999) Twisting of α-quartz tetrahedrons at pressure close to transition to an amorphous state. J Exp Theor Phys 69:431–435

Hazen RM, Finger LW, Hemley RJ, Mao HK (1989) High-pressure crystal chemistry and amorphization of alpha-quartz. Solid State Commun 72:507–511

Kihara K (1990) An X-ray study of the temperature dependence of the quartz structure. Eur J Mineral 2:63–77

Levien L, Prewitt CT, Weidner DJ (1980) Structure and elastic properties of quartz at pressure. Am Miner 65:920–930

Solovyov LA (2004) Full-profile refinement by derivative difference minimization. J Appl Crystallogr 37:743–749

Zhernovsky IV, Kozhukhova NI, Lebedev MS (2018) Crystallochemical aspects of technological typomorphism of quartz geomaterials during mechanoactivation. In: The collection of papers: fundamental and applied aspects of technological mineralogy. Under the editorship of Doctor of Geology and Mineralogy V.V. Shchiptsov. Karelian Research Center of the RAS, Petrozavodsk, pp 97–100

Biomimetic Materials on a Mineral Basis, Biomineralogy

Effect of Earthquake on the Landscape of Jiuzhaigou-Huanglong Travertine and Its Restoration

F. Dong[✉], Q. Dai, Q. Li, F. Wang, and Y. Luo

Key Laboratory of Solid Waste Treatment and Resource Recycle,
School of Environment and Resource, Southwest University of Science
and Technology, Mianyang, People's Republic of China
fqdong2004@163.com

Abstract. The hydrogeology and hydrochemistry-calcium characteristics, microbial community changes, and dam structure geophysical exploration for the core heritage sites of Jiuzhaigou after the "8.8" earthquake were investigated and analyzed. The results indicated that the hydrogeology has undergone significant changes, especially the Nuorilang Waterfall and the Sparkling Lake. The surface hydrodynamic balance has been broken by the earthquake, which may cause a chain reaction such as travertine deposition, travertine erosion, and microbial community change. The Nuorilang Waterfall/Sparkling Lake dam and the cascade waterfall are seriously affected by the earthquake. The results of the profile study and geophysical exploration show that the surface fissures, cracks, subsidence pipes, and other potential geological disasters are significant. The risk of secondary collapse of the dam body and the cascade waterfall is high. It also shows that the plant and microbes in Jiuzhaigou travertine participate in higher deposition than Huanglong. The monitoring results show that both Jiuzhai and Huanglong have extremely high levels of prokaryotic and eukaryotic algae, and Jiuzhaigou is higher than Huanglong. It is suggested that on the basis of scientific argumentation, scientific restoration of the slope and dam break of the Sparkling Lake damage will be carried out so that the original water storage function of the Spark Sea will be restored as soon as possible. Finally, the Nuorilang Waterfall and Sparkling Lake Conservation Remediation Plan were proposed and part of demonstration implementations was carried out.

Keywords: Jiuzhai earthquake · Travertine hydrogeology ·
Algae community · Hydrochemical characteristics · Geophysical exploration

S. Glagolev (Ed.): ICAM 2019, SPEES, pp. 413–414, 2019.
https://doi.org/10.1007/978-3-030-22974-0_100

Microbial Colonies in Renal Stones

A. Izatulina[1(⊠)], M. Zelenskaya[2], and O. Frank-Kamenetskaya[1]

[1] Department of Crystallography, St. Petersburg State University,
St. Petersburg, Russia
alina.izatulina@mail.ru
[2] Department of Botany, St. Petersburg State University, St. Petersburg, Russia

Abstract. The presence and study of the species composition of bacterial and fungal colonies in renal stones was determined. It was shown that the presence of microorganisms depends on the phase composition of the renal stone. No microbial colonies were detected in oxalate stones. Under the action of the acid-producing bacterial and fungal colonies, secondary crystallization of calcium oxalates (whewellite and weddellite) on phosphate aggregates can occur.

Keywords: Renal stones · Calcium oxalates · Crystallization · Phosphate renal stones · Microorganisms

1 Introduction

Interest in pathogenic crystallization is growing every year, which is primarily due to the wide prevalence of diseases associated with stone formation, such as urolithiasis. Very few works are devoted to the influence of bacteria, viruses and micromycetes on stone formation in the human body. Thus, Sagorika et al. (2013) described a patient with renal aspergillosis along with urolithiasis, Zhao et al. (2014) reported on the crystallization of calcium oxalate in the presence of *E. coli*. Other studies have noted the initiation of calcium oxalate crystallization and aggregation in the presence of *E. coli*. Most of the known works are devoted to the so-called infectious renal stones, consisting mainly of struvite, and sometimes containing hydroxylapatite and brushite.

2 Methods and Approaches

The study was conducted using 21 samples of renal stones of different composition: 10 – oxalate, 5 – urate, and 6 – phosphate stones.

The substance of renal stones was sieved on the Czapek-Dox medium, potato glucose agar and Saburo medium to detect the presence and to determine the species composition of microorganisms. Preparation of nutrient media was carried out in accordance with GOST (GOST 9.048-89). The media was sterilized in an autoclave after preparation. After incubation period (20 days, 1 month), the samples were microscoped and examined in accordance with identifier (de Hoog and Guarro 1995).

Powder x-ray diffraction (PXRD) studies were carried out using the Rigaku «MiniFlex II» diffractometer (CuKα radiation of wavelength λ = 1.54178 Å, X-ray

S. Glagolev (Ed.): ICAM 2019, SPEES, pp. 415–418, 2019.
https://doi.org/10.1007/978-3-030-22974-0_101

tube parameters were 30 kV/15 mA; highspeed solid state energy-dispersive detector LYNXEYE was used). X-ray diffraction patterns were collected at room temperature in the range of 2θ = 5–50° with a step of 0.02° 2θ and a counting time of half second per data point, the specimens were rotated 30 times per second during the data collection.

3   Results and Discussion

As the result of urate and oxalate renal stones sieving, the detection of micromycetes on the surface of the nutrient medium was random, and revealed a very small number of fungal colonies; the species of micromycetes were not constant. Abundant growth of fungal and bacterial colonies was detected on the phosphate renal stones: *Cladosporium cladosporioides, Penicillium expansum, Aspergillus niger*, sporiferous light colored fungus, *Geotrichum candidum, Candida sp., Fusarium chlamydosporum, Cladosporium sphaerospermum*, white and pink bacterial colonies (Fig. 1). Secretion of micromycetes on the surface of the nutrient medium was probably accidental, in many cases an insignificant number of fungal colonies was detected (1–3 colonies), the species distribution of micromycetes was not constant. There is a probability that some types of identified micromycetes (*Aspergillus niger, Cladosporium sphaerospermum, Phoma herbarum, Penicillium purpurogemum, Penicillium expansum, Fusarium chlamydosporum*) accidentally hit the test renal stone material (for instance, as a result of transportation or storage). The secretion of bacterial colonies and colonies of the fungus *Candida sp.*, is most likely not accidental, since a number of recent works indicate the possibility of a bacterial biofilm formation on the surface of renal (urinary) stones (Romanova et al. 2015). The possibility of micromycetes detection on the surface of renal stones within the microorganisms' biofilm requires further investigation.

Fig. 1. Phosphate renal stone sample. The growth of bacterial and micromycetes colonies on the surface of the nutrient medium in a Petri dish (∅ 100 mm).

A colony of *Aspergillus niger* micromycetes was found on a nutrient medium on one of the phosphate samples, consisting of hydroxylapatite, struvite and brushite. Aspergillus infections have grown in importance in the last years (Hedayati et al. 2007). Oxalate crystals may be present in clinical samples, due to the high acid-producing ability of this fungus. When *Aspergillus niger* is growing on a liquid nutrient medium, it was found that acidification of the culture fluid begins almost immediately after spore germination and continues during the whole period of mycelium active growth. After a month of incubation (Fig. 2), under the influence of Aspergillus niger culture, the renal stone softens and many small crystals are observed during microscopy. According to the results of PXRD analysis, the observed crystals turned out to be mainly calcium oxalate dihydrate (weddellite); crystals of calcium oxalate monohydrate (whewellite) are also present but in subordinate quantities. Thus, under the influence of the *Aspergillus niger* culture, secondary crystallization of calcium oxalate occurs on a phosphate renal stone. The secondary crystallization of calcium oxalates under the action of microscopic fungi was recently described for monuments of cultural heritage (Rusakov et al. 2016). This phenomenon may be one of the reasons for the frequent presence of a phosphate nucleus in the center of an oxalate renal stones (Izatulina and Yelnikov 2008; Xie et al. 2014).

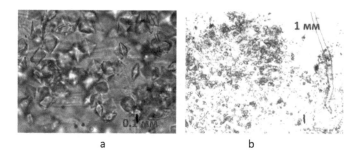

a   b

Fig. 2. The formation of calcium oxalate crystals in the presence of *Aspergillus niger* and a non-disrupting light-colored fungus during growth on a nutrient medium in a Petri dish (potato glucose agar was used as a nutrient medium).

4 Conclusions

Bacterial and fungal colonies were found on the surface of phosphate renal stones; no microbial colonies were found in oxalate and urate stones. It has been shown for the first time that under the influence of the microscopic fungus *Aspergillus niger*, secondary crystallization of calcium oxalates (whewellite and weddellite) can occur on phosphate aggregates. Thus, the possibility of the oxalate stones formation on phosphate nuclei with the participation of acid-producing bacterial and fungal colonies was shown.

Acknowledgements. This work was supported by the Russian Science Foundation (no. 18-77-00026). The XRD studies have been performed at the X-ray Diffraction Centre of St. Petersburg State University.

References

Bonaventura M, Gallo M, Cacchio P, Ercole C, Lepidi A (1999) Microbial formation of oxalate films on monument surfaces: bioprotection or biodeterioration? Geomicrobiology 16:55–64

De Hoog GS, Guarro J (1995) Atlas of clinical fungi, Baarn

Hedayati MT, Pasqualotto AC, Warn PA, Bowyer P, Denning DW (2007) Aspergillus flavus: human pathogen, allergen and mycotoxin. Microbiology 153:1677–1692

Izatulina AR, Yelnikov VY (2008) Structure, chemistry and crystallization conditions of calcium oxalates - the main components of kidney stones. In: Krivovichev SV (ed) Minerals as Advanced Materials I. Springer-Verlag, Heidelberg, pp 231–241

Romanova YuM, Mulabaev NS, Tolordava ER, Seregi AV, Seregin IV, Alexeeva NV, Stepanova TV, Levina GA, Barhatova OI, Gamova NA, Goncharova SA, Didenko LV, Rakovskaya IV (2015) Microbial communities on kidney stones. Molekulyarnaya Genetika, Mikrobiologiya i Virusologiya 33(2):20–25

Rusakov AV, Vlasov AD, Zelenskaya MS, Frank-Kamenetskaya OV (2016) The crystallization of calcium oxalate hydrates formed by interaction between microorganisms and minerals. In: Frank-Kamenetskaya OV, Panova EG, Vlasov DY (eds) Biogenic-Abiogenic Interactions in Natural and Anthropogenic Systems. Springer International Publishing, Switzerland, pp 357–377

Sagorika P, Viswajeet S, Satyanarayan S, Manish G (2013) Renal aspergillosis secondary to renal instrumentation in immunocompetent patient. BMJ Case Rep 2013:bcr2013200306

Xie B, Halter TJ, Borah BM, Nancollas GH (2014) Aggregation of calcium phosphate and oxalate phases in the formation of renal stones. Cryst Growth Des 15(6):3038–3045

Zhao Z, Xia Y, Xue J, Qingsheng Wu (2014) Role of E. coli-secretion and melamine in selective formation of $CaC_2O_4 \cdot H_2O$ and $CaC_2O_4 \cdot 2H_2O$ crystals. Cryst Growth Des 14:450–458

Fabrication of ZnO/Palygorskite Nanocomposites for Antibacterial Application

Y. Kang, A. Hui, and A. Wang$^{(\boxtimes)}$

Key Laboratory of Clay Mineral Applied Research of Gansu Province,
Center of Eco-Material and Green Chemistry, Lanzhou Institute
of Chemical Physics, Chinese Academy of Sciences, Lanzhou, China
aqwang@licp.cas.cn

Abstract. ZnO/palygorskite nanocomposites were synthesized by chemical deposition and calcination process for antibacterial application. The results indicated that ZnO nanoparticles were deposited on the surface of rod-like palygorskite. Antibacterial evaluation confirmed that ZnO/PAL nanocomposites presented the good antibacterial behavior against *Escherichia coli* and *Staphylococcus aureus*, which was mainly attributed to the synergistic effect of ZnO and palygorskite.

Keywords: ZnO · Palygorskite · Nanocomposites · Antibacterial application

1 Introduction

Widespread overuse of the antibiotics on animals has been faulted for creating potentially threat for human beings (Li et al. 2014; Mckenna 2013). Antibiotic-resistant bacterial strains emerge and pose increasing health risks, and thus new antibacterials are urgently needed (Morrison et al. 2014). Inspired by the antibacterial property of nanocomposites in the nature, functional application of natural clay minerals are of great interest in academia and industry (Williams et al. 2011). It is well-known that zinc oxide (ZnO) possesses excellent antibacterial properties against gram-positive bacteria and gram-negative bacteria (Hui et al. 2016; Liu et al. 2019). However, it is difficult to prevent from the aggregation of ZnO nanoparticles during preparation process, which results in the adverse effects of the antibacterial properties. Palygorskite (PAL) as a kind of clay minerals is common in many parts of the world, typically forming in volcanic ash layers as rocks. Interestingly, PAL has one-dimensional rod like morphology, high specific surface area and better ion-exchange capacity, which can be adsorbed onto the bacterial cell by electrostatic adsorption. Therefore, it seems to be a promising attempt to achieve a possible synergistic effect by combine PAL and ZnO after loading ZnO nanoparticles on the surface of PAL. Therefore, ZnO/palygorskite (ZnO/PAL) nanocomposites were synthesized by chemical deposition and calcination process in this study, and the antibacterial properties against *Escherichia coli (E. coli)* and *Staphylococcus aureus (S. aureus)* were also investigated.

S. Glagolev (Ed.): ICAM 2019, SPEES, pp. 419–422, 2019.
https://doi.org/10.1007/978-3-030-22974-0_102

2   Methods and Approaches

Fabrication of ZnO/PAL Nanocomposites: In a typical synthesis of ZnO/PAL nanocomposites, 2 g PAL and 20 wt% $Zn(NO_3)_2 \cdot 6H_2O$ were dissolved into deionized water, and then 10 wt% NaOH solution was added into above solution for 2 h. The mixture was ultrasonically dispersed for 30 min and aged for 24 h. The powder was collected by centrifugation and dried at 80 °C for 6 h, and finally annealed at 400 °C for 3 h in muffle furnace.

Characterization: The products were characterized by X-ray diffractometer (XRD, D/MAX-2200, Rigaku, Japan), transmission electron microscopy (TEM, JEM-1200EX, FEI, USA) and energy dispersive spectroscopy (EDS) elemental composition analyzer. The specific surface area of the samples was evaluated by Brunauer-Emmett-Teller analysis (BET, Micromeritics, Norcross, USA).

Antibacterial Evaluation: *E. coli* and *S. aureus* were tested as a representative culture both Gram-negative and Gram-positive bacteria, which kindly provided by China Veterinary Culture Collection Center. The antibacterial activity of the sample was evaluated by examining the minimum inhibitory concentration (MIC).

3   Results and Discussion

The XRD patterns of PAL and ZnO/PAL nanocomposites were shown in Fig. 1. The diffraction peaks at 2θ of 8.38°, 13.74°, 16.34° and 34.38° were the characteristic peaks of PAL (Wang et al. 2015). The diffraction peaks of the obtained ZnO/PAL were corresponded to a wurtzite ZnO structure (JCPDS standard card 36-1451) (Hui et al. 2016).

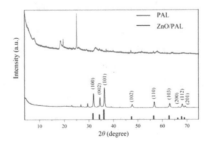

Fig. 1. XRD patterns of PAL and ZnO/PAL nanocomposites

The morphology of ZnO/PAL nanocomposites presented rod-like structure with a rough surface, and the length and the diameter were around $200 \sim 300$ nm and a 30 nm, respectively (Fig. 2a), which indicated that ZnO nanoparticles were successfully loaded onto the rod-like PAL surface to form a heterostructure. The specific surface area of PAL and ZnO/PAL was found to be $173 \ m^2 \cdot g^{-1}$ and $56 \ m^2 \cdot g^{-1}$, respectively. Compared with pure ZnO, this strategy could obviously improve the specific surface area of ZnO, which was favorable to enhance the antibacterial activity of ZnO/PAL nanocomposites. The ring-type selected area electron diffraction pattern indicated the generated ZnO possessed polycrystalline nature (Fig. 2b). What's more, EDS result showed the surface element compositions of the as-prepared nanocomposites were Mg, Al, Si, Fe and Zn (Fig. 2c).

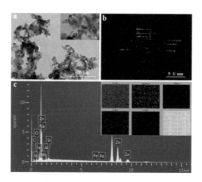

Fig. 2. (a) TEM image (inset is enlarged image), (b) selected area electron diffraction pattern and (c) EDS result of ZnO/PAL nanocomposites

As illustrated in Fig. 3, the microbial colonies of *E. coli* was visible when the concentration of sample was 0.25 mg·mL^{-1}, therefore, the MIC value of sample against *E. coli* was 0.5 mg·mL^{-1}. By contrast, there was small *S. aureus* colonies appeared in the plate when the sample contacts with *S. aureus* (Fig. 3i), and thus the MIC value of sample against *S. aureu* was 2.5 mg·mL^{-1}.

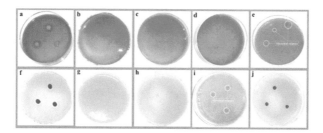

Fig. 3. (a, f) positive control of *E. coli* and *S. aureus*, *E. coli* treated by ZnO/PAL nanocomposites with various concentrations (b) 2.5 mg·mL^{-1}, (c) 1.5 mg·mL^{-1}, (d) 0.5 mg·mL^{-1}, (e) 0.25 mg·mL^{-1} and *S. aureus* (g) 5.0 mg·mL^{-1}, (h) 2.5 mg·mL^{-1}, (i) 1.0 mg·mL^{-1}, (j) 0.5 mg·mL^{-1}, respectively

In fact, PAL as a natural carrier with large special area, could absorb the bacterial cell by electrostatic adsorption. The reason for the synergistic antibacterial effect of ZnO/PAL nanocomposites was mainly due to the active factor of ZnO, which generated reactive oxygen species such as OH, O_2^1, O_2^{2-} and H_2O_2 (Hui et al. 2016; Liu et al. 2019; Ma et al. 2015). Incorporation of ZnO not only enhanced the antibacterial activity of natural PAL, but also reduced the cost of the preparation process, as well as efficiently realized the functional utilization of natural clay mineral resources.

4 Conclusions

In summary, ZnO/PAL nanocomposites were synthesized by chemical deposition and calcination process. The ZnO/PAL nanocomposites exhibited the excellent antibacterial activity against *E. coli* and *S. aureus*, and the MIC values for *E. coli* and *S. aureus* were 0.5 and 2.5 mg·mL^{-1}, respectively.

Acknowledgements. This work was financially supported by the Major Projects of the National Natural Science Foundation of Gansu, China (18JR4RA001).

References

Hui AP, Liu JL, Ma JZ (2016) Synthesis and morphology-dependent antimicrobial activity of cerium doped flower-shaped ZnO crystallites under visible light irradiation. Colloid Surf A: Physicochem Eng Aspects 506:519–525

Li XN, Robinson SM, Gupta A, Saha K, Jiang ZW, Moyano DF, Sahar A, Riley MA, Rotello VM (2014) Functional gold nanoparticles as potent antimicrobial agents against multi-drug-resistant bacteria. ACS Nano 8:10682–10686

Liu JL, Wang YH, Ma JZ, Peng Y, Wang AQ (2019) A review on bidirectional analogies between the photocatalysis and antibacterial properties of ZnO. J Alloy Compd 783:898–918

Ma JZ, Hui AP, Liu JL, Bao Y (2015) Controllable synthesis of highly efficient antimicrobial agent-Fe doped sea urchin-like ZnO nanoparticles. Mater Lett 158:420–423

Mckenna M (2013) Antibiotic resistance: the last resort. Nature 499:394–396

Morrison KD, Underwood JC, Metge DW, Eberl DD, Williams LB (2014) Mineralogical variables that control the antibacterial effectiveness of a natural clay deposit. Environ Geochem Health 36:613–631

Wang WB, Zhang ZF, Tian GY, Wang AQ (2015) From nanorods of palygorskite to nanosheets of smectite via a one-step hydrothermal process. RSC Adv 5:58107–58115

Williams LB, Metge DW, Eberl DD, Harvey RW, Turner AG, Prapaipong P, Poret-Peterson AT (2011) What makes a natural clay antibacterial? Environ Sci Technol 45:3768–3773

Bacterial Oxidation of Pyrite Surface

S. Lipko[1(✉)], I. Lipko[2], K. Arsent'ev[2], and V. Tauson[1]

[1] Vinogradov Institute of Geochemistry SB RAS, Irkutsk, Russia
slipko@yandex.ru
[2] Limnological Institute SB RAS, Irkutsk, Russia

Abstract. The article considers the study of the role of bacteria in the surface oxidation of pyrite. The experiment provided the data on characteristic morphological changes of the surface and the first data on influence of a non-autonomous phase (NP) on bacterial oxidation.

Keywords: Pyrite · Iron-oxidizing bacteria · SEM-EDAX · Non-autonomous phase · Surface

1 Introduction

For many years, extensive studies have been conducted on the processes of diagenetic redistribution of ore-forming components in the Earth's lithosphere and the formation of iron-containing nodules and ores, aimed at the their prospective industrial use. Most researchers believe that the earliest microbial ecosystems were based on sulfur transformations – sulfate-reduction and disproportionation (Wacey et al. 2011).

The choice of pyrite as the object of study is geochemically justified by the close connections of iron sulfides with organic matter in various environments, including hydrothermal conditions (Lindgren et al. 2011).

Although the chemistry of the processes has been studied in principle, there remains a number of unresolved issues. The most important are proof of paleobacterial processes and determination of their role in the formation of mineral deposits. Here the range of opinions is very wide: from complete denial to recognition of their leading character at the sedimentary-hydrothermal stage of ore formation (Vinichenko 2007). The morphological effects of the interaction of mineral surfaces with bacteria have not been sufficiently investigated, which complicates interpretation of natural observations. In particular, it is unclear what effect non-autonomous phases located within the surface layer of the crystal have on the interaction of bacterial communities with the pyrite surface (Tauson et al. 2008, 2009a). Non-autonomous phases (NP) are nanocrystalline objects formed in the surface layer of the crystal through interaction with its surface of the growth medium components or contacting autonomous (classical) phases. The experiment within the framework of present research used specially synthesized pyrite crystals with different degrees of NP development on the surface (Tauson and Lipko 2013), with the aim to study the process of interaction between bacteria and NP and to establish the role of surface phases in the oxidative processes initiated by the acidophilic iron bacteria.

© The Author(s) 2019
S. Glagolev (Ed.): ICAM 2019, SPEES, pp. 423–426, 2019.
https://doi.org/10.1007/978-3-030-22974-0_103

2 Methods and Approaches

The culture of acidophilic iron-oxidizing bacteria isolated from natural habitats (sulfide occurrences) of the Baikal area of the Irkutsk region was used for research on the bio-oxidation of the pyrite surface. This bacterial culture was provided by the laboratory № 7 of Irkutsk scientific-research Institute of rare and precious metals and diamonds, JSC "Irgiredmet". Iron-oxidizing bacteria are used in laboratory tests for bacterial oxidation of resistant iron-sulfide ores containing gold.

The synthesis of pyrite crystals was performed according to the standard technique of hydrothermal thermogradient synthesis in titanium inserts at T = 450 °C and 500 °C and a pressure of 1 kbar (Tauson et al. 2008). In the synthesis of pyrite Fe^+S charge was used, the composition of the surface non-autonomous phase was regulated by the activity of sulfur depending on Fe/S ratio. The obtained crystals were up to 5 mm in size. Pyrite, obtained at high sulfur activity, contains virtually no NP on the surface. At lower sulfur activity, a layer of NP up to ~ 500 nm thick with a base composition similar to pyrrhotite, but with different forms of sulfur, is formed: Fe^{2+} $[S, S_2, Sn]^{2-}$ (Tauson et al. 2008). These surface formations are able to absorb cationic impurities and oxysulfide anions.

To conduct research on pyrite bio-oxidation, a mixture of acidophilic iron bacteria was grown on a liquid 9 K medium at room temperature and constant stirring within 5 days. In eight 250 ml conical flasks with 50 ml of 9 K medium (without $FeSO_4$) there were placed pieces of polished pyrite (4 flasks) and 5–6 pieces of pyrite with nonautonomous phases (4 flasks). The medium and pyrite flasks were sterilized at 0.5 atm. for 10 min to minimize pyrite oxidation. After cooling the medium, 6 flasks were inoculated with 5-day bacterial culture. Previously, the culture was centrifuged and washed from the medium residues with iron in 0.01 m H_2SO_4. The concentration of iron bacteria cells added to the medium with pyrite was about $1*10^7$ cells/ml. The remaining two flasks with polished pyrite and NP on the pyrite surface were used as a control, without bacteria. Cultivation took place at room temperature and constant stirring on a shaker (about 110 rotation/min) for three weeks. Every week 2 flasks with different samples of pyrite were selected for further research. The flasks with control samples were examined after 3 weeks of cultivation. The bacterial film from the pyrite samples were washed with 2% aqueous solution of Polysorbate Tween 80. Pyrite crystals washed after the experiment were dried in air and analyzed on the scanning multi-microscope SMM 2000 in atomic force mode, scanning electron microscope FEI Quanta Company (USA) 200 with energy dispersive device EDAX for X-ray microanalysis.

3 Results and Discussion

The experiment on pyrite bio-oxidation established that the surface of polished pyrite is less susceptible to bacterial oxidation as compared with NP-containing pyrite. Microphotographs show the surfaces of polished pyrite (roughness less than 5 nm) and NP-containing pyrite (roughness more than 300 nm) after two weeks of bacterial cultivation (Fig. 1).

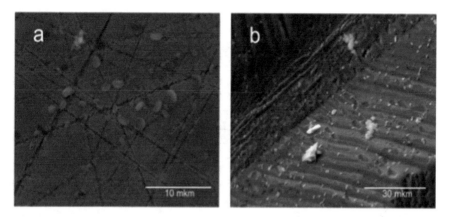

Fig. 1. Surface morphology of pyrite (scanning electron microscopy). a – polished surface, b – NP-containing surface.

The surface with NP exhibits bacteria and characteristic traces of interaction between bacteria and pyrite in the form of holes of different size comparable to the sizes of bacteria. For polished pyrite, these traces are almost absent and are observed only on the borders of scratches left from polishing. A similar result was obtained for the pyrite surface with minimal NP development synthesized at high sulfur activity. Therefore, the activity of bacteria is associated with the structure of pyrite surface. Similar formations, but significantly smaller (nano-holes) were discovered earlier, in the study of pyrites from the Sukhoi Log gold deposit (Irkutsk region) (Tauson et al. 2009b). This confirms the affinity of processes occurring in nature and in the experiment.

4   Conclusions

The research resulted in acquisition of data on pyrite bio-oxidation taking into account the structure of crystal surface under the given conditions. For this purpose, crystals with different degrees of non-autonomous phases development on the surface controlled by growth conditions were synthesized and used in the experiment for the first time. It was found that the surface of polished pyrite is less susceptible to bacterial oxidation, as compared with pyrite containing a non-autonomous phase. The resulting characteristic morphological changes in the surface will further be instrumental in addressing the issues of ore genesis, as well as identifying minerals that were formed at the initial or final stage of growth involving bacteria.

Acknowledgements. We thank Alexandra Mikhailova for suppling of bacterial culture. The research was performed within a state assignment, Project IX.125.3, No. 0350-2016-0025 and was funded by the Federal Agency for Scientific Organizations (FASO) within the framework of State Tasks No. 0345-2016-0003 (AAAAA16- 116122110061-6).

References

Lindgren P, Parnell J, Holm NG, Broman C (2011) A demonstration of the affinity between pyrite and organic matter in a hydrothermal setting. J Geochem Trans 12(3):3–7

Tauson VL, Lipko SV (2013) Pyrite as a concentrator of gold in laboratory and natural systems: a surface-related effect. In: Whitley N, Vinsen PT (eds) Pyrite: Synthesis, Characterization and Uses Chapter 1. Nova Science Publisher Inc., New York, pp 1–40

Tauson VL, Babkin DN, Lustenberg EE, Lipko SV, Parkhomenko IY (2008) Surface typochemistry of hydrothermal pyrite: electron spectroscopic and scanning probe microscopic data I. Synthetic pyrite. J Geochem Int 46(6):615–628

Tauson VL, Kravtsova RG, Grebenshchikova VI, Lustenberg EE, Lipko SV (2009a) Surface typochemistry of hydrothermal pyrite: electron spectroscopic and scanning probe microscopic data II. Natural pyrite. J Geochem Int 47(3):245–258

Tauson VL, Lipko SV, Shchegolkov YuV (2009b) Surface nanoscale relief of mineral crystals and its relation to non-autonomous phase formation. J Crystallogr Rep 54(7):1219–1227

Vinichenko PV (2007) Biogeology and Ore Formation. Izd-e Sosnovgeologiya, Irkutsk

Wacey D, Saunders M, Brasier MD, Kilburn MR (2011) Earliest microbially mediated pyrite oxidation in 3.4 billion-year-old sediments. J Earth Planet Sci Lett 301:393–402

Biomimetic Superhydrophobic Cobalt Blue/Clay Mineral Hybrid Pigments with Self-cleaning Property and Different Colors

B. Mu[1], A. Zhang[1,2], and A. Wang[1(✉)]

[1] Key Laboratory of Clay Mineral Applied Research of Gansu Province,
Center of Eco-material and Green Chemistry,
Lanzhou Institute of Chemical Physics, Chinese Academy of Sciences,
Lanzhou 730000, People's Republic of China
aqwang@licp.cas.cn

[2] Center of Materials Science and Optoelectronics Engineering,
University of Chinese Academy of Sciences, Beijing 100049,
People's Republic of China

Abstract. Inspired by self-cleaning and water-repellent properties of the lotus leaf, biomimetic superhydrophobic cobalt blue/clay mineral hybrid pigments were facilely fabricated based on the rough surface of hybrid pigments and the modification with various organosilanes. The obtained hybrid pigments were characterized using various analytical techniques. Due to the difference in the compositions and morphologies of clay minerals, the obtained cobalt blue/clay mineral hybrid pigments exhibited different color properties. Superhydrophobicity of hybrid pigments was mainly regulated by the types of organosilanes instead of the morphologies of hybrid pigments. The sprayed coating of the superhydrophobic hybrid pigments exhibited the excellent self-cleaning performance with high water contact angle and low sliding angle. The coatings also presented excellent environmental and chemical durability even under harsh conditions. Therefore, the obtained biomimetic superhydrophobic cobalt blue/clay mineral hybrid pigments may be applied in various fields, such as anticorrosion, self-cleaning coating, etc.

Keywords: Biomimetic · Superhydrophobic · Cobalt blue · Clay minerals · Self-cleaning

1 Introduction

Cobalt blue (cobalt aluminate, $CoAl_2O_4$) pigment is a typical eco-friendly blue inorganic pigment. Due to high refractive index, excellent chemical and thermal stability, it can be widely applied in ceramics, paints, engineering plastics, etc. However, the high cost of $CoAl_2O_4$ pigment severely restrains their relevant applications due to the high price of cobalt compounds and the disadvantages of the traditional solid phase method

© The Author(s) 2019
S. Glagolev (Ed.): ICAM 2019, SPEES, pp. 427–431, 2019.
https://doi.org/10.1007/978-3-030-22974-0_104

(Armijo 1969). In addition, most of the common application fields of $CoAl_2O_4$ pigment are hydrophilic, which may be easily contaminated by dark liquid, oil stains, blot in our daily life. Therefore, it is very necessary to develop low-cost $CoAl_2O_4$ pigment with and self-cleaning ability.

Inspired by the unique water-repellent surfaces of the lotus leaf in the natural world (Barthlott and Neinhuis 1997), the design of superhydrophobic surfaces has become the focus in both fundamental research and industrial applications by construction of rough surface structure and modification using materials with low surface free energy. Recently, substrate-based inorganic hybrid pigments composed of an inorganic substrate coated with inorganic pigment nanoparticles have attracted increasingly attention. Our groups successfully prepared $CoAl_2O_4$ hybrid pigment after incorporation of different clay minerals (Mu et al. 2015; Zhang et al. 2017). Incorporation of clay minerals greatly decreased the production cost and calcining temperature for formation of spinel $CoAl_2O_4$, as well as preventing from the aggregation of $CoAl_2O_4$ nanoparticles after being uniformly anchored on the surface of clay minerals. Based on the rough surface of the hybrid pigments, it could realize the superhydrophobic modification of hybrid pigments using materials with low surface free energy. In this study, different $CoAl_2O_4$ hybrid pigments derived from kaoline (Kaol), palygorskite (Pal), halloysite (Hal) and montmorillonite (Mt) were prepared and modified using different organosilanes including octyl triethoxysilane (OTES), dodecyl trimethoxysilane (DTMS), hexadecyltriethoxysilane (HTES) and perfluoroctyl trimethoxysilane (PFOTMS), and the effect of the types organosilanes and the morphologies of hybrid pigments on the superhydrophobic properties was comparatively studied.

2   Methods and Approaches

Cobalt blue/hybrid pigments were prepared according to the similar procedure reported in our previous study (Zhang et al. 2017). Typically, 2.91 g of $Co(NO_3)_2 \cdot 6H_2O$, 7.50 g of $Al(NO_3)_3 \cdot 9H_2O$ and 1.09 g of clay minerals were added to 50 mL of water under magnetic stirring at 150 rpm for 30 min, and then 3 M NaOH was added dropwise into above mixture until the pH was reached to 10. The suspension was continuously stirred for 2 h at room temperature, and then the solid products were collected by centrifugation, washed with water and directly calcined at 1100 °C for 2 h with a rate of 10 °C/min. Next, the superhydrophobic modification of hybrid pigments was conducted in the ammonia saturated ethanol solution (Zhang et al. 2018). 0.54 g of organosilanes and 1.5 g of hybrid pigments with different weight ratios were firstly added into 45 mL above ethanol solution and stirred for 1 h at room temperature, and then 4.00 g of water was injected quickly into the solution and stirred for 24 h at room temperature. The solid products were finally washed with ethanol for three times and dried in an oven at 60 °C. The obtained samples were labeled as clay mineral-HP-organosilanes according to the involved clay minerals and organosilanes.

3   Results and Discussion

Table 1. Color parameters, water contact angle (CA) and sliding angle (SA) of superhydrophobic hybrid pigments before and after being treated at various conditions

Samples	Conditions	L^*	a^*	b^*	CA/°	SA/°
Kal-HP-HTES	-	38.1	2.6	−63.8	164.2	1.0
	98% H_2SO_4	37.4	2.6	−63.6	163.8	1.2
	3 M NaOH	38.1	2.6	−63.8	163.3	1.3
	UV for 3 days	38.0	2.6	−63.9	164.0	1.1
Pal-HP-HTES	-	16.2	−21.1	−23.5	148.1	2.4
	98% H_2SO_4	15.8	−20.8	−26.2	148.2	2.3
	3 M NaOH	16.7	−21.3	−24.8	147.5	2.4
	UV for 3 days	16.9	−22.2	−23.3	148.3	2.4
Hal-HP-HTES	-	55.8	−2.3	−55.0	151.2	1.2
	98% H_2SO_4	55.6	−2.4	−54.6	151.6	1.5
	3 M NaOH	55.8	−2.8	−53.7	151.8	1.8
	UV for 3 days	55.2	−2.6	−54.2	150.8	1.2
Mt-HP-HTES	-	24.5	−21.2	−32.3	156.4	3.7
	98% H_2SO_4	23.2	−21.3	−33.2	155.6	3.4
	3 M NaOH	24.6	−21.4	−32.7	157.5	3.1
	UV for 3 days	21.2	−20.4	−30.5	154.8	3.3
Kal-HP-OTES	-	38.1	2.5	−63.8	135.1	4.0
Kal-HP-DTES	-	38.1	2.6	−63.8	151.2	3.0
Kal-HP-PFOTMS	-	38.0	2.5	−63.7	165.2	2.0

The color parameters, CA and SA of superhydrophobic hybrid pigments before and after being treated at various conditions were summarized in Table 1. It was found that superhydrophobic hybrid pigments derived from different clay minerals presented different color properties (Fig. 1), which might be attributed to the difference in the compositions of clay minerals, especially Fe element. Among the employed clay minerals, the higher content of Fe element was observed in Pal and Mt than Kal and Hal, which decreased the color parameters of hybrid pigments. By contrast, hybrid pigment prepared using Hal exhibited the optimum color properties, and hybrid pigments obtained from Kal came second. In addition, the water contact angle and sliding angle of different hybrid pigments derived from different clay minerals had no obvious difference after being modified using HTES (Fig. 1). Except for Pal-HP-HTES, the values of CA and SA of them were higher than 150° and below 5°, respectively. Although the hybrid pigments were treated at various conditions, the color properties, CA and SA almost kept stable, indicating the excellent environmental and chemical durability. Furthermore, superhydrophobic Kal-HP modified with various organosilanes presented different CA and SA. With the increase in the carbon chain length, the values of SA increased while the SA values decreased. Meanwhile, the incorporation of fluorine atom in organosilanes also favored enhancing the superhydrophobic properties of hybrid pigments.

Fig. 1. Digital photos of (a) Hal-HP-HTES, (b) Mt-HP-HTES, (c) APT-HP-HTES and (d) Kal-HP-HTES

4   Conclusions

Superhydrophobic cobalt blue/clay mineral hybrid pigments with self-cleaning property and different colors were successfully prepared by modifying using organosilanes based on the rough surface. The color and superhydrophobicity were closely related to the compositions of clay minerals and the types of organosilanes.

Acknowledgements. The authors are grateful for financial support of the Major Projects of the National Natural Science Foundation of Gansu, China (18JR4RA001), and the Youth Innovation Promotion Association of CAS (2017458).

References

Armijo JS (1969) The kinetics and mechanism of solid-state spinel formation—a review and critique. Oxid Met 1:171–198

Barthlott W, Neinhuis C (1997) Purity of the sacred lotus, or escape from contamination in biological surfaces. Planta 202:1–8

Mu B, Wang Q, Wang AQ (2015) Effect of different clay minerals and calcination temperature on the morphology and color of clay/CoAl$_2$O$_4$ hybrid pigments. RSC Adv 5:102674–102681

Zhang AJ, Mu B, Luo ZH, Wang AQ (2017) Bright blue halloysite/CoAl$_2$O$_4$ hybrid pigments: preparation, characterization and application in water-based painting. Dyes Pigm 139:473–481

Zhang AJ, Mu B, Hui AP, Wang AQ (2018) A facile approach to fabricate bright blue heat-resisting paint with self-cleaning ability based on CoAl$_2$O$_4$/kaoline hybrid pigment. Appl Clay Sci 160:153–161

Silicon Dioxide in Mineralized Heart Valves

A. Titov[1,2(✉)], V. Zaikovskii[1], and P. M. Larionov[1,3]

[1] National Research University, Novosibirsk, Russia
titov@igm.nsc.ru
[2] Sobolev V.S. Institute of Geology and Mineralogy of SB RAS,
Novosibirsk, Russia
[3] Boreskov Institute of Catalysis of SB RAS, Novosibirsk, Russia

Abstract. This study showed that silicon dioxide of plant origin penetrated into the human body unchanged and was transferred through the blood to the heart, where absorbed by the pathological hydroxyapatite of mineralized heart valves.

Keywords: Silicon dioxide · Scanning and transmission electron microscopy · Mineralized heart valves · Bioavailability

1 Introduction

Over the past decades, numerous studies have shown that Si is an essential element and it affects human health. Silicon is present in the body as a trace element, but so far its biochemical function has not been confirmed by experimental data. Bioavailable silicon usually enters the body from solutions. Most easily it diffuses through the membranes and penetrates into the circulatory system in the form of orthosilicate acid, which is present in water, beer and some beverages. Among food products, the most significant sources of silicon are the products of plant origin: wheat, rice, oats, and barley. Despite the fact that vegetable food has a high content of silicon, its bioavailability is very limited due to the poor solubility of the forms of silicon present in the plants (Farooq 2015).

In this paper we present amorphous silica found in the composition of pathological formations - calcified heart valves. To explain the possible route of ingestion of amorphous silica, we considered one of the most common food crops for silicon consumption - rice.

2 Methods and Approaches

We studied intraoperative material, which included mineralized aortic and mitral valves of the heart, obtained from patients with acquired heart defects of rheumatic and septic genesis (Titov et al. 2016). Siliceous formations found in the calcifications of the heart valves were compared with the siliceous formations of plant origin: rice straw and husk.

The inventory of methods for structural and elemental analyses included high-resolution transmission electron microscopy (HR TEM), scanning electron microscopy (SEM), electron diffraction, and energy-dispersive X-ray spectroscopy (EDX). Electron microscopy was performed using a JEM2010 transmission electron microscope

S. Glagolev (Ed.): ICAM 2019, SPEES, pp. 432–435, 2019.
https://doi.org/10.1007/978-3-030-22974-0_105

(acceleration voltage 200 kV, resolution 1.4 Å) equipped with an EDAX EDS detector (spectral resolution 130 eV) and a TESCAN MIRA3 scanning electron microscope with an Oxford EDS detector (resolution 128 eV) and built-in INKA ENERGY software.

3 Results and Discussion

Our investigation relied on the data obtained in the studies of the calcified formations on heart valves and their bioprostheses (Titov et al. 2016). The presence of silicon at the trace level was observed in the EDS analyzes of calcified heart valves. The studies of the dispersed material of heart valve calcifications by means of transmission electron microscopy revealed amorphous particles of rounded shape about 100 nm in diameter, along with hydroxyapatite nanocrystals (Fig. 1). The EDS spectra of these amorphous particles revealed Si, O, and C (carbon deposition), which corresponded to silicon dioxide (Fig. 1).

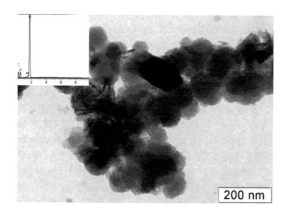

Fig. 1. Electron microscopic image (TEM) of amorphous silica particles among the dispersed calcified mineralized substance of a heart valve. Insert: EDX spectrum from one of the particles of silicon dioxide

As can be seen on the SEM images and EDS spectra of rice straw and husk (Fig. 2.), their surfaces were covered with a thin layer of silicon dioxide. After annealing the rice straw and husk at a temperature of 750 °C, white silica powder remained. The examination of the powder preparation by means of transmission electron microscopy showed that the rice substrate was represented by the rounded particles of amorphous silica with a diameter of about 50 nm (Fig. 3). The similarity of the amorphous particles of the rice substrate with the amorphous particles from heart valve calcifications in chemical composition and structure seemed quite obvious to us.

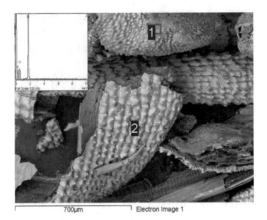

Fig. 2. SEM morphology of rice straw and husk. On the insert is the EDX spectrum from area 2 in the image.

Plants accumulate Si in the form of biogenic (phytolytic) silicon dioxide from soil solutions. It becomes included in plant tissues as a structural component imparting strength and rigidity to the stems. The main route of silicon intake into the organism starts from the gastrointestinal tract. Phytolytic silica is considered to be an insoluble form of Si.

However, most of Si is absorbed from solid products, therefore it is assumed that the phytolytic silicon dioxide is destroyed and absorbed (Jugdaohsingh 2007). Silicon dioxide detected by us in calcifications with nanocrystalline hydroxyapatite, remained unchanged. Apparently silica particles were transferred to the heart valves from the patients' blood and precipitated on hydroxyapatite. Hydroxyapatite is known to have a high sorption capacity (Titov et al. 2013). Perhaps a significant amount of silica can be deposited on the bone tissue as hydroxyapatite is one of its components. Our analysis of the chemical composition of calcified mineralized heart valves using an EDS spectrometer detected silicon at a trace level. The Si content in calcifications is quite significant for this trace element considering the sensitivity level of this method as 0.n weight percent.

500 nm   50 nm

Fig. 3. A and B.Electron microscopic images (TEM) of cytolytic silica obtained from rice straw and husk after annealing at a temperature of 750 °C. B – the image of a fragment of A.

4   Conclusions

To conclude, we suggest that dispersed phytolytic silica may penetrate through the gastric tract into human blood in the unchanged form.

Acknowledgements. This research was carried out within the State Assignment to IGM SB RAS (project 0330-2016-0013).

References

Farooq MA, Dietz K-J (2015) Silicon as versatile player in plant and human biology: overlooked and poorly understood. Front Plant Sci 6:994–1023

Titov AT, Zaikovskii VI, Larionov PM (2016) Bone-like hydroxyapatite formation in human blood. Int J Environ Sci Educ 11(10):3971–3984

Jugdaohsingh R (2007) Silicon and bone health. J Nutr Health Aging 11(2):99–110

Titov AT, Larionov PM, Zaikovskii I (2013) Calcium phosphate mineralization of bacteria. In: Proceedings of the 11th International Congress For Applied Mineralogy (ICAM), pp 9–17

Preparation of Macroporous Adsorbent Based on Montmorillonite Stabilized Pickering Medium Internal Phase Emulsions

F. Wang[1,2], Y. Zhu[1], W. Wang[1], and A. Wang[1(✉)]

[1] Key Laboratory of Clay Mineral Applied Research of Gansu Province,
Center of Eco-Material and Green Chemistry,
Lanzhou Institute of Chemical Physics, Chinese Academy of Sciences,
Lanzhou, People's Republic of China
aqwang@licp.cas.cn
[2] College of Petroleum and Chemical Engineering, Qinzhou University,
Qinzhou, People's Republic of China

Abstract. A macroporous material was prepared using oil-in-water Pickering medium internal phase emulsions (Pickering MIPEs) as template. The obtained macroporous materials with interconnected pore structure exhibited good adsorption capacities to Ce (III) and Gd (III) in water. The adsorption process could be achieved in 30 min, and the maximum adsorption capacities reached 230.64 mg/g for Ce (III) and 240.49 mg/g for Gd (III). Furthermore, the macroporous monolith exhibited excellent reuseability after consecutive adsorption-desorption cycles.

Keywords: Pickering emulsion · Medium internal phase emulsions · Rare metal · Adsorption · Porous material

1 Introduction

Compared with the conventional emulsion, Pickering emulsions exhibit peculiar long-term stability against droplet coalescence, in which the solid particles adsorbed at the oil-water interface act as a mechanical barrier to protect the dispersion phase liquid drops from the coalescence with the continuous phase (Tajik et al. 2017). Due to the unique properties, Pickering emulsions have been used as template to prepare porous polymer monoliths (Briggs et al. 2015). However, the dispersion phase of oil/water Pickering high internal phase emulsions mostly need a large amount of organic solvent, and thus it is indispensable to develop Pickering MIPEs by replacing poisonous organic solvent with low-cost and eco-friendly plant oil and reducing the internal phase volume.

Herein, monolithic macroporous materials were fabricated for adsorption of Ce (III) and Gd (III) based on Pickering MIPEs template, which was composed of the stabilizer of montmorillonite (Mt) and Tween-20 and the continuous phase of flaxseed oil. The effects of adsorption parameters including initial concentration and contact time on the adsorption properties were investigated, and the reusable performance of the adsorbent was also evaluated.

© The Author(s) 2019
S. Glagolev (Ed.): ICAM 2019, SPEES, pp. 436–439, 2019.
https://doi.org/10.1007/978-3-030-22974-0_106

2 Methods and Approaches

Typically, macroporous carboxymethyl cellulose-g-poly(acrylamide)/montmorillonite (CMC-g-PAM/MMT) monolith was prepared based on Pickering MIPEs, which stabilized with 5% of Mt and 4% of Tween-20 (Wang et al. 2017). The obtained monolithic polymers were washed with acetone for 12 h and then immersed into 0.5 M NaOH aqueous alcohol solution ($V_{water}/V_{alcohol}$ = 3/7) for 24 h to transfer the amide group to carboxyl.

The effect of the adsorption time and the initial concentration on the adsorption capacities were conducted according to the following procedure: 20 mg porous adsorbents were added into 25 mL Ce (III) and Gd (III) solution and shocked in a thermostatic shaker at 120 rpm and 30 °C for a given time. After the adsorption, the adsorbents were separated and the concentrations of the Ce (III) and Gd (III) were determined via UV-vis spectrophotometer using the chlorophosphonazo and azo arsine as the complexing agents, respectively. The adsorption capacities qe (mg/g) of the porous monolithic adsorbents were calculated according to the following equation:

$$q_e = \frac{(C_0 - C_e)V}{m} \tag{1}$$

where C_0 and Ce were the initial and equilibrium concentrations of Ce (III) and Gd (III) (mg/L), V (L) was the volume of the Ce(III) and Gd(III) solution.

The reusability studies were performed as follows: The adsorbents were desorbed by immersing 30 mL hydrochloric acid solution (0.5 M) for 2 h after adsorption, and then regenerated with 0.5 M NaOH solution. Finally, the adsorbents were filtered and washed to reach neutral using distilled water before next adsorption process. The adsorption-desorption cycle was repeated five times.

3 Results and Discussion

The representative images of macroporous monolith prepared by Pickering-MIPEs with 5% Mt and 4% Tween-20 as the stabilizer were shown in Fig. 1. The emulsions didn't flow in the inverted plastic centrifuge tube (Fig. 1a), indicating that the oil droplets were closely packed and the formed emulsion was a typical gel emulsion. The macroporous polymer monoliths of CMC-g-PAM/MMT were synthesized by free radical polymerization using APS as the initiator. The prepared wet monoliths were cut into pieces and Soxhlet extracted using acetone to remove the oil phase and surfactant, and then immerged into NaOH alcohol solution to complete hydrolysis of amide groups. Finally, the white monoliths of CMC-g-PAM/MMT were obtained after the dehydrated with acetone and dried in oven at 40 °C (Fig. 1b). The surface morphology of CMC-g-PAM/MMT was shown in Fig. 1c, and it presented a hierarchical pore structure with high connectivity. According to the statistical result of Image-Pro Plus 6.0 software, the average pore size of the macropore and the pore throat were 1.43 μm and 0.39 μm, respectively. Furthermore, the as-prepared porous materials exhibited narrow macropores and pore throats size distribution (Fig. 1d).

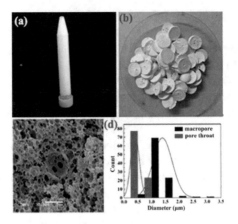

Fig. 1. Digital photographs of (a) the as-prepared Pickering MIPEs, (b) CMC-g-PAM/MMT monolith, (c) SEM image of porous CMC-g-PAM/MMT monolith and (d) pore size distribution of porous CMC-g-PAM/MMT

CMC-g-PAM/MMT monoliths were employed to remove of Ce (III) and Gd (III) from water. As shown in Fig. 2a and b, the adsorption capacities increased with the increase in the initial metal ions concentrations until the adsorption saturation was reached. The maximum adsorption capacities of the macroporous monoliths were 230.64 mg/g for Ce (III) and 240.39 mg/g for Gd (III). The higher adsorption capacity might be due to the sufficient functional groups and the highly interconnected pore structure. The effect of contact time of on the adsorption behavior was depicted in Fig. 2c and d. It was obvious that the porous monolithic adsorbent showed fast adsorption rate for Ce (III) and Gd (III), and the adsorption equilibrium could be reached within 30 min and 25 min Ce (III) and Gd (III), respectively. The macro-pores allowed fast and efficient mass transport, as well as provided sufficient contact between active groups and adsorbents, while the pore throats contributed to a high specific surface area.

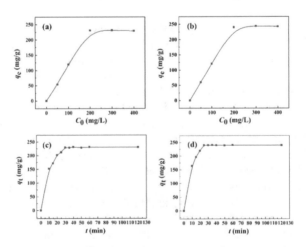

Fig. 2. Effect of the initial concentration of (a) Ce (III) and (b) Gd (III) on the adsorption capacity of porous monolith. Adsorption kinetic curves of the porous monolith for (c) Ce (III) and (d) Gd (III).

4 Conclusions

Macroporous polymer monoliths of CMC-g-PAM/MMT were successfully synthesized by free radical polymerization based on based on Pickering MIPEs stabilized with 5% of Mt and 4% of Tween-20. The as-prepared macroporous polymer monoliths possessed a hierarchical pore structure and highly interconnection, which favored enhancing the adsorption properties to Ce (III) and Gd (III), such as high adsorption capacity, quick adsorption rate, and good reusability.

Acknowledgements. The authors are grateful for financial support of the Major Projects of the National Natural Science Foundation of Gansu, China (18JR4RA001) and the National Natural Science Foundation of China (21706267).

References

Tajik S, Nasernejad B (2017) Surface modification of silica-graphene nanohybrid as a novel stabilizer for oil-water emulsion. Korean J Chem Eng 34:2488–2497

Briggs NM, Weston JS, Li B, Venkataramani D, Aichele CP, Harwell JH, Crossley SP (2015) Multiwalled carbon nanotubes at the interface of Pickering emulsions. Langmuir 31:13077–13084

Wang F, Zhu YF, Wang WB, Zong L, Lu TT, Wang AQ (2017) Fabrication of CMC-g-PAM/Pal superporous polymer monoliths via eco-friendly Pickering-MIPEs for superior adsorption of methyl violet and methylene blue. Front Chem 5:33

Environment and Energy Resources

Depletion of the Land Resources and Its Effect on the Environment

M. Abou Zahr Diaz, M. A. Alawiyeh, and M. Ghaboura$^{(\boxtimes)}$

Department of Mineral Developing and Oil & Gas Engineering,
Engineering Academy, RUDN University, Moscow, Russia
mostafa_ghab@live.com

Abstract. Resources depletion refers to the situation where the consumption of natural resources is faster than it can be replenished. In order to achieve economic growth, developing countries are abusing their lands on the grounds of economic interests. Population Explosion is acting as a catalyst for resources depletion. It seems evident that developing countries pursuing rapid economic growth disregard environmental concerns. The natural resources contribute at large to the economic development of a nation. Consumption pattern if not addressed will lead to irreversible climate change and declined economic growth, as a result of increased social, economic, and environmental costs and decreased productivity. Resource utilization has always been part of human history; however, the acceleration of economic growth activities together with the pursuit of an urgent economic development is the core cause of resources overexploitation. Consumption pattern will lead to irreversible climate change and declined economic growth.

Keywords: Resources · Economics · Population · Utilization ·
Depletion of the soil

1 Introduction

In addition to the problems of each civilization, humanity is faced with the urgent need to solve planetary problems. At the end of XX century, the first terrible signs of the deterioration of the quality of the biosphere had already appeared as a result of the development of man-made civilization and the installation to conquer nature started.

Smog over large cities, deforestation and the onset of deserts, depletion of the soil and basins of many rivers, a decrease in the number of fish and wild animals – all this worried people at the beginning of the twentieth century.

2 Methods and Approaches

A no less formidable problem is the ecological catastrophe approaching the planet. At present, mankind produces organic waste in an amount of two thousand times more than the waste volumes of the rest of the biosphere. Obviously, the violation of this equilibrium caused a whole complex of complex problems. Man, unlike all living things, is not strictly bound by the environmentally friendly conditions of his being, in

S. Glagolev (Ed.): ICAM 2019, SPEES, pp. 443–444, 2019.
https://doi.org/10.1007/978-3-030-22974-0_107

a certain sense he was always going against nature, not adapting to it, but changing it in accordance with its needs (Satterthwaite 2009).

3   Results and Discussion

The demographic problem has become global long ago. In 1987, the five billionth inhabitant of the planet was born, and the growth rate is now such that every second the number of people on Earth is increased by three people (Anderson 2012). According to the figurative expression of scientists, the Earth is now "biting man" and it is quite natural to expect a demographic collapse in the near future, that is, a sufficiently sharp decline in population.

It can be caused by global hunger, depletion of mineral resources and soil, poor drinking water, thermal overheating of the surface, etc.

4   Conclusions

The current generation cannot help thinking about future children and grandchildren, who are to continue to carry the baton of history. Unfortunately, our civilization largely lives at the expense of the future, exhausting irreplaceable resources (oil, gas), polluting water, air and soil with its imperfect technologies, preserving many archaic social structures, sowing seeds of national and religious hatred that will sprout another century.

Based on the above, it can be concluded that only the efforts of the entire world community can prevent an environmental catastrophe that threatens all life on Earth.

References

Anderson R (2012) Resource depletion: opportunity or looming catastrophe? https://www.bbc.com/news/business-16391040
Satterthwaite D (2009) The implications of population growth and urbanization for climate change. Environ Urbanization 21(2):545–567

Geochemical Behavior of Heavy Metals During Treatment by Phosphoric Fertilizer at a Dumping Site in Kabwe, Zambia

H. Kamegamori[1(✉)], K. Lawrence[1], T. Sato[2], and T. Otake[2]

[1] Graduate School of Engineering, Hokkaido University, Sapporo, Japan
skj-9mm@eis.hokudai.ac.jp
[2] Faculty of Engineering, Hokkaido University, Sapporo, Japan

Abstract. Kabwe area in Zambia has been affected by heavy metal contaminations which derived from past mining activities. Particularly, Pb is one of the most concerned elements for human health in Kabwe. In this context, treatment by phosphoric fertilizer was conducted to reduce Pb solubility in soil and slag, limiting their bioavailability. Because leach plant residue in Kabwe contains metal sulfate minerals with high solubility, concentration of heavy metals in groundwater is high. We clarified the geochemical behavior of heavy metals (Pb, Cd, Zn and Cu) after the addition of phosphoric fertilizer (Triple Super Phosphate: TSP) in column experiment. Immobilization of Pb and Cd lowers concentration of the metals in ground below WHO environmental standard.

Keywords: Insolubilization · Soil amendments · Heavy metal contamination · Phosphate mineral · Mine waste

1 Introduction

Kabwe town is the worst polluted place in Africa due to mining and smelting of Pb and Zn ores. Orthophosphate has been receiving a lot of attention as stabilization agent for heavy metals, In order to reduce dispersion and mobility of Pb metal from the slag, we suggests treatment by adding phosphoric fertilizer (Triple super phosphate: TSP) which is effective and locally available.

2 Methods and Approaches

We conducted a series of column experiments in 50 ml of syringe tubes, simulating treatment for stacked slags at a dumping site in Kabwe. The syringes were filled with slags obtained from Kabwe site with 10 g of TSP on the top of slag sample. 6 mL of rain water obtained from the site was added every day, which is consistent with average daily precipitation rate. Infiltrated water was collected at the bottom of syringe and analyzed by ICP-AES and ICP-MS. After the column experiments, the slag samples in the column also investigated to understand geochemical processes occurred during the experiments by SEM/EDS.

© The Author(s) 2019
S. Glagolev (Ed.): ICAM 2019, SPEES, pp. 445–446, 2019.
https://doi.org/10.1007/978-3-030-22974-0_108

3   Results and Discussion

We confirm the reduction in Pb and Cd concentrations in the eluents. Remarkably, the reduction for Pb concentration is 96%. In the infiltrated slags simultaneously, we observe the alteration from $PbSO_4$ to $(Pb, Ca)_5(PO_4)_3Cl$ (Fig. 1), which effects to reduce the mobility of Pb. In contrast, elution of Zn and Cu from the slags are promoted by the presence of TSP. This is due to lowering pH by TSP, desorbed Zn and Cu from amorphous and crystalline iron hydroxides. It suggests to supply orthophosphate at neutral pH range is effective for immobilization of heavy metals in slags.

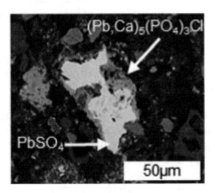

Fig. 1. The alteration from anglesite to pyromorphite

4   Conclusions

We confirmed the behavior of some heavy metals applied TSP in column scale. From the results, TSP could immobilize Pb and Cd, however, it promoted elution of Zn and Cu due to soil acidification. This suggests applying TSP with dolomite to the slag could be a better remediation method.

Acknowledgements. This study is supported by International Collaborative Research Program (SATREPS): Visualization of Impact of Chronic/Latent Chemical Hazard and Geo-Ecological Remediation in Zambia. I'm deeply grateful to Dr. Kasama who taught me how to use SEM/EDS in Center for Electron Nanoscopy, Denmark Technical University.

Murataite-Pyrochlore Ceramics as Complex Matrices for Radioactive Waste Immobilization: Structural and Microstructural Mechanisms of Crystallization

S. Krivovichev[1,2(✉)], S. Yudintsev[3], A. Pakhomova[4], and S. Stefanovsky[5]

[1] Kola Science Center, Russian Academy of Sciences, Apatity, Russia
krivovichev@admksc.apatity.ru
[2] Department of Crystallography, St. Petersburg State University, St. Petersburg, Russia
[3] Institute of Geology of Ore Deposits, Petrography, Mineralogy, and Geochemistry, Russian Academy of Sciences, Moscow, Russia
[4] Deutsches Elektronen-Synchrotron (DESY), Petra III, Hamburg, Germany
[5] Frumkin Institute of Physical Chemistry and Electrochemistry, Russian Academy of Sciences, Moscow, Russia

Abstract. Murataite-pyrochlore titanate ceramics are attractive waste forms capable to immobilize radioactive waste streams of complex compositions, thus eliminating the need for further chemical separation. We have investigated structures of three types of murataite: $3C$, $-5C$, and $-8C$ phases and demonstrate their polysomatic nature and structural complexity. Structurally simple pyrochlore crystallizes first, followed by crystallization of murataite-$5C$ containing pyrochlore cells surrounded by fragments of Keggin clusters. This phase is overgrown by murataite-$8C$ containing both murataite and pyrochlore cells. The crystallization finishes with the formation of murataite-$3C$, which is the most stable and less actinide-rich. The microstructure formed via this mechanism reminds a Russian doll, which creates additional barrier for the actinide leaching from the pyrochlore core. The high chemical and structural complexity of the pyrochlore-murataite series is unparalleled in the world of crystalline materials proposed for the HLRW immobilization, which makes it unique and promising for further exploration.

Keywords: Murataite · Pyrochlore · Crystal structure · Crystallization · Microstructure · Radioactive waste · Actinides

1 Introduction

One of the most important tasks for the advanced nuclear cycle is the elaboration of waste forms capable to immobilize waste streams of complex compositions, thus eliminating the need for further chemical separation. In this regard, the murataite-pyrochlore titanate

© The Author(s) 2019
S. Glagolev (Ed.): ICAM 2019, SPEES, pp. 447–450, 2019.
https://doi.org/10.1007/978-3-030-22974-0_109

ceramics attract considerable attention due to their ability to immobilize radioactive wastes with different and complex chemical compositions, including actinides such as Pu-238. Over last few years, there has been a renewed interest in their synthesis and investigations (Maki et al. 2017; Lizin et al. 2018, etc.).

Murataite-(Y) is a complex titanate mineral first discovered in alkali pegmatites in St. Peters Dome area in Colorado, United States and later found in pegmatites in the Baikal region in Russia. Its crystal structure (cubic, space group $F\text{-}43\ m$, $a = 14.886$ Å) was determined by Ercit and Hawthorne (1995) as based upon a framework of corner-linked α-Keggin clusters hosting a complex metal-oxide substructure. The simplified formula of natural murataite-(Y) can be written as $^{[8]}R_6^{[6]}M1_{12}^{[5]}M2_4^{[4]}TX_{43}$, where R = Y, HREE, Na, Ca, Mn, $M1$ = Ti, Nb, Na, $M2$ = Zn, Fe, Ti, Na, T = Zn, Si and X = O, F, OH. The interest in murataite-(Y) was renewed in 1982, when its synthetic analogue was identified in Synroc-type titanate ceramics with imitators of high-level radioactive waste at the Savannah River nuclear power plant (Morgan and Ryerson 1982). Laverov et al. (1998) reported the formation of murataite-type titanate phase in the uranium-bearing Synroc matrix from the Mayak factory, a radiochemical facility for the reprocessing of nuclear fuel located in Southern Ural, Russian Federation. It was found that five volume percent of synthetic murataite accumulate about 40% of the total amount of uranium present in the sample, which led to follow-up detailed studies of chemistry and properties of this material. Transmission electron studies allowed identification of synthetic varieties of murataite with $3 \times 3 \times 3$, $5 \times 5 \times 5$, $7 \times 7 \times 7$ and $8 \times 8 \times 8$ fluorite-like cubic supercells, referred in the following as murataite-$3C$, -$5C$, -$7C$ and -$8C$ phases (Laverov et al. 2011).

2   Methods and Approaches

We have studied crystal structures of murataite-$3C$, -$5C$ and -$8C$ using single-crystal X-ray diffraction analysis on the samples obtained by melting the mixture of oxides in an electric furnace at 1500 °C with subsequent cooling to the room temperature as described by Laverov et al. (1998). The details of the experimental procedures used to obtain structure models have been described in detail in (Krivovichev et al. 2010; Pakhomova et al. 2013, 2016).

3   Results and Discussion

Urusov et al. (2005) proposed that synthetic murataites can be considered as members of murataite-pyrochlore polysomatic series consisting of different combinations of 2D modules. The structural determination of murataite-$5C$ reported by Krivovichev et al. (2010) confirmed the assumption about the modular nature of the polysomatic series and demonstrated that the murataite- and pyrochlore-type modules are not layers but zero-dimensional blocks (nanoscale clusters), combination of which in a 3-dimensional space generates at least two different derivative structures, which combine structural features of both murataite and pyrochlore. In particular, the crystal structure of murataite-$5C$ can be described as an ordered arrangement of pyrochlore unit cells immersed into the recombined murataite matrix, i.e. a substructure consisting of murataite structure elements.

The crystal structure of murataite-3C was reported by Pakhomova *et al.* (2013), who demonstrated its general identity to natural murataite, with some important chemical and structural modifications. The modular nature of the murataite-pyrochlore polysomatic series was discussed by Laverov et al. (2011).

The crystal structure of murataite-8C was reported by Pakhomova et al. (2016) as based upon a three-dimensional octahedral framework that can be described as an alternation of murataite and pyrochlore modules immersed into transitional substructure that combine elements of the crystal structures of murataite-3C and pyrochlore. The obtained structural model confirmed the polysomatic nature of the pyrochlore-murataite series and illuminated the chemical and structural peculiarities of crystallization of the murataite-type titanate ceramic matrices.

Table 1. Information-based structural complexity parameters for the synthetic members of the pyrochlore-murataite polysomatic series

Material	v [atoms]	I_G [bits/at.]	$I_{G,total}$ [bits/cell]
Pyrochlore	22	1.686	37.088
Murataite-5C	336	4.892	1643.840
Murataite-8C	1387	6.558	9096.031
Murataite-3C	71	3.226	229.044

4 Conclusions

The information-based complexity parameters for the members of the pyrochlore-murataite are listed in Table 1. Both kinds of information-based complexity parameters (per atom and per cell) behave in a similar fashion: they are relatively small for the initial pyrochlore phase, then increase for murataite-5C, reach their maxima for murataite-8C and decrease for the final murataite-3C phase. This trend is also followed in the crystallization of murataite-pyrochlore ceramics: structurally simple and actinide-rich pyrochlore crystallizes first, creating conditions for the saturation of melt with Keggin clusters, which triggers crystallization of murataite-5C containing pyrochlore unit cells surrounded by fragments of Keggin clusters. This pyrochlore-rich phase is overgrown by murataite-8C containing both murataite and pyrochlore unit cells. The crystallization finishes with the formation of pure Keggin phase murataite-3C, which is the most stable and less actinide-rich. The microstructure formed during such a crystallization reminds a Russian doll ('matryoshka'), which creates additional barrier for the actinide leaching from the pyrochlore (or crichtonite) core. The high chemical and structural complexity of the members of the pyrochlore-murataite series is unparalleled in the world of crystalline materials proposed for the high-level radioactive waste immobilization, which makes it unique and promising for further technological and scientific exploration.

Acknowledgements. This work was supported by the President of the Russian Federation grant for leading scientific schools (grant NSh-3079.2018.5 to SVK).

References

Ercit TS, Hawthorne FC (1995) Murataite, a UB_{12} derivative structure with condensed Keggin molecules. Can Mineral 33:1223–1229

Krivovichev SV, Yudintsev SV, Stefanovsky SV, Organova NI, Karimova OV, Urusov VS (2010) Murataite-pyrochlore series: a family of complex oxides with nanoscale pyrochlore clusters. Angew Chem Int Ed 49:9982–9984

Laverov NP, Sobolev IA, Stefanovskii SV, Yudintsev SV, Omel'yanenko BI, Nikonov BS (1998) Synthetic murataite: a new mineral for actinide immobilization. Dokl Earth Sci 363:1104–1106

Laverov NP, Urusov VS, Krivovichev SV, Pakhomova AS, Stefanovsky SV, Yudintsev SV (2011) Modular nature of the polysomatic pyrochlore-murataite series. Geol Ore Dep 53:273–294

Lizin AA, Tomilin SV, Poglyad SS, Pryzhevskaya EA, Yudintsev SV, Stefanovsky SV (2018) Murataite: a matrix for immobilizing waste generated in radiochemical reprocessing of spent nuclear fuel. J Radioanal Nucl Chem 318:2363–2372

Maki RSS, Morgan PED, Suzuki Y (2017) Synthesis and characterization of a simpler Mn-free, Fe-rich M3-type murataite. J Alloys Compd 698:99–102

Morgan PED, Ryerson FJ (1982) A "cubic" crystal compound. J Mater Sci Lett 1:351–352

Pakhomova AS, Krivovichev SV, Yudintsev SV, Stefanovsky SV (2013) Synthetic murataite-3C, a complex form for long-term immobilization of nuclear waste: Crystal structure and its comparison with natural analogues. Z Kristallogr 228:151–156

Pakhomova AS, Krivovichev SV, Yudintsev SV, Stefanovsky SV (2016) Polysomatism and structural complexity: Structure model for Murataite-8C, a complex crystalline matrix for the immobilization of high-level radioactive waste. Eur J Mineral 28:205–214

Urusov VS, Organova NI, Karimova OV, Yudintsev SV, Stefanovskii SV (2005) Synthetic "murataites" as modular members of a pyrochlore-murataite polysomatic series. Dokl Earth Sci 401:319–325

Cs Leaching Behavior During Alteration Process of Calcium Silicate Hydrate and Potassium Alumino Silicate Hydrate

K. Kuroda[1($\boxtimes$)], K. Toda[1], Y. Kobayashi[1], T. Sato[2], and T. Otake[2]

[1] Graduate School of Engineering, Hokkaido University, Hokkaido, Japan
k7927k@eis.hokudai.ac.jp
[2] Faculty of Engineering, Hokkaido University, Hokkaido, Japan

Abstract. Zeolite, used to remove Cs from a contaminated water, would be solidified for the safety disposal. Recently, geopolymer is considered as a new binder for disposal. Geopolymer has an advantage that primary phases such as potassium almino silicate hydrate (K-A-S-H) may sorb radioactive nuclides. In this study, Cs adsorption, co-precipitation and desorption experiment were conducted, and C-S-H, which is primary phases of cement, were also employed for experiments for comparison. From these experiments, it is obtained that K-A-S-H has higher adsorption capacity of Cs than C-S-H. Cs adsorption ratio and co-precipitation ratio by C-S-H were almost same. Cs is likely sorbed by C-S-H thoroughly via ion exchanging. The desorption experiment demonstrated that most Cs was desorbed from C-S-H while 90% of Cs remained in K-A-S-H. Therefore, K-A-S-H has a higher retention capacity than that of C-S-H. Consequently, geopolymer is considered to be a better material in terms of Cs storage.

Keywords: Geopolymer · K-A-S-H · C-S-H · Radioactive waste

1 Introduction

After the accident at the Fukushima Daiichi Nuclear Power Station that occurred due to the The2011 off the Pacific coast of Tohoku Earthquake, contaminated water with radioactive nuclides such as cesium (Cs) have been continuously generated. Zeolite have been used for removing Cs from the contaminated water, and the spent zeolite are currently planned to be solidified for the safety storage and disposal. Recently, geopolymer is considered as a new binder for safety disposal of spent zeolite. Geopolymer has an advantage that primary phases such as potassium almino silicate hydrate (K-A-S-H) may have property for sorbing radioactive nuclides. However, there are few data about the adsorption behavior of Cs by K-A-S-H and the Cs leaching during their alteration.

2 Methods and Approaches

In this study, C-S-H, which is primary phases of cement, were also employed for experiments for comparison. In adsorption experiments, powder C-S-H and K-A-S-H were put into Cs-solution whose concentration is 1.0 mM at 298 K for a week. And in

© The Author(s) 2019
S. Glagolev (Ed.): ICAM 2019, SPEES, pp. 451–452, 2019.
https://doi.org/10.1007/978-3-030-22974-0_110

co-precipitation experiment, the materials to synthesize C-S-H were put. K-A-S-H could not be conducted co-precipitation experiment because water react with materials during synthesize. The solid sample after adsorption experiment were investigated in batch test and flow-through test as desorption experiment of Cs. The period of batch test is 4months and that of flow-through test is a month. And deionized water was used in both of them.

3   Results and Discussion

The adsorption ratio by K-A-S-H is 92%, while the adsorption ratio by C-S-H is 29%. The reason of this is considered that the size of sorption site is based on ionic radius of K or Ca, and that of Cs is similar to K than Ca. The adsorption ratio and co-precipitation ratio by C-S-H is almost same. Cs is likely sorbed by C-S-H thoroughly via ion exchanging, so it may be easy to sorb even after generation. In batch test as desorption experiment, the reaction between solid and water phase became equilibrium in 1month, and Cs concentration were almost stable after that. The desorption ratio from C-S-H was around 20% and from K-A-S-H was around 2%. But in flow through test, C-S-H desorb almost all of Cs in a day. It is considered that Cs sorption by C-S-H is ion exchange, so it is easy to leach by ion exchange too. On the other hand, The desorption ratio from K-A-S-H was almost 1% per day until 1month had past. Cs/Si ratio in each day was constant, and Si concentration is considered to depend on the dissolution amount of K-A-S-H. It is considered that Cs concentration also depended on that. From these results, it can be said that K-A-S-H has higher property to prevent desorption of Cs than C-S-H.

4   Conclusions

Consequently, K-A-S-H has higher retention capacity than that of C-S-H. These results show that geopolymer whose matrix is composed of K-A-S-H is considered to be better in terms of Cs storage.

Acknowledgements. This work was supported by MEXT 8桁の認可番号, Long-term performance of cement disposal systems for synthetic zeolites and titanates arising from reprocessing of contaminated water.

Environmental Pollution Problems in the Mining Regions of Russia

E. Levchenko[(✉)], I. Spiridonov, and D. Klyucharev

FSBI IMGRE, Moscow, Russia
imgre@imgre.ru

Abstract. The main types of environmental impact during exploration, development and mining of mineral deposits are considered. The indicators of the environmental situation caused by the mining and mineral processing in the mining regions, as well as the environmental consequences of accumulated mining and industrial waste are presented. The results of environmental monitoring of the Russian industrial cities are demonstrated.

Keywords: Environmental safety · Mining · Man-made waste ·
Heavy metals · Pollution of the ecosystem

1 Introduction

Intensive economic development due to the steady progress in science and technology entails an inevitable increase in the consumption of minerals. In this regard, the increase in mineral production during the last century, a sharp increase in the mining activities contributed to the accumulation of mining waste and man-made pollution of ecosystems. Besides, despite the obvious benefits of mining for the benefit of man, on the other hand, it is also a powerful source of environmental hazards for biota and humans (Aleksandrova and Nikolaeva 2015).

Many chemical elements contained in waste products, in addition to industrial value, cause toxic effects on the ecosystem.

The mining of mineral deposits leads to a change in the basic physicochemical properties of the lithosphere, including its main functions, i.e., geodynamic, geophysical, resource, and geochemical. The study of changes in the ecosystem's parameters during the life period of a mining enterprise is one of the key goals of an ecological-geochemical assessment.

The high level of the environmental impact is typical of the waste produced by ore processing and metallurgical operations, since their storage requires special engineering structures, and the waste contains chemical components harmful to nature and human health. Their mass is inferior to that of stripped overburden and host rocks, but they affect the environment more perniciously (Spiridonov and Levchenko 2018).

The environmental situation has deteriorated significantly due to the fact that at the end of the last century after the collapse of the USSR, many large mining complexes did not cope with economic difficulties and ceased their activities. The tailings of the enterprises, by majority toxic, have remained uncontrolled. Their conservation and

© The Author(s) 2019
S. Glagolev (Ed.): ICAM 2019, SPEES, pp. 453–456, 2019.
https://doi.org/10.1007/978-3-030-22974-0_111

reclamation have not been carried out timely; hence pollution keeps on growing. In the soils buried under the dumps tangible geochemical transformations occur. The soils buried 20 and more years ago display a strong oxidation over the whole depth of their profile (e.g., pH stays as low as 3.5–4.0), and soil colloids become destroyed. The soil absorbing complex is disturbed, the mobility of organic matter increases, the soil horizons gain ore components, which additionally differentiate due to unequal mobility. These facts testify the mobility of chemical elements in the dumps, and the latter often remain connected to the watercourse systems and can affect the territory of the mining and processing works in the area of air emissions and waste storages.

2   Methods and Approaches

Monitoring of the natural environment should be carried out at all stages of the mining area life, from exploration to mining and further reclamation of disturbed lands and until the site becomes completely stabilized.

The basis of this paper are ecological and geochemical studies, including the identification of areas of environmental pollution by toxic substances, assessment of their extent and composition of their pollution; assessment of potential geochemical endemicity; zoning of the territory according to the pollution level and the degree of environmental danger. identification of pollution sources; identifying areas of potential man-made objects; ecological and geochemical monitoring and forecast of the development of negative processes; development of recommendations for the rehabilitation of areas of poor ecological condition; identification of populations with an increased risk of morbidity. The result of these studies is the compilation of ecological and geochemical maps portraying the ecological status of the territory.

The study of the environmental health is carried out in the following main areas: mapping of the man-made pollution in soil and snow cover; establishing the characteristics of the response of plants to soil pollution; geochemical studies of ground and surface water, and stream sediments; analysis of the chemical composition of atmospheric air, precipitation and aerosols, industrial waste materials as sources of environmental pollution and objects for the extraction of secondary raw materials; relationships of environmental pollution and health indicators of the population living in the pollution hot spots.

3   Results and Discussion

The share of mining industries accounts for 70–80% of the volume of all man-made formations, which have their own characteristics, due to the composition of the feedstock, the technology of extraction, enrichment or processing, and a number of other factors.

As demonstrated by ecological and geochemical studies, the most serious negative effects are related to: the functioning of large industrial hubs (Nizhny Novgorod, Irkutsk-Cheremkhovo, Khabarovsk, Vladivostok, etc.), as well as exploration and development of mineral deposits in active mining areas (Kirovsk, Mama-Bodaibo, Khapcheranga, Dalnegorsk-Kavalerovsk, Norilsk, and other areas of similar profile). On the basis of the analysis of the updated database of available technogenic objects, including rare metal deposits, the allocation of 576 technogenic formations on the territory of the Russian Federation is analyzed.

Relevant location maps were compiled, and ranking of technogenic deposits and formations was carried out using the following parameters: areal extent, storage type(s), type(s) of technogenic formations, hazardousness classes, and level of environmental impact. The man-made deposits and formations were ranked by their effect on the elements of the environment.

The analysis of the hottest spots suggests that a series of the factors provokes the deterioration of the ecological situation in the territories.

Of particular concern is the ore processing plant waste, since it requires special engineering structures, and the waste itself contains chemical elements and compounds harmful to nature and human health. Their amounts are inferior relatively to the masses of stripped barren overburden and hosting rocks, but they affect the ecological situation more perniciously. For example, the environmental situation caused by the extraction of mineral raw materials and the disposal of waste on more than 25% of the territory of the Urals economic region is estimated as a crisis. Slightly less than the area of such lands in the south of the Russian Far East, Khanty-Mansi Autonomous Area, Tyumen Region, Krasnoyarsk Territory and other areas of intensive mining and processing of mineral resources.

According to the environment impact degree, the highly hazardous objects list is as follows: apatite concentrates of the Khibiny apatite-nepheline deposits (TR, Sr, F), enrichment tails of the eudialyte lujavrites of the Lovozero GOK (TR, Th), tailings of enrichment of baddeleyite-apatite-magnetite ores of the Kovdor Mining apatite, baddeleyite ZrO_2). Medium-level objects are waste storages accumulated from the apatite concentrate processing in the Khibiny group deposits (phosphogypsum) containing rare earth metals and gypsum (Bykhovskiy et al. 2016; Karnachev et al. 2011).

The toxicity of mining products depends on their physical condition and chemistries. Understanding the mechanisms of the action of chemical elements and compounds on the environment and public health makes it possible to optimize medical consequences and to carry out acceptable mining and processing of mineral raw materials. At the same time, it is necessary to take into account the whole range of sources and objects of impact in order to create a system of medical and environmental safety of the work areas.

The problems of the urbanized environment as a human habitat become similar to those experienced by geologists, representatives of related professions and the population of geological exploration, mining, oil and gas, and metallurgical enterprises.

Three indicators are accepted in Russia as measures of the soil chemical pollution in Russia; these are the maximum acceptable content (MAC), the background content (Z_b) and crustal abundance/clarke (Zc). We analyzed the weighted average bulk content distribution of heavy metals (the hazardousness classes 1 and 2): Pb, Cd, Hg, Zn, Ni, and Cu. By the above mentioned three evaluation criteria, the cities falling into the 1st (highly dangerous) category are Irkutsk, Penza, Saratov, Chelyabinsk, Yekaterinburg, the 2nd (dangerous) are Perm and St. Petersburg, and the 3rd (moderately dangerous) include Blagoveshchensk and Vologda.

4   Conclusions

The extent of the loss of land, water, forest, recreational and other resources from subsoil use in general and from unused waste in particular places these processes on a par with negative factors that pose a threat to the country's security.

The environmental consequences of accumulated mining and industrial waste are larger than it is declared in various publications concerning the problem under consideration and are of a global scale.

References

Aleksandrova TN, Nikolaeva NV (2015) Ecological-geochemical estimate of the Russian mining and metallurgy waste. Polytech University Publishers, St. Petersburg (in Russian)

Bykhovskiy LZ, Potanin SD, Kotelnikov EI, Anufrieva SI et al (2016) Rare earths and Sc-bearing man-made formations and deposits in Russia. In: Rare earth and Sc minerals in Russia: Mineral commodities, VIMS Economic Geology Series, No. 31, pp 112–120 (in Russian)

Karnachev IP, Zhirov VK et al (2011) Ecological and sanitary estimate of the Khibiny mining area, Murmansk oblast. Vestnik MGU, vol 14, no 3, pp 552–560 (in Russian, with English abstract)

Spiridonov IG, Levchenko EN (2018) Mining waste and ecological safety. Prospect and protection of mineral resources, no 10, pp 15–24 (in Russian, with English abstract)

Environmental Solutions for the Disposal of Fine White Marble Waste

I. Shadrunova, T. Chekushina$^{(\boxtimes)}$, and A. Proshlyakov

Academic N.V. Melnikov Institute of Problems of Comprehensive
Exploitation of Mineral Resources, Russian Academy of Sciences,
Moscow, Russia
tvche.2016@gmail.com

Abstract. The article deals with environmental problems of formation of fine white marble wastes on the territory of Koelga deposit and total mining complex. An inventory analysis of marble waste was carried out, environmental assessment of fine marble waste and their impact on the ecology of the complex territory was carried out and theoretically justified. Planned and scientifically justified ways of large-scale utilization in the production of ceramic bricks.

Keywords: Formation of fine marble waste · Environmental assessment · Amount of waste · Waste disposal

1 Introduction

The growth of industrial and mining production, the progress of civilization increase environmental problems due to the increasing consumption of mineral and other resources from the bowels of the Earth, due to the rapid rise in the number of solid man-made wastes of different productions. These wastes can be used for the production of building materials and to improve the environmental safety of mining regions. Abandoned lands are exempted from waste dumps and territories have environmental and economic benefit (Oreshkin 2017).

2 Methods and Approaches

In the world and in the Russian Federation there are the technologies of extraction and processing of non-metallic mineral resources. During these technological processes man-made wastes are formed. Their queries and the surrounding areas are withdrawn from economic circulation, violate natural landscapes - their man-made options are created. It destroys the soil, changes modes of rivers, lakes, reservoirs, underground and surface groundwater and causes great damage to the environment (Meshheryakov et al. 2009).

Koelga deposit of white marble began the work from 1924. During this time in the dumps huge amounts of fine wastes of extraction and processing marble were accumulated (Tseytlin 2012).

© The Author(s) 2019
S. Glagolev (Ed.): ICAM 2019, SPEES, pp. 457–460, 2019.
https://doi.org/10.1007/978-3-030-22974-0_112

Almost all kinds of new productions require new construction, materials and mineral resources. Therefore, to improve the environmental safety of the regions it is necessary to carry out comprehensive development of deposits, and also to utilize man-made wastes in the production of building materials. To solve the above problems a comprehensive environmental assessment of man-made waste requires. The assessment should include amount of accumulated volumes for large-scale utilization of man-made waste in the production of building materials, products (Khokhryakov et al. 2013).

The purpose of the article is the ecological assessment of formation of man-made waste products of mining production in the form of white marble with a decrease in the available subsoil mineral resources for the production of building materials and products.

To achieve the goal, it is necessary to justify the use of these wastes as raw materials for the production of building materials and products. This will simultaneously improve the environmental safety of the territories due to large-scale utilization of man-made waste marble and will free up the areas occupied by dumps.

3   Results and Discussion

It was calculated that in 2018 the total mass in the dumps is more than 25 million tons of fine marble wastes, and the area of dumps - more than 20 hectares. An important task was also the calculation of the environmental damage from the abandoned territories under the dumps of fine marble waste.

To calculate it was analyzed the environmental effect from their utilization by reducing the area under the dumps, and the pollution of the environment of mining complex territory (Fadeichev et al. 2012).

When calculating it was determined that for the Chelyabinsk region the damage to the environment from storage of fine marble waste in dumps is about 500 thousand rubles a year (in the prices of 2018). Taking into account the amount of wastes already placed in dumps of JSC "Koelgamramor", the environmental damage will amount to over 30 million rubles.

According to calculations the utilization of fine marble waste in brick production will significantly reduce the environmental load on the environment. It will take place by reducing waste mass in dumps, that will allow to reduce the abandoned areas under dumps and to return the land to use.

In the articles it was determined the amount of recyclable fine marble waste at 1 m^3 of molding mixture for the production of ceramic bricks of multiple colors: terracotta or dark brown; light red or pinkish; fawn or straw.

The analysis of the results of technical tests showed that, on the basis of fine marble wastes, it is possible to obtain ceramic bricks of danger class 4, which corresponds to state standard GOST of the Russian Federation. It was found that burning of over-moulded ceramic raw makes at temperature 850… 900 °C. It is proved that at that temperature, the particles of marble are not affected by the process of decarbonization. Therefore, there is no greenhouse gas emissions - carbon dioxide, i.e. ecology of this mining territory is not the subject to harmful effects. Moreover, as above stated, the level of danger of marble wastes was higher (class 3) by one step, than the level of danger of the ceramic bricks produced (4 class) on the basis of these wastes. Also amounts of energy for manufacture of these ceramic bricks reduced significantly as compared to common ceramic brick (Moumouni et al. 2016).

4 Conclusions

Thus, the total mass accumulated fine marble wastes and environmental damage to mining area of Koelgo deposit were determined. The technology of improving the ecology of the region due to large-scale utilization of the above marble wastes in the production of ceramic bricks was elaborated. The possible number of bricks of different colors at full disposal of accumulated marble wastes was determined. Using environmental life cycle assessment of finished products based on fine marble wastes the possibility of obtaining an environmentally safe effective bricks was theoretically justified and the technology of their production was elaborated. The dependence of color products from fine marble waste was defined. So, at an amount of 20% of fine marble waste in the mixture by mass of clay rocks the ceramic brick has a dark brown color, and at 40% - has straw color. The influence of the elemental composition of the mixture on the color of the brick was determined (Merem et al. 2017).

It was proved that the most environmentally safe, resource-saving way of man-made waste disposal is their utilization in the production of building materials and products. This method releases territories abandoned for storing waste and provides environmental and economic effects from the elimination of dumps (Bilgin et al. 2012).

Thus, the environmental problems of the Russian Federation connected with rise of man-made waste of white marble with a decrease in available reserves of mineral resources for the production of building materials and products were specified. The scientific foundations of the integrated environmental assessment methodology of man-made wastes and their large-scale utilization in the production of building materials and products were elaborated (Hebhoub et al. 2011). The possibility to use these wastes as raw components for their production, while solving environmental problems of the territories due to large-scale utilization of man-made waste was justified. This extends the raw material base and contributes to the integrated development of bowels, their mineral and man-made resources. Utilization of man-made waste allows to get a huge environmental and economic effects on the territory of the Russian Federation.

References

Bilgin N, Yeprem HA, Arslan S, Bilgin A, Günay E, Mars MO (2012) Use of waste marble powder in brick industry. Constr Build Mater 29:449–457

Fadeichev AF, Khokhryakov AV, Grevcev NV, Cejtlin EM (2012) Dynamics of negative impact on the environment at different stages of mining development. News High Educ Inst Mountain Mag 1:39–46

Hebhoub H, Aoun H, Belachia M, Houari H Ghorbel E (2011) Use of waste marble aggregates in concrete. Constr Build Mater 25(3):1167–1171

Khokhryakov AV, Fadeichev AF, Cejtlin EM (2013) Application of an integral criterion for determining the environmental hazard of mining enterprises. News Ural State Mining Univ 1:25–31

Merem EC et al (2017) Assessing the ecological effects of mining in West Africa: the case of Nigeria. Int J Mining Eng Mineral Process 6(1):1–19

Meshheryakov YuG, Kolev NA, Fedorov CV, Suchkov VP (2009) Stroymaterialy Production of granulated phosphogypsum for the cement industry and building products, vol 5, pp 104–106

Moumouni A, Goki NG, Chaanda MS (2016) Natural Resources Geological exploration of marble deposits in Toto Area, Nasarawa State, Nigeria, vol 7, pp 83–92

Oreshkin DV (2017) StroymaterialyInvironmental problems of integrated development of mineral resources in the large-scale utilization of man-made mineral resources and waste in the production of building materials, vol 8, pp 55–63

Tseytlin EM (2012) Features of environmental hazard assessment of mining enterprises Theses of the report of VII Krakow conference of young scientists. AGH University of Science and Technology, Krakow, pp 809–819

Security Test of New Technology in View of Increased Performance of Oil Platforms Without Increasing Environmental Risks

E. M. Tanoh Boguy[(⊠)] and T. Chekushina

Department of Mineral Developing and Oil & Gas Engineering,
Engineering Academy, RUDN University, Moscow, Russia
boguymartialeddy@gmail.com

Abstract. In this article, it will be important to note the context of the gradual depletion of existing fields, which are pushing to expand research and exploitation of new fossil fuel resources in order to meet the growing demand for fuel and, despite international regulations to combat global warming. Consequently, an increase in offshore platforms in global hydrocarbon production to compensate for the depletion of the earth's reserves is becoming a major problem for the oil industry. Given the financial unforeseenness that is represented, and the energy autonomy which is provided, marine exploitation has become a problem for states with a large sea area and, therefore, an environmental.

Keywords: Security test · New technology · Increased performance · Environmental risks

1 Introduction

The use of new reserves, in economically viable conditions depends on the available technologies. The development of deep and ultra-deep offshores requires considerable research and development efforts. Progress has also been made in managing the multiple risks associated with this activity. A disaster like «Deepwater Horizon» led to a detailed analysis and sharing of findings by industry experts.

Despite the security rules on the platforms, in fact, some major incidents are revealed, the causes of which are multifactorial in nature and which have dire consequences for both humans and the environment.

Legislation forces organizations to take responsibility for dealing with disasters, which has been developed over time and in different ways in different countries.

Our analysis of how environmental risks are taken into account by various subjects and offers development prospects to ensure better safety of offshore activities.

2 Methods and Approaches

Some of the accidents at oil rigs, such as the Deepwater Horizon, in the spring of 2010 caused a shock wave in their magnitude and severity that convinced that such accidents could occur. In fact, some states have taken steps to raise the level of security. In fact,

© The Author(s) 2019
S. Glagolev (Ed.): ICAM 2019, SPEES, pp. 461–462, 2019.
https://doi.org/10.1007/978-3-030-22974-0_113

some states have taken upon themselves the task of "solving the problem of providing security on the shelf." Global hydrocarbon production is becoming more and more offshore, accounting for more than 35% of oil and 19% of gas. Since the deposits are located at great depths, states and companies must develop the potential for their use, seeking to control the risks inherent in this activity carried out under extreme conditions. In order not to have restrictions in the conditions in which exploration, drilling and mining operations are carried out, by more and more complex and risky methods. In addition, oil companies are well aware of their interest in investing in the development of new technologies. This allows them to gain an industrial competitive advantage in strategic areas of deepwater exploitation. Significant progress has been made in managing the multiple risks inherent in offshore operations. Oil companies put prevention at the level of operating conditions.

3 Results and Discussion

Regardless of the achievements observed, it is obvious that the safety rules applied on the platforms guarantee greater efficiency than environmental protection, and that more and more risks are encountered. Increasing risks to humans and the environment is inextricably linked with the complexity of drilling operations. Working platforms continue to find solutions that completely avoid any potential risks in the protected areas. Advanced technology and security measures suggest that there is a clear improvement, but the limit between politically correct and pollution is quickly exceeded when it comes to such profits.

4 Conclusions

Finally, it is important to include risk management in determining policies, procedures and plans, as well as specific risk mitigation measures that will be taken to manage security risks. The environment is associated with all sorts of accidents, while drilling and operating the platform. The accident, which is a major problem for these exploitation will be a hydrocarbon spill, which is highly unlikely and will be limited to pumping oil and fuel stored on support vessels in the event of a tank failure or reloading pipe.

Calcite Mineral Generation in Cold-Water Travertine Huanglong, China

F. Wang, F. Dong[(⊠)], X. Zhao, Q. Dai, Q. Li, Y. Luo, and S. Deng

School of Environment and Resource,
Southwest University of Science and Technology, Mianyang, China
fqdong@swust.edu.cn

Abstract. Mineral generations could help us to understand the physical, chemical and biological processes within their formation, and then to reconstruct the sedimentary paleo-environment and paleo-climate. The calcite in the Huanglong cold-water travertine can be divided into three mineral generations, which reveal two different sedimentary environment systems respectively. In the calcium cycle, calcite mineral generation exposes a step in recycling marine matter to the land, and it also allows the land to proliferate, which mainly manifeste in the addition of plant debris, algae and microbial residues, so that the topography has been accumulating.

Keywords: Cold-water travertine · Mineral generation · Paleo-environment · Huanglong

1 Introduction

Calcite is the main mineral component of travertine/tufa, and it plays a decisive role in the sedimentary evolution of travertine, whether inorganic or bio-organic (Pentecost 1995). The size of the calcite in travertine is a reflection of the deposition rate and can therefore be used to characterize its sedimentary environment, which is the result of physical, chemical and biological synergy during the deposition process. Herein, we divide the calcite in the Huanglong cold-water travertine into different mineral generations according to the sedimentary environment and evolution time series, i.e., from the generation of the parents to the descendants. The classification of these mineral generations helps to understand the physical, chemical and biological processes within their formation, and then to reconstruct the sedimentary paleo-environment and paleo-climate. On the other hand, the mineral generation of calcite will help to understand the architecture of travertine landscape (Wang et al. 2018), so that they can be better protected and leave more natural heritage of travertine for human beings.

2 Methods and Approaches

A comprehensive field geological survey of rocks consisting of calcite was performed, mainly from sedimentary rocks, and systematic sample collection based on the geological background of these rocks was carried out. The mineralogy studies of calcite

© The Author(s) 2019
S. Glagolev (Ed.): ICAM 2019, SPEES, pp. 463–465, 2019.
https://doi.org/10.1007/978-3-030-22974-0_114

were carried out by polarized light microscopy, XRD and SEM to determine their generational relationship.

3 Results and Discussion

From the diagenetic time series of calcite, the types of rocks are Mesozoic limestone and dolomite, and the travertine deposited since the Late Cenozoic. The calcite in travertine is further divided into two mineral generations, namely calcite and secondary in primary travertine travertine. The calcite in travertine is further divided into two mineral generations, namely calcite in primary travertine and calcite in secondary travertine. Therefore, calcite is divided into three generations from the generation of the parent to the descendants, i.e., calcite in the Mesozoic carbonate rock, calcite in the Late Cenozoic travertine, and calcite re-precipitated after travertine leaching.

The calcite in the Mesozoic limestone is micritic, microcrystal and sparry calcite, ranging in size from centimeters to micrometers. CaO and MgO in the rocks composed of these calcite are close to the theoretical value, and the other components are very low, which belong to the soluble carbonates. These calcite became the parent generation in the whole calcite evolution sequence, and they provided the material source for the calcite of later generations after being leached. The calcites of the descendants form the different morphologies of the cold-water travertine. During the formation process, physical, chemical, biological and other factors participate in the diagenesis. Among these travertines, no matter what color, except for calcite, other minerals hardly develop. The calcites of the descendants of travertine are very numerous and complex. Here, we mainly listed two of them, which are calcite in the laminal travertine and calcite in the porous travertine, because these are the main components of most travertines. The calcite of the laminal travertine is long columnar and slablike. The brown and white calcite is continuously growing without interruption. These characteristics are very different from those observed on the eye assay, which indicates that the calcite growth in the dry and cold seasons is continuous (Wang et al. 2014). On the other hand, it reflects that the hydrodynamic conditions are very stable, and the water layer is very thin with little or no biological involvement. The calcite in the porous travertine tends to be granular, and the particle size is much smaller than that of the laminal travertine, and its particle size is generally less than 100 μm. These characteristics reflect the rapid crystallization of calcite, Due to the strong hydrodynamics and the participation of biological effects, calcite cannot be continuously grown, but suddenly nucleates and grows to a certain extent then no longer grows.

The last generation of calcite is the secondary calcite in travertine. The ancestral body of this type of calcite is the deposited travertines, which are dissolved in the water by weathering and leaching, then the calcite re-precipitates through the deposition of a parent-like travertine. These calcites will adhere to the cracks, edges and even the surface of the primary calcite.

4 Conclusions

The calcite in the Huanglong cold-water travertine can be divided into three mineral generations, which reveal two different sedimentary environment systems respectively. They are the generations of the marine carbonate rock diagenesis system, and the descendant generation is the continental freshwater karst sedimentary system. Unlike conventional weathering, which converts terrestrial carbonate rocks to the ocean phase, this is done in the opposite direction. The study of different calcite mineral generation can reconstruct the paleo-environment and paleo-climate of its sedimentation.

Acknowledgements. This research was supported by National Natural Science Foundation of China (Grants nos. 41572035, 41603041 and 41877288), the Open Funds of Key laboratory of mountain hazards and surface processes (grant No. 19zd310501) and Longshan Talents program of Southwest University of Science and Technology (18lzx663).

References

Pentecost A (1995) The quaternary travertine deposits of Europe and Asia Minor. Quaternary Sci Rev 14(10):1005–1028

Wang HJ, Yan H, Liu ZH (2014) Contrasts in variations of the carbon and oxygen isotopic composition of travertines formed in pools and a ramp stream at Huanglong Ravine, China: implications for paleoclimatic interpretations. Geochimica et Cosmochimica Acta 125:34–48

Wang FD, Dong FQ, Zhao XQ (2018) The large dendritic fissures of travertine dam exposed by Jiuzhaigou earthquake, Sichuan, southwestern China. Int J Earth Sci 107(8):2785–2786

Optimization of the Natural-Technical System "Iron Ore Quarry" Management Based on the Algorithm of the Rock Mass Stability Ensuring

L. Yarg[✉], I. Fomenko, and D. Gorobtsov

Department of Engineering Geology,
Russian State Geological Prospecting University (MGRI), Moscow, Russia
ifolga@gmail.com

Abstract. The method of natural-technical system (NTS) "Iron ore deposits" optimal control in terms of the pit walls stability is based on two-level systems with cross-links. The algorithm for optimizing the pit walls angles designed values includes the following steps: separation of rock massif into engineering-geological complexes (EGC), typing of the pit walls within the EGC, substantiation of the calculation geomechanical models and stability analysis of the pit walls based on mathematical modeling. Based on the results of the calculations the maximum angle of the pit wall is determined at which it remains stable. As minimized performance criteria the deviations of the stability factors current state from the maximum allowable values are considered. The proposed approach is one of the ways to ensure the stability of the deep-pit quarries walls during their long-term development.

Keywords: Natural-technical system (NTS) · Open pit · Stability assessment · Optimization of pit walls angles · Control of NTS

1 Introduction

The development of iron ore deposits and permanent deepening of the open pit leads to changes in the stress state, decompaction of rocks, an increase in massif fracture, weathering rates and a decrease in the strength properties of rocks that form the open pit, activation of geological processes: debris, rock falls, landslips and landslides, suffusion, surface erosion.

The considered natural-technical system "Ore deposits of KMA" is a complex system of the local level. The functioning of the local NTS "Iron Ore Quarry" is characterized by: a certain set of processes developing permanently without the stabilization stage under the influence of long-term man-made interactions which form the basis of the NTS operation. Reduction of negative consequences is possible only with a clear understanding of the processes developing in the field of interaction of natural-technical systems (NTS) "mining and processing plant (GOK")" (Yarg et al. 2018).

Research objective: optimization of the NTS "iron ore quarry" management based on the algorithm of the rock mass stability ensuring.

S. Glagolev (Ed.): ICAM 2019, SPEES, pp. 466–470, 2019.
https://doi.org/10.1007/978-3-030-22974-0_115

2 Methods and Approaches

The processes development initiated by technological work is progressive in space and time. Long-term exploration of deposits leads to the changes of boundaries, mode and set of processes (Bondarik and Yarg 2015).

The system of engineering geological support in the quarry areas includes a range of work and research aimed to obtain the information about engineering geological conditions during the entire life of the quarry, assessment and forecast of the slope stability at various stages of their construction to achieve the technical, economic and environmental safety of mining work.

Effective management of the natural-technical system "Iron Ore Deposit KMA" should be carried out taking into account both local and global stability factors of the pit walls.

Separation of the Rock Massif to EGC. Features of engineering-geological conditions including lithologic-petrographic composition, physical and mechanical properties, structural disturbance, parameters of the natural stress field require an individual approach to the process of predicting the behavior of an array of rocks. This becomes possible only on the basis of correct engineering and geological research data.

Elementary NTS "Stoilensky Quarry" is divided into two engineering-geological complexes (Yarg et al. 2018):

- The upper one is composed of loose and semi-rock soils with a thickness up to 90 m. The sedimentary cover is typified taking into account the geological structure, hydrogeological conditions (the water inflow along the open-pit contour water permeability) and the physical and mechanical properties of the soils.
- The lower EGC is represented by rocks with a thickness of up to 600 m. The main stability determining factors are: anisotropy of the massif properties due to its fracturing and spatial orientation of the cracks.

Engineering geological processes developing during the operation of the elementary NTS "Open pit" of the iron ore deposits of Stoilensky and Lebedinsky GOK are: scree formation; collapse; landslides; surface erosion; suffusion: mechanical, chemical; filtration deformations.

Tiping of the Pit Walls Within the EGC. A "bowl" of a quarry with a simple structural plan of the rock mass (Fomenko et al. 2016, Hoek and Bray 1981, Wyllie and Mah 2010) (i.e. assuming that the direction of weak zones and fracturing within the pit remains constant) can be divided into zones of conditional stability and potential instability of the walls (Fomenko et al. 2016).

In accordance with these factors, three types of pit walls quarrying were identified: relatively difficult, difficult and very difficult.

3 Results and Discussion

Optimization of the NTS "Mineral Deposit" functioning is based on a modern methodology for stability calculation (Pendin and Fomenko 2015, Bar et al. 2018).

For potentially unstable pit wall the probable collapse can occur according to the following schemes:

1. The azimuth of crack systems fall coincides with the azimuth of the pit wall fall. In this case a flat problem can be solved.
2. The azimuth of crack systems fall does not coincide with the azimuth of the pit wall fall, but at the same time according to the kinematic analysis results the formation of wedge-type collapses is likely. In this case the pit wall stability problem is solved in a three-dimensional formulation, for example using the method of volume blocks.

Based on the results of the calculations the maximum pit wall angle is determined at which it remains stable.

In accordance with the "large-scale interconnected" theory (Tsurkov and Litvinchev 1994), the management of local NTS "Iron Ore Quarry" in terms of the pit walls stability can be based on two-level systems with cross-links. As minimized performance criteria the deviations of the stability factors current state from the maximum allowable values are considered. As optimized parameters the following were taken: the level of the upper Jurassic aquifer, the strength properties of Alb-Cenomanian sands and Devonian clays, fracturing and blockiness of the Precambrian massif.

The graphs (Figs. 1, 2 and 3) of the relation between safety factor and the dynamics of aquifer, blockiness and strength properties of rocks allow setting the limit values of the system coordinates at which the system does not leave the zone of admissible states.

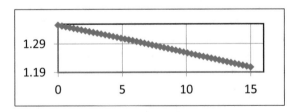

Fig. 1. The effect of groundwater level rise (horizontal axis) on the global safety factor of the pit walls (vertical axis).

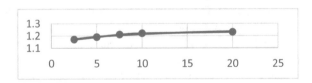

Fig. 2. The relation between Ku (vertical axis) and the blockiness of the rock massif (distance between cracks), (horizontal axis)

Fig. 3. The effect of the ore-crystalline rocks strength (horizontal axis) in the massif on the safety factor (vertical axis): 1 – densely fractured rocks; adhesion C = 690 kPa, angle of internal friction φ = 32°; 2 — moderately fractured, adhesion C = 1130 kPa, angle of internal friction φ = 36°; 3 — weakly fractured, adhesion C = 3140 kPa, angle of internal friction φ = 39°

4 Conclusions

The obtained results are in good agreement with the generally accepted ideas about the causes of stability infraction in the massif zones near the pit walls. At the same time the obtained data advantage is the possibility of establishing the limiting values of changes in the safety factors on graphs. These safety factors determine both local and global stability of the pit walls at which the system will not leave the zone of admissible states.

Control solutions that ensure the safe work performance should include: adjustment of the drainage system and water supply process, taking into account the position of the GWL; interception and organized disposal of surface and seepage water into the circulating system; maintaining the water level in the settling ponds, excluding flooding of the territory; adjustment of the blasting operations technology during the quarrying of deep horizons taking into account the stress state of the massif and the occurrence of the rock technogenic fracturing during drilling and blasting operations.

The obtained predictive estimates of the pit walls stability can be used in the design and development of fields with similar engineering geological conditions.

References

Bar N, Weekes G, Welideniya S (2018). Benefits and limitations of applying directional shear strengths in 2D and 3D limit equilibrium models to predict slope stability in highly anisotropic rock masses. https://www.researchgate.net/profile/Neil_Bar/research

Bondarik GK, Yarg LA (2015) Engineering geology. Questions of theory and practice. Philosophical and methodological foundations of geology. KDU (in Russian)

Fomenko IK, Pendin VV, Gorobtsov DN (2016) Estimation of the stability of quarries of quarries in rocky soils. Min Sci Technol (3), 10–21 (in Russian)

Hoek E, Bray JW (1981) Rock slope engineering, 3rd edn. Institution of Mining and Metallurgy, London

Pendin VV, Fomenko IK (2015) Methodology of landslide hazard assessment and forecast. Publishing House of the Russian Federation Lenand, Moscow (in Russian)

Tsurkov VI, Litvinchev IS (1994) Decomposition in dynamic problems with cross-links. Science: Physics and Mathematics, Moscow (in Russian)

Wyllie DC, Mah CW (2010) Rock slope engineering: civil and mining, 4rd edn. Spon Press/Taylor&Francis Group, London

Yarg LA, Fomenko IK, Zhitinskaya OM (2018) Evaluation of slope optimization factors for long-term operating open pit mines (in terms of the Stoilensky iron ore deposit of the Kursk Magnetic Anomaly). Gornyi Zhurnal (11), 76–81

Utilization of Associated Oil Gas: Geo-ecological Problems and Modernization of the State

L. Z. Zhang[1,2] and H. Y. Sun[1,3(✉)]

[1] Department of Mineral Developing and Oil & Gas Engineering,
RUDN University, Moscow, Russia
657273629@qq.com
[2] Liaoning Shihua University, Fushun, China
[3] Qinhuangdao Experimental Middle School, Qinhuangdao, China

Abstract. In the world vast of oil is extracted, especially in China. Respectively produce associated petroleum gas is in a large volume. There are geo-ecological problems in the utilization of associated petroleum gas. In connection with the increasing requirements for the preservation of the state of the biosphere in China, the process of modernization was begun. Chinese modernization of associated petroleum gas utilization is presented.

Keywords: Associated petroleum gas · Modernization · Technology · Geo-ecological problems

1 Introduction

Associated petroleum gas (after this APG) is a mixture of various gaseous hydrocarbons dissolved in oil and released in the process of extraction and preparation of oil. The oil gases also include gases released in the operations of thermal processing of oil (cracking, reforming, hydrotreating, etc.), consisting of saturated and unsaturated (methane, ethylene) hydrocarbons.

From geology, APG is often formed during the Ordovician and Silurian periods. Sometimes it is built late to the Cretaceous. (Vorobiev and Zhang 2018).

2 Methods and Approaches

The PRC's world ranking in oil production is quite high and, accordingly, the volumes of simultaneously produced associated gas are very significant. In 2011–2013 in China, APG was provided in the amount of 27.3, 28.9, 30.2 billion m^3. In addition, the share of production took more than 5.1% in the world.

Previously, in China, APG was traditionally considered not as a valuable resource, but as a by-product of oil production, the simplest method of which utilization is flaring in many fields, especially in Northeast China.

S. Glagolev (Ed.): ICAM 2019, SPEES, pp. 471–472, 2019.
https://doi.org/10.1007/978-3-030-22974-0_116

3   Results and Discussion

Associated gas recovery technology using membrane separation is based on the following steps:

- removing micro solid particles, crude oil and heavy hydrocarbon emulsion contained in associated gas;
- after preliminary impurity removal, heating up to 590 °C in a heat exchanger;
- introducing into a liquid rotary compressor;
- introducing the heated gas into a desulfurization tank, and desulfurizing;
- introducing the desulfurized gas into a membrane separator, and separating;
- introducing the gas from the membrane separator to a molecular sieve tank, and performing deep desulfurization and decarburization;
- cooling to obtain the product (Mo 2013).

This technological process is quite simple and convenient in industrial operation, and besides, it is characterized by low operating costs, high recovery rate and can be, after a little adaptation, widely circulated.

4   Conclusions

Associated petroleum gas will become a valuable raw material for further processing. China's economy needs to use APG to reduce greenhouse gas emissions.

Modernization of processing of APG in China and its prospects lie in the area of increasing the efficiency of processing of associated gas, reducing energy consumption in the course of processing, flexible operation, convenient installation and operation.

References

Mo JL (2013) Recovery process for petroleum associated gas. China Patent CN102994180A, 27 March 2013
Vorobiev AE, Zhang LZ (2018) Apply innovative technologies for processing of associated gas in China. Eurasian Sci. J. 10(2)

Cultural Heritage, Artifacts and Their Preservation

Identifying the Decorative Stone Samples from the Mining Museum's Collection: First Results

N. Borovkova[1(✉)] and M. Machevariani[2]

[1] Mining Museum, St. Petersburg Mining University, St. Petersburg, Russia
borovkova_nv@pers.spmi.ru
[2] Assistant of the Department of Mineralogy, Crystallography and Petrography, St. Petersburg Mining University, St. Petersburg, Russia

Abstract. The report presents the primary results of a study of a unique collection of polished decorative stone samples belonging to Empress Catherine the Great. Primary macroscopic analysis of 83 plates, divided into 13 groups according to similar features, was performed. The bulk chemical composition of rocks was estimated on the basis of XRF- analysis data, performed using a Delta Olympus XRF portable analyzer. Preliminary studies allowed to outline the characteristic fields of the studied samples of decorative rocks on the ternary plots of their bulk composition. In the future, it is planned to perform Raman spectral imaging to generate detailed maps of the mineralogical composition of the decorative stone samples.

Keywords: Polished decorative stones ·
Collections of Empress Catherine the great · Handheld XRF analyzer ·
Mining museum

1 Introduction

Natural stone serves as a unique raw material for the objects of decorative and applied arts, as well as architectural monuments. Museums and monuments of St. Petersburg store rare objects made of natural stone. The problems of their preservation are increasingly forcing restorers and art researchers to turn to geologists to identify various types of gemstone materials needed for restoration. This raises a number of problems, primarily related to the lack of reliable information about the origin of various types of such materials, as well as their accurate identification in art objects. Their study is complicated by the need to use exclusively non-destructive analytical techniques, which greatly complicates the task. Unlike European museums, in Russia, there is no complex reference collection of natural decorative (ornamental) stone, supported by current results of laboratory research, reliable information on the location, and a complete catalog of art object made of such materials. Obviously, the need for such data is highly in demand not only among restorers and art historians but also among geologists, whose research interests include the preservation of the diversity of gemstone raw materials and objects of cultural heritage.

© The Author(s) 2019
S. Glagolev (Ed.): ICAM 2019, SPEES, pp. 475–478, 2019.
https://doi.org/10.1007/978-3-030-22974-0_117

The object of this study was one of the oldest collections of polished flat samples of natural decorative (ornamental) stone of the XVIII century, previously owned by Empress Catherine the Great (Borovkova 2017). The research contributes to the development of a methodology for evaluating historical gemstone materials of considerable cultural, museum and scientific value. In 1816, a collection was transferred from the Imperial Hermitage to the Museum of the Mining Cadet Corps (now the Mining Museum), which included collections of marbles and «rock sampled as polished plates». In the inventory of this collection, which is stored in the archives of the State Hermitage Museum, there is the following note: «Jasper and solid rocks found in the Ural Mountains, starting from Tura River by noon over the rivers Uyu and Ural» [State Hermitage archives. F. 1. O. 6l. D. 1 a-c]. The total number of such samples is not indicated, but in the catalog (1798) they are recorded from №1525 to №1725; and in the next catalog (1811), there are a number of additional samples. Thus, with the same description, samples from № 2052 to 2330 are recorded in a later catalog.

2 Methods and Approaches

Currently, more than 100 items of such decorative rocks are found in the Mining Museum. The primary macroscopic evaluation of the samples suggested wider geography of their origin. The need for reliable authentication of historical samples necessitated their thorough review.

As the first phase of the study, 83 plates were selected and broadly classified into 13 groups. The groups were formed on the basis of visual estimation of similar characteristics: the rocks structure, texture, and color. Eight of the thirteen selected groups were pre-diagnosed as known geo-referenced decorative rock types that received code names corresponding to their regional and historical affiliation, namely: Korgon porphyry; Tigiretsky breccia; Tigiretsky quartz; Kalkan, Kushkuldinskaya, Nikolaevskaya, Urazovskaya, and Surguchnaya jasper. The remaining groups require further diagnostics and are pre-defined as andesite, hornfels, marble, and green marble.

The planned research method involves the use of known geo-referenced samples. Comparative analysis of the studied samples and reference rocks will be divided into three steps: visual comparison, comparison of chemical and mineral compositions.

As mentioned above, the study of museum objects requires a special approach and the use of non-destructive techniques. In this regard, the samples chemical composition analysis was performed using a Delta Olympus XRF handheld analyzer. The measurements were carried out in the Mining mode with preliminary calibration. There was threefold spectra collection from each point with a 30 s acquisition time. In general, 50 bulk chemical composition analysis of rocks belonging to seven previously selected groups (Korgon porphyry, Tigiretsky quartz, Urazovskaya jasper, green marble, marble, andesite, and hornfels) were made.

3   Results and Discussion

The absence of reference samples let us consider the obtained results as preliminary. Nevertheless, on the ternary plots of the rock samples bulk composition, it is possible to outline the characteristic fields corresponding to certain samples groups.

The Ca-Fe-Si diagram illustrates obvious trends: visible outlining of marble, Urazovskaya jasper, and andesite characteristic fields. The bimodal distribution of points corresponding to green marble possibly occurred due to the presence of large calcite phenocrysts, the content of which varies not only during the transition from sample to sample, but also unevenly distributed over the analyzed area of the sample. On the Ca-Fe-K triangular plot, the characteristic field of Urazovskaya jasper shows that it is ferruginous which is also expressed in its characteristic purple-red color (Fig. 1). The field corresponding to the decorative andesites and the bimodality of the green marble analysis distribution are presented on all charts and can later serve as a diagnostic feature.

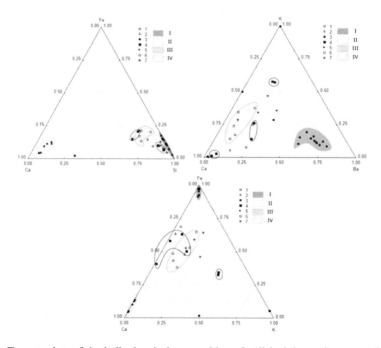

Fig. 1. Ternary plots of the bulk chemical composition of polished decorative stones from the collection of Empress Catherine the Great, plotted according to the data acquired by handheld XRF analyzer Delta Olympus. Rock types and characteristic fields: 1 - Korogon porphyry; 2 - Tigiretsky quartz; 3, I - Urazovskaya jasper; 4, II - green marble; 5, III - marble; 6 - andesite; 7 - hornfels

4 Conclusions

Thus, preliminary studies allowed us to outline the characteristic fields of the studied decorative rocks samples on their bulk composition plots. To obtain accurate results, it is necessary to expand the number of samples under study and add reference samples to the diagram. The lack of variation in the petrogenic elements content level in decorative rocks of different types and geographic reference also requires the analysis of not only chemical but also mineralogical composition of rocks. Driven by the need in non-destructive techniques, it is planned to use Raman spectroscopy to make mineralogical composition maps of the samples.

Reference

Borovkova NV (2017) Personal mineralogical collection of Empress Catherine the great in the mining museum collection. Bulletin of St. Petersburg State University of Technology and Design, № 1, Series 2. Arts Philology 8–15

Monitoring of the State of St. Petersburg Stone Monuments and the Strategy of Their Preservation

O. Frank-Kamenetskaya[1](✉), D. Vlasov[1], V. Rytikova[2],
V. Parfenov[3], V. Manurtdinova[2], and M. Zelenskaya[1]

[1] St. Petersburg State University, St. Petersburg, Russia
ofrank-kam@mail.ru
[2] State Museum of Urban Sculpture, St. Petersburg, Russia
[3] St. Petersburg Electrotechnical University «LETI», St. Petersburg, Russia

Abstract. The results of the multi-year monitoring of the state of Saint Petersburg stone monuments are summarized. The unique collection of decorative stones in museum Necropolis and the deposits that were most likely used to create them are studied. The processes of stone monuments' degradation in response to physical, chemical and biogenic influences are discussed. Special attention is paid to describing the monitoring methodology and the structure of the monitoring information database. Drawing on received results, the strategy for the conservation and restoration of monuments are discussed. The obtained data are of exceptional scientific interest in studying the processes of stone deterioration under the impact of the environment.

Keywords: Cultural heritage · Monitoring · Stone deterioration · Anthropogenic weathering · Restoration and conservation works

1 Introduction

Preservation of the monuments of cultural heritage is one of the priorities of the modern society. This problem becomes especially acute where the monuments are exhibited in the open air and subjected to destructive effects of the environment. In large cities, such as St. Petersburg, the deterioration of natural stone is notably fast, which is primarily due to the influence of the anthropogenic factor (The Effect 2019). Now we present the results of a multi-year, comprehensive study of the state of historical stone monuments of St. Petersburg, which are exposed to the destructive impact of the urban environment. The obtained data are of exceptional scientific interest for studying the processes of stone deterioration under the impact of the environment.

2 Methods and Approaches

Monitoring studies have been carried out in the Historical Necropoleis of the Museum of Urban Sculpture since 1998, where on a small square there is a unique collection of decorative and facing stone. The stone is intensively destroyed due to destructive

© The Author(s) 2019
S. Glagolev (Ed.): ICAM 2019, SPEES, pp. 479–482, 2019.
https://doi.org/10.1007/978-3-030-22974-0_118

influence of the volatile and humid Petersburg climate and unfavorable ecological situation. In this work, in addition to the Saint Petersburg scientists, museum staff and restorers, post-graduate students and students of the St. Petersburg State University, the Russian and Herzen State Pedagogical University took part. Over the past years, more than 1300 monuments of the Museum Necropolis have been examined (some of them several times), Based on the results obtained, a methodology for monitoring studies of stone materials of monuments was developed, which included the following steps: 1. Visual inspection of the object. Photographic documentation. Sampling. 2. Qualimetric evaluation of the integral state of the monument material (performed for 348 monuments). 3. Mapping of the types of material deterioration. 4. Examination of the samples of material and products of its deterioration by instrumental procedures (petrographic description of thin sections under a polarizing microscope, SE microscopy with EDX, X-ray phase analysis, biological methods). 5. Examination of the species composition of the microbial community on the surface of the monument. 6. Developing a 3D model of the monument and a quantitative estimate of the types of destruction of its material by the results of laser scanning. 7. Study of the local corrosivity of the air environment near the monuments. 8. Archival research. 9. Creating and maintaining a database on the state of the sculptural monuments in St. Petersburg.

3   Results and Discussion

Stone Material of Monuments. The diverse stone material in the museum Necropoleis is represented by marbles, limestones, granites and other hard rocks (gneisses, gabbroids, amphibolites, quartzites). The museum Necropoleis are not inferior to the historical center of St. Petersburg in the variety of stone. Basically, the stone came from Italy and the areas close to St. Petersburg (from the territory of the present Leningrad region, Karelia and Finland).

Qualimetric Evaluation of the Integral State of the Monument Material. The technique was developed jointly with V.M. Marugin (VITU, SPb). It was shown that the degree of stone destruction in the museum Necropoleis varies from 2 to 51%. In most cases, the extent of carbonate rock deterioration does not exceed 25% and that of granite and other hard silicate rocks - 10%. This is due to the considerable contribution of chemical weathering (formation of gypsum-enriched patina) in the deterioration of memorials of marble and limestone. Cracks occur on the surface of carbonate rocks that are heterogeneous in composition and structure (Ruskeala, Italian breccia and brecciated marbles, Pudost and Putilovo limestones) at least 10% more often than on other denser and more homogeneous marbles and limestones. But on denser solid silicate rocks (granites, etc.), cracks occur no less frequently than on carbonate rocks. At the same time, they are much more common (found on 80% of monuments) on such dense homogeneous rocks as Serdobol granite and Shokshinsky quartzite, which indicates a possibility of their anthropogenic or constructional origin. The incidence of the primary gypsum crust on the surface of limestones is more frequent than on the surface of marbles. Among limestones, the gypsum-rich patina is most often found on the surface of the porous Pudost travertine (on the surfaces of 50% of examined monuments). Among marbles, it is most often seen on the homogeneous Carrara marble (on the surfaces of 26% of the surveyed monuments). Its detachment together with marble and the formation of a secondary gypsum crust are observed only on the monuments with a complex surface relief made of dense homogeneous marble: (white Carrara and light gray Bardiglio). In fouling, biofilms with dominant fungi are widespread on the surface of all rocks The input of microorganisms (fungi, algae, lichens) in rock deterioration varies from 2 to 10%. The degree of manifestation of various types of stone destruction significantly varies depending on the exhibiting conditions of the monument, the characteristics of the stone material, as well as the timing and effectiveness of work on the care.

Mapping of Deterioration Forms. Ultrasonic sounding was used to detect hetero-geneities of the rock material invisible from the surface. Method for monitoring the biofouling of cultural heritage sites using computer technology allowing to register the areas of the most threatened biodeterioration sites was developed (Fig. 1). Beside 3D laser scanning method was used for to create 3D computer models and to carry out the quantitative measurements of various kinds of damage of the monument materials: cracks, chips, scratches, gypsum crusts (Fig. 2) and others.

Database on the State of the Sculptural Monuments of St. Petersburg. One of the most important stages in monitoring the state of the monuments was creating and populating a specialized database used to store, analyze and structure the accumulated factual information Currently, the database includes characteristics of the state of 650 stone monuments in the Necropoleis of the Museum of Urban Sculpture and in other parts of St. Petersburg.

Approaches and Methods of Monument Protection from Damage. To assess the effectiveness and safety of different approaches when removing biofilms, mud buildups and gypsum crusts from the surface of stone monuments, a comparative analysis was made of the potential of various chemical biocidal treatments and of laser cleaning options. The results of the experiments showed that the laser cleaning technology for removal of biofilms from the surface of the stone is comparable, and in some cases even superior to chemical treatment with hydrogen peroxide and kaolin. In the case of intensive development of biofouling, containing mosses and lichens, the efficiency of laser cleaning is significantly higher than the efficiency of chemical biocidal treatment. The use of laser cleaning to remove gypsum-rich patina is also effective.

Fig. 1. Biofilms with dominant algae (color yellow) and dark-colored micromycetes (color brown): a-photograph, b-map

Fig. 2. Electronic 3D-model of fragment of the mourner sculpture, on which the area of the gypsum crust is highlighted

4   Conclusion

Integrated monitoring the state of St. Petersburg monuments provided an objective picture of the state of the their materials, makes it possible to take timely interventions for the restoration and conservation of works of art, to plan the necessary measures to protect the stone from deterioration and in result make it possible to preserve and adequately exhibit the works of monumental sculpture and memorial art of St. Petersburg, which are an impressive, imaginative part of the world history and culture.

Acknowledgements. This study was supported by RSF project no 19-17-00141 and performed using the equipment of the SPBU resource centers "X-Ray Diffraction Methods for Studying Matter," "Nanotechnologies," and "Geomodel".

Reference

Frank-Kamenetskaya OV, Vlasov DY, Rytikova VV (Ed) (2019) The Effect of the Environment on Saint Petersburg's Cultural Heritage. Springer, Switzerland

Ceramics Sugar Jars Pieces from Aveiro Production

S. Moutinho[1(✉)], C. Costa[1,2], Â. Cerqueira[2], C. Sequeira[2],
D. Terroso[2], J. Nobre[1], P. Morgado[2], A. Velosa[1,2], and F. Rocha[1,2]

[1] RISCO, Civil Engineering Department, University of Aveiro,
3810-193 Aveiro, Portugal
sara.moutinho@ua.pt
[2] Geobiotec, Geosciences Department, University of Aveiro,
3810-193 Aveiro, Portugal

Abstract. Ceramics sugar jars pieces are, from morphological point of view, conical containers of fired clay with a hole in the vertex, which were used to sugar cane pulp maturation into sugar cake. These ceramic materials were produced in Aveiro given the existence of local raw material. Also, this occurrence of geological deposits exploited for red clays allowed the local development of strong pottery production center, transforming the city of Aveiro into one of the major Portuguese cultural heritage sites very rich in traditional ceramic tiles (*azulejos*) and other ceramic products. After local manufacture, the ceramic sugar jars pieces were exported as sugar production devices for Madeira island, Cape Verde archipelago and later, for Brazil. Also, these materials were found in buildings construction. So, this work focuses on the characterization of ceramic sugar jars produced in Aveiro and its construction use comparing with properties of other ceramics, justifying their preference for export to several countries of the world.

Keywords: Ceramic products · Sugar jars pieces · Properties · Construction elements

1 Introduction

The use of ceramic sugar jars pieces in ancient masonry walls in the Aveiro district reflects the use of these materials in construction beyond the production and transportation of sugar (Nobre 2017). Ceramic sugar jars materials were produced in Aveiro given the existence of raw material in abundance and of very good quality. In the fiftieth century the production center of Barreiro (close to Lisbon) would have ceased production and until the independence of Brazil, in the early nineteenth century, Aveiro would have been the only producer of these ceramic materials in Portugal (Morgado 2014). This production will have provided intense trade with the major sugar producing centers. Due to the local absence of natural stone for construction, the rejected/surplus ceramics were used as building material on the walls. Recently, following old house demolitions in the city of Aveiro, whole walls have been discovered with these ceramic

© The Author(s) 2019
S. Glagolev (Ed.): ICAM 2019, SPEES, pp. 483–484, 2019.
https://doi.org/10.1007/978-3-030-22974-0_119

materials, many of which were practically intact, which allowed the development of this comparative study.

2 Methods and Approaches

Mineralogical analysis was carried out by X-Ray diffraction, using a Panalytical X'Pert-Pro MPD, $K\alpha$ Cu ($\lambda = 1,5405$ Å) radiation on random-oriented powders; chemical composition was assessed by X-Ray Fluorescence using a Panalytical Axios PW4400/40 X-Ray Fluorescence spectrometer for major and trace elements and Lost on Ignition (LOI) was also determined. Compressive strength was assessed by a Shimadzu: AG-IC equipment. TGA analysis was also performed.

3 Results and Conclusions

The chemical and mineralogical properties of ceramics were similar, pointing to local production using only local raw materials. Quartz is present in all samples. The phyllosilicates are not present in any sample of the sugar ceramics but are present in all the remaining ceramic samples. The presence/absence of phyllosilicates is an indicator of the heating process temperature, higher on the case of the sugar ceramic jars. The compressive strength analysis of the ceramics sugar jars pieces shows higher values (mean 9.5 MPa) than other ceramics (mean 8.0 MPa).

References

Morgado P, Rocha F (2014) Produção da cerâmica do açúcar em Aveiro pode explicar origem dos Ovos moles. Univ Aveiro J
Nobre J, Faria P, Velosa AL (2017) Paredes pão-de-açúcar em edifícios de Aveiro Evolução, materiais e características. Master thesis, New University of Lisbon

Author Index

A

Abarzúa, G., 3
Abou Zahr Diaz, M., 443
Afanasiev, V., 179
Agapov, I., 15, 19, 24, 29
Akhmetzyanova, M., 323
Akpınar, İ., 7
Alashanov, A., 385
Alaskhanov, A., 360, 365, 369
Alawiyeh, M. A., 443
Aleksandrova, T., 59
Aliev, S., 335
Anchugova, E., 281
Andreicheva, L., 11
Andrejkovičová, S., 381
Anisimov, I., 15, 19, 24, 29
Arduin, D., 313
Arsent'ev, K., 423
Arsentyev, V., 66
Artamonov, A., 323
Askhabov, A., 33
Astakhova, Yu., 119
Aupova, N., 107

B

Babaev, V., 389
Balykov, A., 307, 372
Bedina, V., 277
Belyanin, D., 195
Berkh, K., 37
Bilskaya, I., 140
Bolatov, A., 267
Bondarenko, D., 286

Borovkova, N., 475
Brodskaya, R., 140
Bulaev, A., 189
Burtsev, I., 41, 192

C

Cerqueira, Â., 311, 313, 381, 483
Chanturiya, V., 45
Chekushina, T., 457, 461
Chernysheva, N., 315
Chikisheva, T., 49
Chulkova, I., 319
Çiftçi, E., 7, 53, 84, 183
Costa, C., 311, 313, 381, 483

D

Dai, Q., 413, 463
Danyushevsky, L., 107
Demyanova, V., 348
Deng, S., 463
Denisova, Y., 235
Deryabin, P., 319
Diarra, K., 53
Dobrinskaya, O., 272
Dolotova, A., 15, 24, 29
Dong, F., 290, 413, 463
Doroganov, V., 296
Drobe, M., 37

E

Elbendari, A., 59
Eliseev, A., 179
Elistratkin, M., 315

© The Editor(s) (if applicable) and The Author(s) 2019
S. Glagolev (Ed.): ICAM 2019, SPEES, pp. 485–488, 2019.
https://doi.org/10.1007/978-3-030-22974-0

Elkina, Yu., 189
Evtushenko, E., 277, 292, 296

F
Fedorov, A., 327
Fischer, H.-B., 344
Fomenko, I., 466
Frank-Kamenetskaya, O., 415, 479

G
Galdina, V., 319
Garkavi, M., 323
Gavshina, O., 296
Gerasimov, A., 66, 70
Ghaboura, M., 443
Glagolev, E., 315, 344, 395
Golubeva, I., 192
Gomes, A., 231
Gomze, L. A., 133
Goncharov, A., 327
Gorbatova, E., 167
Gorobtsov, D., 466
Grakova, O., 235
Gromilov, S., 179
Gubareva, E., 376
Gurova, E., 319
Gusev, V., 195
Gutiérrez, L., 3
Gzogyan, S., 75, 80
Gzogyan, T., 75, 80

H
Hadisov, V., 369
Hajjaji, W., 381
Hardaev, P., 315
Harja, M., 255
Huang, J., 290
Huang, Y., 290
Hubaev, M., 335
Hui, A., 419
Huo, T., 290

I
Idir, R., 339
Ignatyev, G., 281
Ilyina, V., 259
Iospa, A., 119
Iskandarov, N., 99
Ismailova, Z., 385
Izatulina, A., 415

J
Javid, F., 84

K
Kamashev, D., 263
Kamegamori, H., 445
Kameneva, E., 146
Kang, Y., 419
Karabaev, S., 219
Karmanov, N., 208
Karpova, A., 49
Kazakov, A., 393
Kazanov, O., 198
Kelm, U., 3
Khalmatov, R., 99
Kharitonova, M., 15, 24, 29
Khatkova, A., 90
Khubaev, M., 360
Kilin, V., 49
Klimenko, T., 94
Klimenko, V., 94
Klimova, L., 300
Klyucharev, D., 453
Kobayashi, Y., 451
Kobzeva, Yu., 140
Kokh, A., 267
Kokh, K., 267
Kolesov, E., 49
Kolodezhnaya, E., 323
Koneev, R., 99
Kononov, O., 152
Kononova, N., 267
Konovalov, V., 327
Korneev, S., 198
Korovkin, D., 307, 372
Kotova, E., 70
Kotova, I., 127
Kotova, O., 103, 133, 255, 281
Koulibaly, M., 143
Kovalev, S., 356
Kovalevski, V., 201
Kovalyov, S., 352
Kozhukhova, N., 331
Krivosheeva, A., 99
Krivovichev, S., 447
Kuladzhi, T., 335
Kuroda, K., 451
Kuzmin, D., 41
Kuznetsov, A., 267

L
Labuzova, M., 376
Large, R., 107
Larionov, P. M., 432
Lashina, I., 344
Lawrence, K., 445

Lazareva, E., 208
Lazareva, V., 66
Le Saout, G., 339
Lesovik, V., 315, 344, 395
Levchenko, E., 453
Li, Q., 413, 463
Likhnikevich, E., 119
Lipko, I., 423
Lipko, S., 423
Loganina, V., 348
Logovskaya, G., 198
Loktionov, V., 292
Luo, Y., 413, 463
Lyahnitskaya, V., 140

M
Ma, B., 290
Machevariani, M., 475
Malyukov, V., 221
Manurtdinova, V., 479
Markarova, M., 281
Masanin, O., 389
Maslennikov, V., 107
Maslennikova, S., 107
Matveeva, T., 45
Matyukhin, P., 239
Mazhitov, E., 348
Meima, J., 37
Melamud, V., 189
Melnikova, E., 189
Milman, B., 15
Min'ko, N., 272
Minibaev, A., 205
Mirsamiev, N., 219
Mishin, D., 352, 356
Morales, J., 3
Moreva, I., 277, 292
Morgado, P., 483
Moutinho, S., 311, 313, 381, 483
Mu, B., 427
Mugisho, J., 219
Murtazaev, S.-A., 335, 360, 365, 385
Murtazaeva, T., 369

N
Nakhaev, M., 365
Nelyubova, V., 389, 403
Nesterenko, G., 195
Nikiforova, Z., 111, 115
Nikitina, L., 90
Nikolaeva, N., 59
Nikonow, W., 244
Nizina, T., 307, 372

Nobre, J., 483
Novikova, S., 172

O
Ogurtsova, Y., 376
Ogurtsova, Yu., 403
Olmaskhanov, N., 219
Onosov, D., 248
Otake, T., 445, 451
Ozhogina, E., 103, 119, 167

P
Pakhomova, A., 447
Parfenov, V., 479
Pateyuk, S., 90
Pavlenko, A., 123
Pavlenko, V., 94
Pavlov, V., 136
Perovskiy, I., 41
Piryaev, A., 208
Podolian, E., 127
Pokhilenko, N., 179
Ponaryadov, A., 192, 255
Ponomarchuk, V., 208
Potemkin, V., 59
Prokopyev, E., 49
Prokopyev, S., 49
Proshlyakov, A., 457
Pursheva, A., 323
Pystin, A., 129
Pystina, Y., 129

R
Rammlmair, D., 37, 244
Rashchupkina, M., 319
Razmyslov, I., 133
Rocha, F., 311, 313, 381, 483
Rogozhin, A., 119, 167
Roux, J.-C., 339
Ryabova, A., 300
Rykunova, M., 403
Rytikova, V., 479

S
Sagitova, A., 15, 19, 24, 29
Saidumov, M., 369
Salamanova, M., 360, 385
Sangu, E., 53
Sato, T., 445, 451
Saydumov, M., 360, 365
Sazonov, A., 136
Selim, H., 183
Sendir, H., 183

Senyut, V., 179
Sequeira, C., 311, 313, 483
Shadrunova, I., 457
Shchemelinina, T., 255, 281
Shchesnyak, E., 227
Shchiptsov, V., 201
Shelukhina, I., 127
Shevchenko, S., 140
Shevchenko, V., 267
Shmakova, A., 192, 214
Shushkov, D., 281
Sigida, A., 99
Silaev, V., 133
Silyanov, S., 136
Smetannikov, A., 248
Sobolev, K., 331
Sokolov, S., 212
Spiridonov, I., 453
Stefanovsky, S., 447
Strokova, V., 286, 331, 376, 389, 399, 403,
    407
Sun, H. Y., 471
Sun, S., 255, 290
Svetlichnyi, V., 267
Sycheva, N., 119
Sysa, E., 296
Sysa, O., 277, 292

T
Tanoh Boguy, E. M., 461
Tauson, V., 423
Tcharo, H., 143
Tchibozo, F. K. N., 143
Terroso, D., 311, 313, 483
Timonina, N., 223
Titov, A., 432
Titov, S., 393
Toda, K., 451
Tolstoy, A., 395
Travin, A., 172
Troshina, O., 19
Tseluyko, A., 107
Tukuser, V., 49
Tursunkulov, O., 99

U
Ugapieva, S., 179
Uljasheva, N., 235

Uralbekov, B., 267
Usikov, S., 389
Ustinov, I., 70, 152

V
Vaisberg, L., 146, 152
Vdovina, I., 157
Velosa, A., 313, 483
Vikhot, A., 235
Vlasov, D., 479
Volodchenko, A., 344, 399
Volodin, V., 307, 372
Vorobev, A., 227
Vorobyev, K., 221, 231
Vorobyov, N., 214
Voytekhovsky, Yu., 162

W
Wang, A., 419, 427, 436
Wang, F., 413, 436, 463
Wang, K., 290
Wang, W., 436

Y
Yakushina, O., 119, 167
Yarg, L., 466
Yashkina, S., 296
Yastrebinskiy, R., 123
Yatimov, U., 107
Yatsenko, E., 300
Yu, H., 290
Yudin, D., 172
Yudintsev, S., 447
Yusupov, T., 172

Z
Zagorodniuk, L., 395
Zaikovskii, V., 432
Zelenskaya, M., 415, 479
Zhang, A., 427
Zhang, L. Z., 471
Zhao, X., 463
Zhao, Y., 290
Zhernovsky, I., 331, 389, 403, 407
Zhmodik, S., 195, 208
Zhu, Y., 436
Zhukova, V., 119
Zvyagina, E., 136

Printed in the United States
By Bookmasters